MACLEAN LIBRARY, SIERRA NEVADA COLLEGE
3 7338 00019 3657

W9-AUI-663

# WHO'S AFRAID OF SCHRÖDINGER'S CAT?

ALSO BY IAN MARSHALL & DANAH ZOHAR

*The Quantum Self*
*The Quantum Society*

# WHO'S AFRAID OF SCHRÖDINGER'S CAT?

*All the New Science Ideas You Need to Keep Up with the New Thinking*

IAN MARSHALL
& DANAH ZOHAR

*with contributions by* F. DAVID PEAT

WILLIAM MORROW AND COMPANY, INC.

*New York*

It is the policy of William Morrow and Company, Inc., and its imprints and affiliates, recognizing the importance of preserving what has been written, to print the books we publish on acid-free paper, and we exert our best efforts to that end.

Library of Congress Cataloging-in-Publication Data

Marshall, I. N.
Who's afraid of Schrödinger's cat? : all the new science ideas you need to keep up with the new thinking / by Ian Marshall and Danah Zohar, with contributions by F. David Peat.—1st ed.
p. cm.
ISBN 0-688-11865-8
1. Science. I. Zohar, Danah, 1945– II. Peat, F. David, 1938– . III. Title.
Q158.5.M356 1997
500—dc20 96-20769
CIP

Printed in the United States of America

*First Edition*

2 3 4 5 6 7 8 9 10

BOOK DESIGN BY CATHRYN S. ALISON

# CONTENTS

# PROLOGUE

**Quantum Theory:** The hypothesis accounting for the stability of the atom and other phenomena that in radiation the energy of electrons is discharged not continuously but in discrete amounts of quanta.

—*The Oxford English Dictionary*

Schrödinger's cat is the mascot of the new physics. Conceived by quantum theorist Erwin Schrödinger to illustrate some of the apparently impossible conundrums associated with quantum reality, he has become a symbol of much that is "mind boggling" about twentieth-century science.

Schrödinger's cat lives in an opaque box. The fact that we can't see what he gets up to inside the box is part of the story. Inside with the cat is a fiendish device, triggered by the random decay of a radioactive sample that determines whether he is fed good food or poison. If a decay particle hits one switch on the device, he gets food. If it hits the other, he gets poison.

In the everyday world of common sense and the old physics, one switch or the other would be triggered, and the cat would eat *either* good food *or* poison; consequently he would be *either* alive *or* dead. But Schrödinger's cat is a quantum cat, so things don't work that way for him. In the quantum world, all possibilities—even mutually contradictory ones—coexist and have a reality of their own. These coexisting quantum possibilities ensure that Schrödinger's cat is fed *both* food *and* poison simultaneously. Consequently he is *both* alive *and* dead at the same time. Both possible realities coexist.

Of course we never see alive/dead cats, and we can never catch Schrödinger's cat in his double act. If we open the box to look at him, we will see that he is *either* alive *or* dead. But it is our looking that has saved or lost the cat. The observer is part of what he or she observes; observation changes things, and the world of observation is a world of either/or. The quantum world of both/and is always at one tantalizing remove.

In this simple story of a curious cat, we see the first outlines of the larger story that this book sets out to tell, the story of worlds within and beyond the everyday world of common sense, of the new science

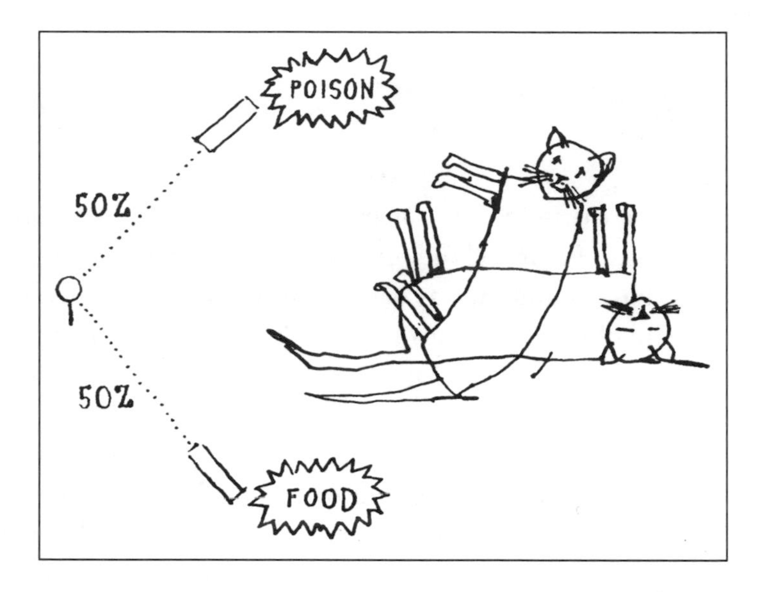


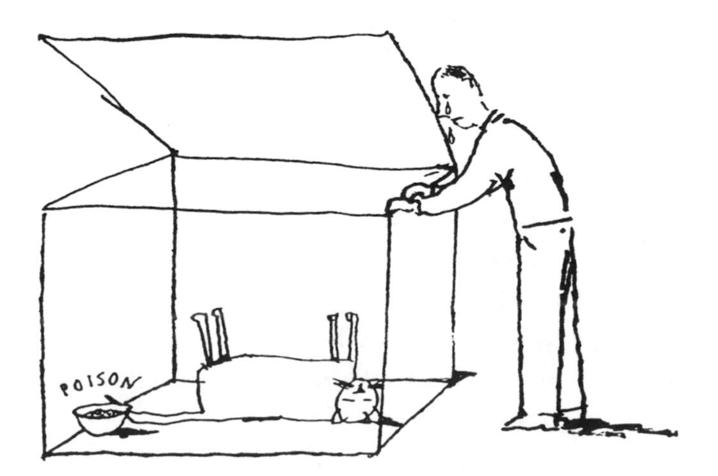

that studies these worlds, and of the new kinds of thinking required to do this science. It is also the story of new ways of looking at ourselves, of understanding the mind and its capacities and our human place in the larger scheme of things. It may even be the story of how we must change common sense itself.

# INTRODUCTION

# *The New Science and the New Thinking*

Niels Bohr, one of the founding fathers of quantum physics, was often asked to speak to lay audiences about the new science. He would begin by telling a story in which a young rabbinical student goes to three lectures by a very famous rabbi. Afterward, he describes these to his friends. The first lecture, says the student, was very good—he understood everything. The second lecture was much better—the student did not understand it, but the rabbi understood everything. The third lecture, however, was the best of all, very subtle and very deep—it was so good that even the rabbi did not understand it.

Bohr, like the rabbi of his story, never understood the science he had helped to create. Nor did Albert Einstein, who didn't even like quantum physics. Even today, some seventy years after it was fully formulated and long since it was joined by the further "bizarre" sciences of chaos and complexity, many scientists have trouble coming to terms with the central concepts of the new physics—its indeterminism, nonlinearity, and acausality; its fractals and wave/particle duality; and its cats that are alive and dead at the same time. Laymen can be forgiven for being largely unaware of how profoundly different the science of the twentieth century really is from past science, and how this difference may bear on their own lives and thinking. Science is often the harbinger of great changes that overtake human thinking. It draws its inspiration from often vague and tentative but wider cultural shifts and transmutes them into highly focused, rigorous, clear language, and into powerful images and metaphors.

A radically new way of thinking runs through the scientific work of the twentieth century. New concepts, new categories, a wholly new vision of physical and biological reality mark a sharp break with nearly everything that science held dear or certain in earlier centuries. The transition to this new thinking has been so profound and so abrupt that it constitutes a second scientific revolution calling, perhaps, for a new scientific method.

One purpose of this book is to present the new scientific thinking, to articulate what is new about it, and to outline in accessible form the major ideas born of it. What are relativity theory and quantum theory? What are their major concepts; what do they have in common? What do both share with chaos and complexity theory? What are the new sciences of mind, the new biology, and the new cosmology? What new vision of the natural world do all these entail?

The book also has a wider purpose: to show that in these new scientific ideas lies a rich repository of language, metaphor, and allusion, a whole new set of images with exciting applications in the realm of daily experience. It reveals how scientific ideas can fire the imagination, and how the new science can become a powerful model for new thinking in many areas of our personal, intellectual, artistic, and business lives.

We arc all surrounded by the new scientific thinking. Articles in the press, television programs, references in conference speeches, even comments at dinner parties constantly mention new scientific concepts like chaos, the butterfly effect, catastrophe theory, quantum leaps, and Heisenberg's Uncertainty Principle. We are told repeatedly that this new scientific thinking suggests a new "paradigm" or "world view" that can enlighten our thinking. Yet the average person with only the most elementary science background often feels this is beyond him or her, or "too clever." For many, it is like trying to read shop fronts in a foreign country, written in an alien script. The temptation not to try is great, but many increasingly feel that to ignore science or to remain ignorant of the big changes it is undergoing is to deny oneself an important new perspective.

If we are to make sense of this new science and be in a position to decide whether or how its ideas have relevance to our own lives, we must become literate about them. We must recognize that there *is* a "new science," how it differs from the old, and why this difference matters. We must appreciate the extent to which the old science has influenced our ways of thinking, and see that the new science offers a whole new way of looking at ourselves and our relationships, our business affairs and management techniques, our theories of mind and organization, global political and economic trends, the meaning of being human, and our place in the larger scheme of things.

Many good popular books introduce some area of the new science to the popular reader—quantum mechanics, the new biology, or "Gaia." But many would-be readers protest, "What I really need is an 'idiot's guide' to all the new science." This book attempts to be that guide—not for idiots, but simply for those who want to know.

## The "Old" Science

The scientific revolution—what we now call the *first* scientific revolution—began in the fifteenth and sixteenth centuries with the work of men like Copernicus and Galileo, who discovered that the earth revolves around the sun; Johannes Kepler, who established the laws of planetary motion; and Francis Bacon, who determined early scientific method. But the publication of Isaac Newton's *Principia* in 1687 marked the true advent of a revolution in thought about the physical world. Newton's three laws of motion synthesized the first gropings toward a scientific view and laid the foundations of the new classical mechanics that was to dominate scientific thought for two centuries.

Classical or "Newtonian" physics had a strong set of principles that distinguished it from anything that had gone before. Philosophy and religion based their truths on reflection or revelation; Newtonian science was based on observation. The world was thought to consist of many observable data that could be analyzed and reduced to a few simple laws and principles, or to a few basic components. The laws and principles became the basis for all-embracing general theories and sets of predictions that could be tested through experiments, which were conducted strictly in accordance with a new scientific method that viewed systems in isolation from their environments, breaking them down into their simplest component parts and using the behavior of these parts to predict the unfolding future of the system.

Simplicity, determinism, and predictability were the cornerstones of the Newtonian approach. Any system or object starting from some given state or position and acted upon by some given force would always behave in exactly the same way. Cause and effect reigned supreme, and there was always a direct, linear relationship between the force acting upon a body (the cause) and the deflection of that body from its original course (the effect).

Newtonian physics also developed the dualism that had run through Western culture for 2,000 years. In Greek philosophy, Plato distinguished between thinking and experience. The Christian Church later emphasized a dualism between man and God, and between body and soul, and the seventeenth-century philosopher René Descartes stressed a sharp division between mind and body. In a similar vein, the Newtonian scientists stressed the distinction between the observer and the observed. Scientists detached themselves from the data, just as they isolated the data from its surrounding environment. Detached observation became the new criterion of objectivity. The scientist qua scientist, above involvement, separate from all that he or she observed, viewed the physical world from an ivory tower.

## The Newtonian World View

Because of the sheer power and beauty of Newton's simple schema and the radically different perspective on truth and experience offered by the new scientific method, Newtonian science had an almost instant impact on the wider culture, well beyond the world of physics. Its ethos, methodology, values, and the vast new technology to which it gave rise exercised a hold on the whole Western imagination. To a very great extent, science replaced the Christian religion as the creative focus for Western civilization.

John Locke's individualism, Adam Smith's self-interest economics, Karl Marx's determinist laws of history, Charles Darwin's reductionist biology based on a blind evolutionary struggle, and the tempestuous forces of Sigmund Freud's dark psyche all, to some extent, owed their inspiration to Newtonian physical theory. Countless other thinkers in economics, politics, sociology, and psychology took Newton as their model in articulating their own theories. The new concept of force, the reduction of any whole to a few simple and separate parts (ATOMISM and REDUCTIONISM), and the sense that events are determined by rigid laws of cause and effect (DETERMINISM) provided the central images and categories in terms of which people viewed their experience. They came to see themselves as separate, isolated islands connected to each other only by force or influence, their behavior all too often determined by biology, background, or conditioning. Newton's description of the universe as a vast machine, and the role of the machine in the new industrial revolution, gave rise to mechanism as the dominant metaphor for how people and organizations function, and these images colored speech and thinking in nearly all aspects of life.

In architecture, Le Corbusier called his new concrete high-rise blocks "machines for living." In politics, we still speak of the cogs and wheels of government and the machinery of power. In psychology, many think of a self governed by the push and pull of determinist forces—sex and aggression, the life instinct, the death wish. In sport, we describe a proficient athlete as a well-oiled machine, a poor one as firing on one cylinder. In business, Frederick W. Taylor's scientific management took Newtonian mechanics onto the shop floor with a stress on efficiency, the division of labor, and the promotion of the cult of the expert—the observer who is detached from what he or she observes. The fragmentation of intellectual life into highly specialized fields of expertise is modeled on the scientific method's isolation of data and their analysis into separate parts or facts. Even in our personal lives, we speak of the dynamics of a relationship or the mechanics of a situation, and we describe ourselves in the language of the new computing machines. We are "mind machines," programmed for success or failure; we switch on and off; we blow our fuses.

Such images, analogies, and metaphors help us to focus, to articulate our experience and bind it into a coherent unity. They are not in themselves thinking, but they are the foundation on which thinking rests. The mechanistic images gave rise to a mechanistic world view—an overarching picture of how the physical world and human nature and affairs are structured, a set of categories in terms of which people could understand themselves and their experience. This world view dominated thought from the seventeenth century until well into the twentieth. It made possible the technological progress of the industrial revolution, and it was compatible with the flowering of Western individualism and free-enterprise economics. It led to the miracle of modern medicine and to a kind of critical and empirical thinking that freed many people from ignorance and superstition. But the extension of mechanistic, atomistic, and reductionist thinking to all areas of human life and experience has also had consequences that we are now beginning to question.

## The Need for New Thinking

An essential step in scientific method is the formation of a theory. The scientist observes the data and tries to fit them into an overall picture that explains what he or she has observed. The theory is then tested against experiment and, if validated, used to predict the pattern of future observations, or to determine responses to certain situations.

Scientists seeking the cause of malaria, for example, noted that its sufferers usually lived near or had recently visited swampy ground, where blankets of heavy mist hung over stagnant water. These observations led to the theory that the disease was caused by breathing in something contained in the misty air. Preventive measures concentrated on avoiding contaminated air. The theory saved some people, who stayed away from swamps, but not everyone who breathed "bad air" got the disease, and not everyone who avoided it was safe. In time, insect-borne disease and the breeding and feeding habits of mosquitoes were better understood, and the theory of bad air was abandoned in favor of a new theory linking malaria to mosquito bites.

World views, like scientific theories, make sense of our knowledge and experience. Theories are tested against experiment; a world view, against experience. Does this solve our problems and help us deal effectively with the challenges that arise? Does it fit the facts as we know them? Does it allow us to make sense of others' behavior and find meaning in our own lives? When the answers to these questions become negative, the world view feels inadequate, and we begin the search for a new and better perspective.

During the 1880s, the German philosopher Friedrich Nietzsche was writing *Thus Spake Zarathustra*, in which he announced that "God is dead." He didn't mean that some old man in the sky, or indeed religion itself had died, but rather that the whole framework of our culture had ceased to retain meaning. Our world view had "died."

Nietzsche, like all good philosophers, was ahead of his time. It is only now, as we approach the millennium, that we begin to realize the full weight of his insight. But very shortly after *Zarathustra*'s madman ran screaming through the streets with his nightmare tale, a more general strain of the mechanistic, or Newtonian, world view began to be felt in the arts and literature. While Dadaism celebrated the negative destruction of form and tradition, Cubism in a more positive vein brought an end to the single perspective—"Newton's single vision," as described by the poet William Blake. The First World War destroyed deeply held assumptions about the nature and stability of European culture, and about the wholly positive effects of technology. T. S. Eliot's *The Waste Land,* published in 1922, replaced glorification of technology and the machine with a bleak vision of natural and spiritual aridity: "What are the roots that clutch . . . ?" Twenty years later, in the midst of another great war, philosopher Susanne Langer wrote, "The springs of European thought have run dry." Across continental Europe, the new existentialist philosophers questioned the value and efficacy of reason and objectivity, and the whole vision of Enlightenment man that had accompanied the scientific revolution came into question.

To a large extent, twentieth-century culture has seen the slow unraveling of and disillusionment with the values and categories of thinking that were the mainstay for nearly three centuries. But it is only now, at the end of our century, that this disillusionment has become obvious in nearly every aspect of our personal, intellectual, political, and business lives. Politics has lost its meaning for large numbers of people in Western democracies, and in business the old models for successfully functioning organizations no longer match the realities of worker or market expectation. In many areas of thinking, we are experiencing what some have described as a starvation of the contemporary imagination.

The scientific revolution promised to rid humans of ignorance and superstition. The mechanists' science succeeded in undermining many of the central beliefs of traditional Western religion, but it left nothing in its place. There is no role for life or consciousness in Newton's vision of a clockwork, determinist universe, no perspective on human struggle, and no scope for human initiative and responsibility. This science qua science is value-neutral. It offers no guide to behavior. Today we are free *from* a great deal, but we have very little idea of what we are free *for*.

The sharp divide between observer and observed in mechanistic science, and the accompanying picture of a physical world composed of lifeless, brute matter, places human beings and their projects outside the context of nature. Nature becomes an object, something to be observed, conquered, and used. Technology is a means to this end. Today's ecological crisis is in large part the product of such thinking, but we have no new overall model of nature, nor of a relationship between the human and the natural, from which we might derive new thinking.

Everything in Newton's universe is ultimately reducible to so many individual atoms and the forces acting between them. Today, largely because of the successes of technology, we live in a world of growing social, political, and economic interdependence. Electronically, we live in "the global village." Political and economic models built on the assumption that people, governments, and companies are separate units, each of which can act most successfully in pursuit of isolated self-interest, have become unwieldy and unstable. Yet we lack coherent new models for a more creative relationship between parts and the whole, individuals and groups. We lack new models for how parts or individual members of any system or organization can function together to enhance its potentiality.

Newtonian science is hierarchical. The physical world is structured in ever-descending units of analysis: Molecules are more basic than complex compounds; atoms are more basic than molecules. Newtonian models of relationship and organization structure power and efficacy on the same ladder of ascending and descending authority. Power radiates out from the center or down from the top. We lack coherent new models in which decision making and power can spread sideways throughout an organization. We lack good models for "lateral thinking."

Newtonian science stresses the absolute, the certain, and the unchanging. Rigid laws of cause and effect lead to predictable, linear change. Organizational models that rely on fixed hierarchies and rigid roles, and that base their strategies on *b* always following *a*, are inflexible and unresponsive. They can't cope with rapid or abrupt change, with which we now find ourselves constantly faced.

In Newton's fixed space-time framework, there is only one way of looking at any situation. Newtonian truth is a truth of either/or. But either/or thinking can't cope with paradox and ambiguity. It is unwieldy in the face of diversity. Yet we lack any coherent models on which to base a new kind of both/and thinking.

## The Second Scientific Revolution

At the end of the nineteenth century, the British physicist Lord Kelvin advised his best students to avoid a career in physics. "All the interesting work has been done here," he told them. Classical physics had apparently solved all the enigmas of the physical universe. Everything was understood and summed up in neat, determinist equations, with the exception of two small experimental anomalies—experimental results that didn't fit into the classical picture. Kelvin, like all his colleagues, believed that these anomalies presented only a temporary problem and would soon be fitted into the general scheme of things.

The first curious result arose from an experiment conducted by the physicists Albert Michelson and Edward Morley to measure the velocity of the earth as it traveled through the universal ether, within which classical scientists believed all solid bodies like planets were suspended. The experiment seemed to indicate that earth's velocity was zero, which made no sense. The second had to do with experiments on black-body radiation. A "black body" is any hot body that radiates its energy uniformly without preference for a particular color. In experiments, a bell-shaped curve was observed while measuring the color of such radiation against its brightness or intensity. The equations of classical physics predicted that the brightness should become infinite when the color reached ultraviolet. Something was wrong.

It was from attempts to understand these two small experimental "blips" that two major pillars of twentieth-century science were born.

The Michelson-Morley experiment gave rise to Einstein's theory of relativity, and the mystery of the black-body radiation led to the first steps in formulating quantum theory. Some sixty years later, toward the end of the twentieth century, another "small anomaly" in the way that physical systems were expected to display order in response to predicted changes gave rise to what we now call the chaos-and-complexity theory. None of these new theories was simply an extension of the old physics. None could be *understood* in the framework of Newtonian or classical science. Each contained radically new and apparently bizarre assumptions about the nature and behavior of physical reality. Each required a new kind of thinking, a new set of categories for structuring how one event follows another or for how things are made, and a new set of mathematical ideas and principles to describe these categories.

This book presents the main ideas of these new twentieth-century sciences and shows how each in its own way challenges the old thinking. We can get some feel for the enormity of the revolution they have brought about by contrasting the broad features of the new science with those of the old. The old science emphasized the distinction between observer and observed. In all the new sciences, this distinction is blurred, sometimes even meaningless. The new science is *interactive*—the scientist interacts with or participates in the system under study. Rather than standing back and observing in a detached way, the new scientist becomes *part* of the process he or she is studying. The scientist's observation—its vantage point, style, intent—is itself part of the data that must be understood. This is true in relativity theory, quantum theory, and chaos-and-complexity theory. In some of the new sciences of the mind, the question is raised as to whether the observer and the observed are not merely two different aspects of the same thing.

Where the old science stressed continuity and continuous, linear change, the new science is about abrupt movements and rapid, dramatic change that is off the scale from what came before. The pages that follow are about quantum leaps, catastrophes, and sudden surges into chaos. Where the old science saw change as determinate, promising certain predictability, the new science stresses indeterminacy and the foibles of prediction. In the hands of quantum or chaos-and-complexity theorists, Newton's clockwork universe becomes a gambling casino where scientific method must sometimes give way to the Monte Carlo method, a computerized rolling of dice and the calculation of odds.

The old science portrayed a physical universe of separate parts bound to each other by rigid laws of cause and effect, a universe of things related by force and influence. The new science gives us the vision of an entangled universe where everything is subtly connected to everything else. Influences are felt in the absence of force or signal; correlations develop spontaneously; patterns emerge from some order within. Where the Newtonian scientist reduced everything to its component parts and a few simple forces acting on them, the quantum or chaos scientist focuses on the new properties or patterns that emerge when parts *combine* to form wholes. A universe where nothing new or surprising ever happens is replaced by a self-organizing universe of constant invention. The scientist learns that this fact or that part cannot be isolated from its overall environment or context, the way holism replaces reductionism and wholes are seen as greater than the sum of their parts. In the new science, organized simplicity gives way to self-organized complexity.

## A "Paradigm Shift"

Many people today speak of paradigms and paradigm shifts. What does it mean to say that the new science represents a new paradigm, or that twentieth-century thinking is undergoing a more general paradigm shift as we move toward the new millennium?

The word *paradigm* was first made prominent by philosopher Thomas Kuhn in his now classic *The Structure of Scientific Revolutions*. He used it to describe the overall framework of basic assumptions used by scientists as they analyze and interpret their data. The assumption that, if a body moves, it was impelled to do so by the application of some force is couched within the larger paradigm of Newtonian mechanics and its assumption that all movement is governed by laws of cause and effect. If a classical scientist observes some movement for which there is no obvious cause, he or she nevertheless assumes that there is one and sets out to find it. In many cases, this kind of thinking leads to further discoveries that themselves support the original paradigm.

Newton's laws of motion, plus his law of gravitation, predicted the movements of the known planets. Sometime in the middle of the nineteenth century, small discrepancies were noticed in the observed orbit of Uranus. Based on the assumption (the paradigm) that some other body must be exerting a force on Uranus's movements, astronomers looked for, and in 1846 discovered, an additional planet—Neptune. The overall paradigm of Newtonian science easily absorbed the new discovery. A new scientific fact was added, but scientific assumptions remained unchanged.

Kuhn demonstrated that, in case after case, scientists cling to their paradigm, their deep, underlying philosophical assumptions, against insuperable odds. Data that challenge the paradigm are ignored or explained away as "experimental anomalies," until the evidence for some new perspective becomes overwhelming. Thus, in the case of planetary orbit abnormalities, small discrepancies in the orbit of Mercury could not be explained in the same way as similar discrepancies in Uranus's orbit. No new planet was exerting gravitational force on Mercury. The discrepancies remained an anomaly until Einstein's theory of General Relativity made sense of them. Relativity theory represented a paradigm shift—it required scientists to adopt a radically new set of assumptions to understand their data. (See SPECIAL RELATIVITY; GENERAL RELATIVITY.)

The "second scientific revolution," which we have outlined briefly, represents such a paradigm shift across the board. It required scientists to give up their deeply held philosophical assumptions and familiar categories of analysis to adopt an entirely new way of looking at the physical world. The scientific ideas described in the pages that follow issue from several apparently different major theories and fields of study—relativity, quantum mechanics, chaos-and-complexity theory, new theories of mind, and the new cosmology—yet all fit harmoniously into a single new paradigm that displays the broad general features of holism, acausality, observer participation (contra dualism), and nonlinearity (discontinuity). As the details of this scientific paradigm shift become clearer, the reader will see how they relate to and substantially underpin a larger shift in the general world view of our culture, which has been taking place simultaneously.

## How to Use This Book

This book is designed to be used in many different ways. Its main text, which consists of short essays on specific ideas, can be used as an "encyclopedia" of the new sciences. The many cross-references and the table of contents allow a reader to go more deeply into a subject.

The book begins with four main essays making up an "Overview of the New Sciences." Each essay defines the main themes and concepts in one area of science and gives some historical background. A reader could begin by reading all four of these essays to get a panoramic understanding of what is new and exciting about twentieth-century science. The main text could then be used for further detail on specific ideas. Alternately, one could read one overview essay at a time, accompanied by its "dictionary" entries. The overview essays are:

A. "Kinds of Being," a general introduction to ancient, classical, and quantum physics. The emphasis is on the building blocks of ordinary matter.

B. "Order in Science and Thought," a survey of ideas having to do with complexity, such as chaos, nonlinear systems, evolution, and the theory of games.

C. "The New Sciences of the Mind," which deals with as-yet-incomplete efforts to understand human thinking and consciousness, and with attempts to articulate a new scientific paradigm for subjective experience via psychology, neuroscience, and artificial intelligence.

D. "The Cosmic Canopy," which concerns very high energy phenomena in cosmology and particle physics. Here, investigations into the largest and the smallest scales of our universe are coming together into a newly unified field of high-energy physics. The thinking here, too, has the new conceptual twist characteristic of twentieth-century science.

An ambitious reader can read the whole book and gain a fairly detailed knowledge of the main ideas of the new science and how all these are related parts of an across-the-board paradigm shift in our culture at large.

To make the location of any specific idea easy to find, the essays in the main text have been alphabetized. "A" and "The" have been disregarded in the ordering process. Cross-references to other essays are indicated by SMALL CAPITALS.

# WHO'S AFRAID OF SCHRÖDINGER'S CAT?

# Overview of the New Sciences

## A. KINDS OF BEING

What are the building blocks out of which everything that exists is made? What forces or interactions or "fit" binds them into the familiar objects we see around us every day?

Every attempt at a systematic view of the physical world faces these questions. Twentieth-century science grew out of earlier, classical attempts to answer them, and classical science in turn evolved in the context of ancient Greek and Christian views on these questions. The answers proposed by each of these three developments in physical thinking remain strands in our larger culture today. To grasp the basic assumptions—the paradigm—of the new science, we need to understand how it arose, and how it differs from what came before.

In ancient Greek views about being, there was a tension between a philosophical urge for simplicity and a wish to include all the world's commonsense features. Democritus' view that all that exists is "atoms in the void" was an ancestor of modern atomic theory, although unrelated to either observation or experiment. Greek atomism was pleasingly simple, but it left out a great deal of everyday experience. The early atomists had nothing to say about human purposiveness, perception, or thought.

Plato added an immaterial world of pure forms that acted as models or archetypes for the beings and qualities we observe around us. Aristotle thought that, in addition to mechanistic processes, the world contained goal-directed processes, such as the growth of organisms, or the plans of human beings. In his scheme, each being had an inner blueprint, its potential, which it would manifest if it could. The overall, mature Greek view of the world was one of many different kinds of processes interacting in complex ways. It was "messy" but close to experience.

The Greek thinkers had something to say about the major topics of this book, but they did not discriminate sharply between the kinds of questions asked. It was only after the flowering of classical physics, with its firm division between the mental and physical worlds and its total preoccupation with the physical, that people could weigh the successes and failures of such an approach and learn how to ask more penetrating questions.

The first scientific revolution emphasized the need for observation and experiment and the search for simple, universal laws of matter. After Copernicus, it was no longer plausible to suppose that the earth was the center of the universe, or that it was made of material different from that of the planets and stars. Scientists came to see that our existence is part of a much vaster picture. After Kepler, it was more attractive to describe the motion of the planets around the sun as forming elliptical orbits, rather than as compounded of complicated epicycles on circles. Earlier Greek and medieval expectations about "perfect" laws of nature—the existence of perfect circles, four elements, or five planets to complement the existence of five regular solids—could not be met by the universe as it really is. But, after Newton, human beings could take pride in being able to predict and partially control large segments of the natural world through the knowledge of a few simple laws.

In Newtonian mechanics, the material world consists of indestructible particles in an infinite, unchanging space, moving and interacting over time via forces. The scientist is distinct from what he or she observes. All other material events can be described in the same objective terms; for example, waves, heat, or energy. Physics is the only fundamental science. Given an accurate description of the fundamental particles and forces and enough computing ability, everything in the physical world is predictable. Classical physics in the nineteenth century now included electromagnetism and thermodynamics, but it retained its general ethos: It was reductionist, determinist, and objectivist. Its categories, unlike those of twentieth-century science, are easily grasped by common sense. Its essential achievement is mathematical precision and specialization.

Classical physics offers a good approximation of the behavior of ordinary material objects. It accurately predicts the motions of the planets and enables us to build bridges and airplanes. It is still used for all such purposes because it is easier to calculate with than the new physics is. But it has definite limits where it breaks down, roughly at the boundaries of the very small, the very large, the very complex, and the subjective. The themes of this book are the province of the new science, the areas where classical physics not only makes the wrong predictions but seems to be using the wrong concepts. It is a matter not of making minor corrections, but of adjusting to a major conceptual revolution.

Classical mechanics considers the motions of indestructible particles or atoms, but it has no way of coming to grips with the origins or large-scale structure of the universe as a whole, or with the high energy processes by which atoms themselves can be created or destroyed. These more ultimate questions are the subject matter of relativity theory, particle physics, and the extension of elementary quantum theory into quantum field theory.

Quantum physics is one of the most successful physical theories. Where its predictions have been tested, they have been found to be accurate to many decimal places. This is believed to apply to all matter, small and large. Without quantum mechanics we cannot begin to explain everyday phenomena like the strengths of solids, the ductility (bendability) of metals, and the colors of chemical compounds, as well as more exotic phenomena like superconductors and laser light.

Quantum physics' basic concepts are not at all like those of classical physics. Quantum physics is not reductionist, determinist, or objectivist; the observer is not separate from what he or she observes. In classical physics, the behavior of a system of particles is the sum of the behavior of its parts and their interactions. In quantum physics, the behavior of the whole system may have emergent properties that were not deducible from the properties of the parts. Such HOLISM seems to bring us closer to ancient or commonsense thinking than we were in the heyday of classical physics.

Quantum physics has as much mathematical elegance as classical physics and a rich set of existing entities that have two levels of being—actuality and potentiality. What we observe at any given time is only one aspect of what a system is. There is also what it might become. To describe such a two-level system fully, we need a range of complementary concepts: waves and particles, wholes and parts, position and momentum (COMPLEMENTARITY). The underlying reality of a quantum system is more elusive and interesting, and perhaps more like our picture of ourselves, than the predictable and somewhat bleak view portrayed by the science of the past two or three centuries.

## Ancient and Medieval Views of Being

The ancient Greeks were chiefly concerned with understanding the underlying nature or substance common to everything. Views on this were numerous but united by a common philosophical question: What universal, permanent substance underlies the variety and change of the physical world?

The pre-Socratics did not see matter as inert or lifeless or in any essential way distinct from the concerns and purposes that order the world of the mind. Some even wondered whether mind or spirit might provide the most basic substance of the world. No sharp distinction was made between the material substance of a thing and its form; both were assumed to follow from the nature of whatever the underlying substance was. Heraclitus' principles of strife and tension in the physical realm were applied to relations between people in society. The vortex theories of Democritus and Epicurus, an early Greek attempt at thermodynamics, became a central explanatory principle for social change.

Most of the early Greeks were atomists. (See ATOMISM.) They believed that everything in the physical world, including minds and souls, could be reduced to tiny indestructible particles whose shapes, sizes, and possible combinations varied. Some pre-Socratics believed that the world was made of innumerable kinds of atoms; others, that everything was a modification of one essential substance, like water (Thales), fire (Heraclitus), air (Anaximenes), or perhaps even some substance beyond our perceptions, like Aristotle's "prime matter." A few early philosophers rejected the atomic hypothesis, arguing that substances could always be broken down still further, that there was no limit to matter's divisibility. This thinking in some ways foreshadowed contemporary developments in quantum field theory, which holds that both discrete particles and continuous fields have roles to play in physical reality.

Mathematician and mystic, Pythagoras represented an important strand in pre-Socratic thinking. He believed that numbers were the building blocks of the universe, a view later developed by Plato's doctrine of eternal "forms." In our own time, some mathematicians and physicists similarly believe that mathematical truths exist in an immaterial world of their own, that physical laws may predate the Big Bang and thus the physical universe in which they are expressed.

Empedocles propounded the view of being closest of any early Greek thinking to modern chemistry. He believed that the world was made up of four indestructible elements—earth, water, air, and fire—which probably represented solids, liquids, gases, and energy. He demonstrated that air was a substance, by placing an upturned cup in water and showing that the cup did not become full because the air kept the water out. He even suggested a primitive theory of evolution. The love and strife that either bound together or separated his four elements were precursors of the attractive and repulsive forces described by modern science. Although he was religious, he denied that there was any purpose behind the world's pattern of events. Everything was due to chance or necessity. (See CAUSALITY; DETERMINISM.)

Plato and Aristotle, unlike their pre-Socratic predecessors, draw a distinction between matter and form and went some way toward seeing the material world as different from more highly organized, organic being. Aristotle emphasized the distinction between matter and form and doubted that simple atomic theory could account for the differences between material substances. He believed that the basic substance, "prime matter," was without shape or essence, which was conferred upon matter by the form it took, and he associated the form with the purpose of the substance: The growth of an acorn into an oak tree was the acorn's striving to realize its inner potential. The whole notion of purpose, the TELEOLOGY, dropped out of the mechanistic classical view of the material world, but the idea that there is a form or pattern "sleeping" within matter, and giving it shape, is coming back with the new sciences of chaos (see CHAOS AND SELF-ORGANIZATION; COMPLEXITY).

Both Plato and Aristotle adopted a version of Empedocles' four elements, and Aristotle suggested a fifth: the ether of which the heavens were composed. The only earthly substances composed of ether were the souls of rational beings. Unlike Empedocles, Aristotle did not see the four earthly elements as primary. He believed that they could transform themselves into one another.

Throughout the Greek period, pre-Socratic and Socratic, there was no distinction between science and philosophy, nor any really firm distinction between philosophy and religion. The nature of the material world and the nature of souls were considered the same *kind* of question, even if the philosophers argued that matter and form, or matter and souls, were made of different stuff. Socrates later introduced a second set of humanistic or ethical concerns: What is a good man or a good society? How can we judge whether our beliefs are well founded, or know the nature of truth and beauty? Plato and Aristotle continued to develop these questions along with their scientific interests.

In the Middle Ages, Christianity added a third concern: the mediation between God and man, sin and salvation, through Jesus Christ. All three strands of thought—the nature of being, how it is good to be, and being in God—came together in "the medieval synthesis." During this period, science, philosophy, and religion were bound together, and Aristotle's distinction between matter and form became a Christian preoccupation, with the sinfulness of the material world opposed to the holiness of the heavenly realm.

The medieval concern closest to something like modern science was alchemy, which was known in ancient Greece, India, and China as well as the medieval West. The alchemists relied largely on experimental methods in their attempts to isolate and identify the qualities of different substances. They attempted to separate the "light" or "active" properties of a substance from its "heavier" solid properties, the goal being to transform one substance, like lead, into another, like gold. Like all medieval thinkers, they mixed human and religious concerns with their "scientific" ones.

The alchemists also adopted the Aristotelian system of five elements—the four earthly elements of earth, air, fire, and water and a fifth, which for them represented spirit. The base metals they wanted to transform were made of earthly elements, whereas the gold that it was hoped would result was associated with spirit, the finest element. Our word *quintessence*, meaning the innermost essence of a thing, is an alchemical term derived from the name of the fifth element. The transformation of one substance into another was associated with the transformation of a human being into a long-lived or immortal creature. The soul was thought of as a chemical or quasi-chemical, and its departure from the body at death was likened to the separation of a gas from a more earthly compound.

In the decline of Greek civilization after Aristotle, little creative thinking occurred. It was only with the rise of modern science, beginning in the fifteenth century, that observation and experiment transformed the whole debate.

## Views of Matter in Classical Physics

The political, moral, and intellectual authority of the medieval church declined gradually over three centuries. The creative anarchy of the Italian Renaissance was followed by the gradual growth of individualism, tolerance, and diversity after the Reformation. Then came the seventeenth-century scientific and philosophical revolution and the birth of the modern outlook.

Inevitably there was tension between religion and science, the rival claimants to intellectual authority. Early on, Copernicus and Galileo challenged the view that the earth and its inhabitants were at the center of the universe. The scientific attitude that theory and experiment are equally important gradually replaced the approaches of philosophy and religion, which rely heavily on reflection or revelation. Truth, according to the scientist, could now be observed, weighed, and measured. In the seventeenth century, new instruments were invented—the telescope, the compound microscope, the thermometer, more accurate clocks. Important work was done piece by piece in many fields—astronomy, magnetism, chemistry, physiology, and the study of microorganisms. But the greatest achievement of seventeenth-century science was classical mechanics.

Mechanics, the science of how bodies move, regards the universe, or some aspect of it, as a vast machine with a set of interacting parts like the inner works of a clock, each part with only a few simple properties and movements determined by its mass and the forces acting on it. This view was developed philosophically by Descartes and Locke, and scientifically by Galileo, before Newton's completed system was presented in his *Principia* in 1687. The key concepts in the mechanistic vision are space, time, mass, forces, and particles. Other properties, such as shape and color, are either reducible to these or, like consciousness, outside physics altogether.

Newton's grand synthesis of the mechanistic vision had three laws of motion (see Box One), a universal force (gravitation), and a new mathematical tool (the calculus). Newton knew that there were other forces—friction and impact, for example—but he had no neat mathematical description for them, and this simple basis was sufficient to explain accurately a vast range of phenomena, including planetary motions, tidal movements, and the paths of falling bodies.

In the nineteenth century, a second force, electromagnetism, became clearly understood. James Clerk Maxwell united work on electricity, magnetism, and light into a single theory. Electric and magnetic forces were seen as two aspects of a field—a state of stress in an all-pervasive, jellylike medium known as the ether. The theory predicted electromagnetic waves traveling at the speed of light and including both visible light waves and other frequencies. Radio waves, discovered soon after, confirmed the theory.

After Maxwell's work, two forces, gravitational and electromagnetic, were known. All others, such as friction, were believed to be reducible to these two. The conceptual framework of classical physics now included the new idea of fields and waves within fields, but this was not too puzzling so long as the universal ether was considered a mechanical substance. (See SPECIAL RELATIVITY.)

Classical physics had one more brilliant success, the theory of heat and statistical mechanics (THERMODYNAMICS). But although thermodynamics extended the range of classical theory, it did not strain existing conceptual categories, except for the reintroduction of PROCESS and BECOMING into physics. Classical mechanics, electromagnetism, and classical thermodynamics can easily be seen from today's perspective as aspects of a single mechanical paradigm. (See ATOMISM; CAUSALITY; DETERMINISM; REDUCTIONISM.)

## Quantum Physics

The main concepts of elementary quantum physics are described in QUANTUM PHYSICS and its cross-references. Higher-energy processes of

BOX ONE

**NEWTON'S THREE LAWS OF MOTION**

1. *The Law of Inertia.* If undisturbed, a material body will continue to move in a straight line at a constant speed.

2. *The Law of Acceleration.* Alterations in speed and direction are caused by and proportional to applied forces. Acceleration is inversely proportional to mass; i.e., it is more difficult to change the course of a heavy moving body than a light one.

3. *The Law of Action and Reaction.* For every action there is an equal and opposite reaction (e.g., if I push something, it pushes back equally on me).

The first two laws, which were anticipated by Galileo, broke with the Greek idea that on earth any moving body will "naturally" halt unless it is alive. In Newton's scheme, the halting is attributed to the force of friction. Since stars and planets do not halt, both the ancients and medieval people thought them to be alive, or at least inhabited by gods or celestial beings. After the conceptual revolution brought about by classical mechanics, motion could be studied within science, but the material world seemed a much bleaker, deader place. However, Newton himself, both a Christian and an alchemist, did not have to worry about this consequence.

the sort described by SPECIAL RELATIVITY, in which particles can be created or destroyed, require an extension into QUANTUM FIELD THEORY. Such violent processes occur in nuclear reactions on earth or inside STARS, in particle accelerators, or not long after THE BIG BANG. They suggest that physics implies a PROCESS philosophy of BECOMING that is at least as fundamental as being. Still higher energies would require still further concepts of material existence, and a new, more comprehensive theory. Attempts at this are not yet complete. (See overview essay D, THE COSMIC CANOPY; THE PLANCK ERA; QUANTUM GRAVITY; SUPERSTRINGS; THEORIES OF EVERYTHING.)

There are also frontiers where quantum physics meets with other domains of new science, and these remain incompletely mapped. For instance, does quantum chaos exist? (See overview essay B, ORDER IN SCIENCE AND THOUGHT.) Or, do quantum phenomena relate to consciousness, and if so, how? (See CONSCIOUSNESS, TOWARD A SCIENCE OF.)

## B. ORDER IN SCIENCE AND THOUGHT

In the Book of Genesis and in Plato's *Timaeus*, both important sources of the Western creation myth, we are told that the world began as a formless void. Eastern religions teach that the primary reality is without features. It is a nothing about which nothing can be said. Modern QUANTUM FIELD THEORY, too, describes THE QUANTUM VACUUM, the first thing created after THE BIG BANG, as a featureless ground state of unformed energy. Why then is there "something rather than nothing"? Why is there any order in the world rather than merely the formless flux with which it began? What brought about this order? Is it essential to the structure of the universe, or just a random happening? Or, worse still, is order a mere product of human perception? Such questions have plagued philosophers and theologians since ancient times and are today a central preoccupation of science.

Human beings seek constantly to order the world and make sense of our experiences. Families, clans, crop plantings, stock markets, and international communication systems are as sophisticated as any theory emerging out of theoretical physics, and are all attempts to structure some aspect of the world. We cannot escape the appeal of order.

We define order roughly as the meaningful or nonrandom arrangement of parts within a structure. In this sense, it is closely related to the concept of information. Some examples of such order are the mathematical forms found in arithmetic, geometry, and calculus; the forms of musical, architectural, and artistic compositions; the natural forms of crystals, plants, and animals; and the very basic laws of physics themselves. Anything that we can perceive or say about the world is necessarily couched within some frameworked way of looking at it, and any framework implies a structure. It is one of the major new insights of twentieth-century science that we are always within a framework, and that therefore no God's-eye view of order is possible. GÖDEL'S THEOREM tells us that we cannot even describe all possible types of order within the limited context of our arithmetic. Where there is no order, we are left with nothing at all to say.

Our own brains are "hardwired" for order. Vision is a good example, although any other sense would observe the same principles. Individual groups of cells in the visual cortex respond to specific stimuli. One cell is excited by an object's boundary or contour, another by a moving edge, a region of color, or an area in motion. In this way, a visual scene is broken down into a variety of codes, which are then reintegrated at the higher centers of the brain to produce what we see. (See VISUAL PERCEPTION.)

Naked vision, like trying to describe a world without order, is impossible. We are born into the world with strategies for seeing and ordering. "Reality," as we perceive it, is a construct that grows out of the intersection of the brain's ordering strategies with incoming sense data. It is probably true that about 50 percent of what we "see" is already present in the brain in the form of processing strategies, visual memories, recognition systems, and so on.

The brain's compulsion to order extends outward into the social group by means of patterned gestures, ritual actions, exchanges, and, above all, human language. Even the simplest and most ancient of societies order their world in highly complex ways. A coherent order enables the group to survive, to hunt or farm, to communicate, to bring up children, to resolve internal differences, and to contain tensions between neighboring groups.

Through creation stories, ceremonies, and acts of renewal, the order of the group is maintained. An important aspect of this overall process is the superposition of order onto the external world. The seasons and the movements of the sky are ordered by means of creation stories, songs, calendars, and periodic ceremonies to facilitate hunting and the planting of crops. Clocks are devised to order "time" and rulers to order distance. Myths embracing our place in the world order the natural world of plants and animals. Very postmodern thinkers suggest that the physical laws of science are themselves one more set of categories imposed by humans on the world, although this is a minority opinion.

Whether founded in nature itself or simply in the wiring of our brains, our perception of order is clearly essential to our ability to make sense of the world, providing our rather slow conscious processes with a crucial backup facility. Nonetheless, the brain's built-in strategy for perceiving order also has its drawbacks. Too often we fall back on stereotypes or clichés rather than forming original thought. This can result in misconceptions and even confusion when the world fails to fit our preconceptions. Such crises, or simple boredom, may force us to think more creatively, which requires more time and energy. Tests have shown that human beings actually do best psychologically in an environment poised between too much order (boredom) and too little order (confusion). The same is true of self-organizing physical systems, particularly biological ones poised at the edge of chaos. Survival requires regularity in the climate, food supply, and so on, but evolution is spurred by the challenge of circumstances to which living systems are not fully adapted.

We are left with a family of related insights and questions about order. Some order is necessary; too much is uncreative. Some order is wired into our perceptual categories, but some seems to exist in Nature herself. There appear to be testable physical laws. Why? And why are the physical laws so simple that we have been able to understand them partially? Might we one day be capable of understanding all physical laws totally? (See THEORIES OF EVERYTHING.) We may need a wholly new philosophical and scientific approach. In the meantime, it can be useful to map out the fundamental kinds of order that have been actually observed in the world.

## Historical Types of Order

Any individual or society uses only a selection from an infinite set of possible types of order. The most fundamental, all-embracing selection describes a society's paradigm, the set of categories in terms of which all experience is interpreted. Paradigms vary from one society to another, as two contrasting studies of the perceptual capacities of North Americans reveal. Psychologists studied the mainstream population and found that its members could perceive horizontal or vertical lines more easily than diagonal ones. They concluded this must be an innate ability. But other psychologists repeated the experiments with Native American subjects who had been raised in wigwams. It turned out they could perceive diagonal lines just as well as horizontal and vertical ones.

Our current Western society and much of the rest of the world now live in terms of the paradigms and precepts of Western science to such an extent that the scientific perception of reality now seems to be inevitable, anything else now being rejected as myth, superstition, or illusion. But at other times societies have been effectively ordered according to quite different precepts. The Shang civilization of ancient China, for example, flourished for some seven hundred years, creating great art and elaborate buildings and maintaining social stability—all based on the emperor's daily consultation of the tortoise oracle!

Medieval Western society, perceiving the world as static and hierarchical, assumed it had always been much the same. Everything had its nature, its purpose, and its proper place. Disruptions to this order—illnesses or disasters—had entered the Garden of Eden via human frailty or sin. Ptolemaic and medieval cosmology, as expressed, for example, in Dante's *Divine Comedy*, flourished for more than a thousand years. The planets, stars, and heavens, each with its own inhabitants, revolved in concentric circles around the earth. This appealing vision was overthrown by the Copernican revolution and the rise of the scientific method.

The seventeenth-century scientific revolution, part of a gradual shift from the medieval to the modern world, was preceded by the

Renaissance and the Protestant Reformation. These, like science, emphasized individual observation and judgment rather than submission to authority. Scientific belief in atoms and mechanical forces, which replaced the earlier vitalism, was paralleled by a similar belief in individuals as the "atoms" of society. The political philosopher John Locke admired and consciously followed Newton's thinking in physics. In Adam Smith's liberal economics, society was held together by market forces, just as physical and chemical forces held atoms together. The general temper was reductionist. (See REDUCTIONISM.)

Though a belief in God and human freedom originally tempered this clockwork vision, religious belief gradually faded and psychology, derived from Newton's model, replaced freedom with DETERMINISM. In its heyday, modernist culture was articulated by materialist and positivist philosophers, who took mechanistic science as their ideal; by barren, concrete housing developments; and by art, music, and literature based on concepts divorced from tradition and from most human feelings.

Modernism was one-sided and thus inherently unstable. Resembling Democritus' bleak vision of "atoms in the void," it left no room for the consciousness, values, triumphs, and tragedies of the scientists and philosophers who had created it. The twentieth century has witnessed a widespread reaction in a family of movements and attitudes known as postmodern, which hold that mechanistic science and rationality are only one way of looking at the world. Cubist art, the later writings of James Joyce, and the philosophies of Henri Bergson, the existentialists, and the later Wittgenstein have all demonstrated an alternative to rationalism and the single point of view. At the same time, interest in non-Western perspectives has increased.

The postmodern developments in the humanities were paralleled by corresponding developments in science. SPECIAL RELATIVITY (1905) introduced multiple frames of reference for space-time, QUANTUM PHYSICS gave us complementary but equally valid pairs of opposites (e.g., waves and particles), HEISENBERG'S UNCERTAINTY PRINCIPLE presented a dependence on context, and HOLISM overthrew the older atomistic viewpoint. Fractals and chaos theory gave us a vision of inexhaustible complexity, far beyond the simple geometrical shapes of Euclid. Science ceased to be a quest for organized simplicity and became an effort to describe randomness, INDETERMINACY, and self-organized complexity.

These developments in science are part of the postmodern paradigm. Being more clearly thought out, articulated, and experimented on than their relatives in philosophy and literature, they help to illuminate the broader cultural situation. If we equate modernism with traditional logic and rationality, postmodernism has two strands, two different directions in which it can go. It can regress to prerational or antirational positions, like Dadaism in art, some aspects of New Age thinking, deconstruction in philosophy and literature, and various kinds of hedonism or superstition. Alternatively, it can attempt a progression to transrational forms of discourse, resembling Zen koans or mystical paradoxes, which do not deny the value of rationality but seek to supplement it.

Quantum theory has this supplementary, extending relation to the simpler, more rational Newtonian mechanics. Quantum paradoxes do not deny the value of Newtonian perceptions; they simply place them in a larger context. In the same way, the different frames of reference of different observers are all parts of a larger whole—a four-dimensional God's-eye point of view that cannot be perceived by any human being but follows its own laws. (See RELATIVITY AND RELATIVISM.) An exploration of the new types of order that twentieth-century science and mathematics have evolved can enrich the imagination and perhaps clarify the outlines of the emerging new paradigm in culture as a whole.

## Main Themes to Be Explored

### *1. Thermodynamics*

Nineteenth-century THERMODYNAMICS had two great laws: the conservation of energy (THE FIRST LAW OF THERMODYNAMICS) and the nondecrease of ENTROPY (THE SECOND LAW OF THERMODYNAMICS). These were underpinned by STATISTICAL MECHANICS.

The Second Law suggested to many people that the universe is gradually running down or cooling with a clockwork regularity. Today we realize that this law applies only to closed systems, where neither energy nor matter can enter or leave. In OPEN SYSTEMS, such as living organisms or the weather, new patterns (i.e., order) can be created. (See CHAOS AND SELF-ORGANIZATION; DISSIPATIVE STRUCTURES.)

### 2. *Biology*

Since the seventeenth-century scientific revolution, the body has been thought of by most scientists as a machine consisting of separate working parts. How, then, does it possess such abilities as growth and repair, reproduction and instinct? VITALISM was a rearguard action proposing that such capacities resided not in the mechanical body but in some special vital force. But the mechanical view of the body won the day, and scientists sought models that could accurately describe it. (See ARTIFICIAL LIFE; CONSTRUCTION COPIER MACHINES.)

DARWINIAN EVOLUTION became the generally accepted conceptual framework for mechanistic biology. It held that life propagated through chance mutations that could be inherited, a competitive struggle between organisms, and survival of the fittest. The discovery of DNA shed light on known genetic laws. COEVOLUTION, yet another mechanistic theory, emphasizes cooperation rather than competition. It has become obvious that the mutation of a single gene can transform not just one characteristic, such as eye color, but also a larger-scale pattern. With the articulation of CHAOS AND SELF-ORGANIZATION, biology was extended into the new paradigm. It is now recognized that not all biological traits can be reduced to genetic behavior. (See EMERGENCE.) The whole field is much richer and more open today despite many controversies and large gaps in understanding.

### 3. *Systems Theory*

In the 1940s, general SYSTEMS THEORY reacted against excessive specialization in the sciences and emphasized a more holistic approach. Systems theory influenced many disciplines—ecology, sociology, engineering, and some aspects of management thinking. It embodied the concepts of positive and negative FEEDBACK between parts of a system called black boxes. (See THE BLACK BOX; CYBERNETICS; INFORMATION; NONLINEARITY.)

Today it is still popular to analyze a system by drawing a succession of black boxes interconnected by arrows (causal loops) representing stimulation and inhibition. Such a schema models the broad, qualitative performance of a given system, but it is doubtful that it is sufficient to understand all the underlying features of a system's behavior. Systems theory remains steeped in the mechanistic paradigm and thus cannot model a kind of holism in which the *parts* of a system change *through their participation in the system*. (See CONTEXTUALISM.) This more fundamental holism is characteristic of new-paradigm or quantum thinking. Systems theory's ambition to recognize more universal patterns and apply them to systems has not been achieved. Since the 1960s, research interest has shifted in other directions, such as complexity, although systems theory remains a useful tool.

### 4. *Other Models of Complexity*

In the 1960s, CATASTROPHE THEORY became popular. Where previous physics had focused on linearity and steady step-by-step change, catastrophe theory took nonlinearity and sudden change as its focus. Catastrophe theorists studied how systems undergo discontinuous change, such as a beam breaking, water freezing, or even a couple having an argument. Their work was appealing as a mathematical classification, and it focused interest on a neglected area of behavior, but it led to few further insights or predictions.

Chaos theory focuses on yet another sort of nonlinear order. It demonstrates how quite simple deterministic systems can spin off into chaotic unpredictability when a very small fluctuation of the initial conditions in some process is repeated many times. (See ITERATION.) Thus animal populations can fluctuate wildly in response to very small changes in breeding patterns from year to year, or water can rush into turbulence with very small changes in the flow patterns of individual molecules. (See ATTRACTORS; THE BUTTERFLY EFFECT; CHAOS AND SELF-ORGANIZATION; NONLINEARITY; THE THREE-BODY PROBLEM.) The mathematical study of FRACTALS, such as THE MANDELBROT SET, is closely connected to these themes. In fractals, no amount of even the most microscopic detail is sufficient to get to the heart of the structure, they are essentially irreducible. Consequently, fractal geometry is wholly different from Euclidean geometry, which focuses on smooth lines, planes, and circles.

Chaos is only one sort of complex behavior. COMPLEXITY theory has also attracted a great deal of interest, even though the complexity involved has no general definition and depends on a handful of suggested models. The most exciting definition of complexity relates to THE EDGE OF CHAOS, where systems are delicately poised between order and chaos and at their most creative.

### 5. *Coherence*

COHERENCE, a specifically quantum type of ordering, has no equivalent in classical physics. Macroscopic examples of coherence include SUPERCONDUCTORS, SUPERFLUIDS, and LASERS. All these structures are examples of BOSE-EINSTEIN CONDENSATION, in which the PHASE of the quantum wave function becomes fixed throughout the system, so that individual elements become completely delocalized and have no individually discernible identity, or position in place and time. Analogies of coherent quantum systems are a group of soldiers marching in step and the many voices of a choir sounding as one voice. Herbert Fröhlich proposed one sort of Bose-Einstein condensate that theoretically might exist in living cells. (See FRÖHLICH SYSTEMS.) Others have suggested that such a system in the brain might underlie the unity of consciousness. (See CONSCIOUSNESS, TOWARD A THEORY OF.)

### 6. *Symmetry*

Symmetry, an important type of order, will be discussed in overview essay D, THE COSMIC CANOPY, in conjunction with elementary particles.

## C. THE NEW SCIENCES OF THE MIND

What is a mind? What can minds do? Are minds the products of brains? How does the brain give rise to the mind? Do animals have minds? Do computers? Are computers good models for brains? Is biology important in studying the brain? What is awareness? Consciousness? Must a mind, to be a mind, be conscious? Can biology or physics or theories of computation explain consciousness?

These are just a few of the questions posed by the new twentieth-century mind sciences: psychology, neuroscience, and cognitive science. Most of them have no definite, universally agreed-upon answer. They are a minefield of uncertainty and speculation. In this field, at this time, the questions are almost more exciting than the answers. All good science begins with questions, and the sciences of the mind are just beginning.

Psychology, a study of the mind or of behavior, is the oldest of the new mind sciences. What are its roots? To what extent is it a science? Psychology differs from art and literature in that it looks for general rules and principles of human behavior. But how does it differ from physics and chemistry? Will it ever be possible to bring objective study to bear on subjective experience?

Much of what we mean by psychology is implicit in our lives as social beings—we all try to relate to and understand others, and in doing so we try to understand their "psychology." This is more a skill than a theory. Such implicit views of mind appear in all human narratives from the earliest times, including myth, novels, and biography. More explicit views are parts of other disciplines. In education we ask how children can best be taught. In law we define when people are and are not responsible for their actions. In medicine we seek the causes and best treatments for disturbed states of mind. In politics we ask what form of government people will best cooperate with. In philosophy we question to what extent people's experience is conditioned by their brains and how far they can achieve accurate knowledge of the world.

The first systematic study of psychology began in ancient Greece with the work of Plato and Aristotle. Both placed great emphasis on the nature of the soul and its relation to the body, and hence to the world of experience. Plato believed the soul had three parts—reason or mind, spirit, and desire. For him, the rational aspect was the most important. It alone was immortal and equated with the soul's ability to grasp mathematical truth. Plato's clear judgment that rational mind was totally distinct from the irrational experience of the senses divorced his psychology from its biological roots.

Aristotle, by contrast, placed great emphasis on the relation between mind and body, and the importance of the soul's "lower" functions (perceiving, feeding, reproducing) for shaping its higher ones. He distinguished four functions of the soul, having to do with intellect, sensation, nutrition, and motion. The son of a doctor and himself a marine biologist, he took his categories of explanation from the realm of living things. This led him to root his psychology firmly in biology and to explain the whole of the natural world in similar terms. Where twentieth-century psychologists were to apply mechanical principles from the physical world to the soul or psyche, Aristotle applied soul, or purposive principles, to the whole of physical nature: "Nature, like mind, always does what it does for the sake of something, which something is its end." Both Plato and Aristotle formed quite systematic views about the nature of perceiving, thinking, and imagining.

In Eastern cultures, a systematic study of consciousness has long been a part of religious tradition, but in the West psychology did not become a subject in its own right until the seventeenth-century scientific revolution, when John Locke and other philosophers popularized a new view. Rather than simply talking about mental faculties such as memory, understanding, and will, Locke attempted to analyze these into smaller units, which he called "ideas." These ideas (qualities like redness or roundness) resembled the atoms of the new scientific physics. Ideas were formed by perception, stored in memory, and later formed new combinations. They were associated with each other by similarity, contrast, or contiguity. This approach represented progress toward a systematic study of mental experience, but it was a philosophy more than a science.

Most of the founders of the scientific revolution, including René Descartes and Isaac Newton, were believing Christians. They had a mechanical model of the physical world, but they believed in a separate and immortal soul. Though they were fierce individualists, they subscribed to the view that everyone should live by God's laws. As a result, psychology was more or less unnecessary in the West so long as Christianity remained strong. In Hinduism and Buddhism, by contrast, spiritual progress is a matter of enlightened vision rather than obedience, making psychology an integral part of religion.

While Descartes rejected all human authority and belief in human habits of thought, he turned to faith in God as an indirect way of sustaining faith in the human world. Descartes believed that God gave him his perceptions and ideas, and that God would not deceive him. Two centuries later, Friedrich Nietzsche began in the same way by rejecting all authority and received truth but, lacking faith in God, was forced to explore the nuances of human psychology. His attempt to fill the spiritual vacuum felt toward the end of the nineteenth century with an emphasis on psychological development was a powerful influence on twentieth-century thinking. It was only when concentration on human nature could be free of religious influence that psychology came into its own. (See PSYCHOLOGY IN THE TWENTIETH CENTURY and its cross-references.)

Experimental psychology is the laboratory study of human beings or animals. It is now more than a hundred years old, but it has limitations. If experimental psychology restricts itself to publicly observable behavior (see BEHAVIORISM), it leaves out most of what interests us about people. If it relies on people's *experiences*, the data are not objectively checkable in the same way as the data of physics and chemistry. This same dilemma affects the scope of clinical and social psychology.

A second approach to the sciences of the mind is through NEUROSCIENCE. Because neuroscience gives us objective data about brain processes that happen between a stimulus and its response, it goes beyond the more narrow behaviorist paradigm. Today, we have greatly improved METHODS OF STUDYING THE BRAIN. Progress in neuroscience has proceeded at an almost dizzying pace. Any detailed knowledge of the brain's vital role in our mental life is less than 150 years old. It was only in 1906 that Camillo Golgi and S. Ramón y Cajal won the Nobel Prize for their discovery of the neuron—the basic cellular structure of which most brain matter is composed. (see NEURONS.)

Today, an explosion of information about the anatomy and physiology of the brain, made possible by the past century's rapid advances in technology, has given us a complex array of facts. We now know a great deal about neurons, neuron bundles, their connections, their molecular structure, and which areas of the brain are associated with various mental faculties. We understand a great deal about brain injuries and something of the role specific brain chemicals play in psychiatric disorders. But so far all these facts—and the facts that we know are only a tiny fraction of what we *could* know—fit into no comprehensive theory about mind and brain. What is more, neuroscience covers many orders of magnitude, from molecules and subcellular structures (see

NANOBIOLOGY) to large systems (see MEMORY; PERCEPTION; PSYCHIATRY; THINKING), yet the discoveries of research workers at these different levels have so far not been integrated. For this to happen, we need not just more *facts*, but a more accurate overall *model*. This would require a conceptual breakthrough.

A revolution in computer technology and the tantalizing analogies between how computers work and some facets of human thinking have led to a powerful computationalist model of mind. (See COMPUTATIONAL PSYCHOLOGY.) Most cognitive scientists have adopted this model in some form and use it to guide their research questions and philosophical reflections. Research into ARTIFICIAL INTELLIGENCE is based on similarities between human thinking and computer capabilities. Brains are conceived of as "mind machines." But computers are not conscious: They feel neither joy nor pain; they do not act with intention. Critics wonder how anyone seriously could believe computers are good models for the brain. Supporters claim it is only a matter of time until computer technology will catch up with any observed differences. (See EXPERT SYSTEMS.)

The new COGNITIVE SCIENCE attempts to integrate the questions raised by philosophers, the data collected by psychologists, and the knowledge of internal brain function gained by neuroscience. The goal of cognitive science is a unified picture of mental ability together with its physiological basis, but as practiced so far it concentrates only on limited aspects of our experience—cognition, or thinking—to the virtual exclusion of emotion, consciousness, and creativity. Cognitive science is also held in thrall to the revolution in computer science, in consequence adopting the Newtonian paradigm of mechanism and computability almost exclusively. (See THE CHURCH-TURING THESIS; FORMAL COMPUTATION; NEURAL NETWORKS.) Human thinking is modeled on what machines can do, or on what can be simulated by machines. The mental capacities and underlying neural bases that are best understood most resemble computer circuits. The result has been a powerful model of some aspects of human thinking, combined with easy neglect of whatever mental experience cannot be squeezed into a reductionist scheme.

While a clear new paradigm is emerging in the other sciences of this century, the sciences of the mind offer an array of models or paradigms, no one of which has clearly taken hold. Some are twentieth-century variations of age-old dualist models, viewing "mind stuff" and brain matter as essentially distinct. Many are reductive models, still steeped in mechanism and its modern computer-age equivalent; these assume that one day all mental function will be explained in terms of neural function. A few models look to Darwinian biology or to the new sciences of quantum theory, chaos, and complexity. Some observers have compared the field to the proverbial blindmen and their elephant, each discipline or research group grabbing hold of some aspect of experience or some part of the brain and saying, in its ignorance, "Ah! *That* is what mind is like and how it must work." But no one has yet seen the elephant.

Indeed, each of us *is* the elephant. We live with and relate to other "elephants" as a matter of daily experience. Our common sense and intuition tell us a great deal about the nature of mind, the kinds of things people who have minds are likely to do, and perhaps even why they do them. But how far can this direct experience be translated into scientific knowledge? Our experience alone can never tell us what neural or subneural activity accompanies something like falling in love or becoming angry.

Consciousness and its attendant creativity are the most glaring omissions in the existing neurosciences and in the dominant cognitive science models of mind. (See CONSCIOUSNESS, TOWARD A SCIENCE OF.) This omission has led to interesting, though not yet mainstream, investigations and speculations at the edge of mind science. Does quantum activity at subneuron level give rise to conscious experience and its unity? Does patterned activity in the firing of neural synapses account for persistent visual patterns or even for well-worn attitudes? Does this lead to CHAOS THEORIES OF MIND? Is the brain creative because its neural structure is constantly evolving in dialogue with experience? Is creative thinking noncomputable (i.e., not able to be simulated by computers), and possibly quantum in origin? These are some of the questions being asked at the leading edge of the mind sciences. We still lack an adequate understanding of THE MIND-BODY PROBLEM, without which our sciences of the mind cannot be solidly founded.

## D. THE COSMIC CANOPY

The heavens have always been a central focus of human inquiry and philosophizing. Since our species' childhood, people have looked up into the night sky with awe, wonder, and questions. What is the universe like as a whole? When did it begin? How did it come to be? What will become of it? Of what are the stars and planets made, and what holds them in the sky? How does the earth, and we who inhabit it, relate to the general universal scheme? Is there intelligence behind it all; are other intelligent beings occupying other regions?

Such questions have always been at the heart of human mythology and religion. In ancient Greek philosophy and early science, they were the subject of COSMOLOGY, "study of the cosmos." In more modern times, they began to blend with the factual data collected by OBSERVATIONAL ASTRONOMY and with physicists' realization that the same physical laws govern both heaven and earth, until, by the twentieth century, the modern science of cosmology was born. It is here that we begin to see the integration of much of our science—the large-scale relativistic equations of Einstein, Newton's gravitational work, Maxwell's wave equations for light, and the infinitesimally small worlds of quantum theory and particle physics. Through the new sciences of QUANTUM GRAVITY and QUANTUM FIELD THEORY, attempts are being made to unite the four fundamental forces of the universe—gravitation, electromagnetism, the strong and weak nuclear forces—in one all-encompassing Theory of Everything. (See THEORIES OF EVERYTHING.)

Observational astronomy is humankind's oldest exact science. From the very beginning, our study of the heavens has been based on direct observation and the careful collection of data that could be used for practical purposes, though in earlier, prescientific eras, such observations were wedded to myth and religion. Prehistoric people saw gods and heroes in the heavens' constellations, but used them nonetheless as aids to navigation and timekeeping. The days, months, and seasons of the year could be measured by the visible movements of the sun, moon, and stars. The stars appeared to be fixed in their relative positions, as a background canopy for seven wandering objects: the sun, the moon, and five visible planets—Mercury, Venus, Mars, Jupiter, and

Saturn. The Greek word for planets means "wanderers." The Babylonians associated the seven wanderers with gods, each ruling over a day of the week. Because of the mythical element, astronomy and astrology arose together.

Some early Greeks developed the scientific side of astronomical observation. During lunar eclipses, the earth invariably casts a curved shadow on the moon, leading to the conclusion that the earth is a sphere approximately three times larger than the moon. Eratosthenes observed that noon, when the sun is at its highest, occurs at different times in different longitudes. From this he calculated the size of the earth fairly accurately. Some early Greeks thought that the sun was considerably larger than the earth, but they had no accurate way of measuring the relative sizes because they lacked a way to measure the distance between the two bodies.

The early Greeks believed that the earth revolved around the sun, but later religious motivations and faulty physical reasoning led to the dominance of Aristotle's and Ptolemy's geocentric theories. The earth was believed to be flat and at rest in the center of the universe while the sun, moon, planets, and stars revolved around the earth. The celestial spheres were thought to be made of a different substance, the ether. Heaven was supposedly located beyond the stars, and hell under the earth. This vision, combining features of common sense, the old astrological religions, and Christianity, survived for over 1,500 years, well into the Middle Ages.

In the sixteenth century, Copernicus demonstrated a much simpler model of the solar system and its movements by assuming that the earth and the five visible planets all revolve around the sun. Based on evidence from the newly invented telescope, Galileo and Kepler argued in favor of this view. The Church forced Galileo to recant. (Copernicus escaped censure by dying the year his work was published.) In the end, however, the Copernican view won the day. People realized that the heavenly bodies were made of the same stuff, and followed the same physical laws, as bodies on the earth.

The Copernican victory was a fundamental paradigm shift in cosmology. Astronomy was separated from philosophy and religion. The medieval view that truth rests on revelation or tradition gave way to the scientific trust in reason and observation. The picture of the earth as the centerpiece of a creation watched over by a loving God gave way to the bleaker view that the earth is a rather small and ordinary piece within an immense universe.

According to the early scientific paradigm, the universe was thought to be static. Either it always had existed and always would exist in more or less its present condition (see THE PERFECT COSMOLOGICAL PRINCIPLE), or it had been created at some time in the past but would remain the same until the Last Judgment. This static universe was infinite and fairly uniform in composition. (See THE COSMOLOGICAL PRINCIPLE.) Newton accepted the static universe theory but argued that the stars must be infinite in number, because a finite array in infinite space and time would ultimately clump together under the pull of universal gravitation. In the nineteenth century, the law of increasing ENTROPY added the gloomy prediction that the universe was gradually running down.

The twentieth century has witnessed yet another revolutionary paradigm shift in cosmology. Einstein's GENERAL RELATIVITY theory (1915) provided a new and subtle picture of space-time and its connection with matter and gravitation. According to Einstein's equations, now well supported by experimental evidence, the universe as a whole cannot be static. It is either expanding or contracting. Ironically, Einstein disliked the implications of his own work and attempted to reestablish the possibility of a static universe by fiddling with his own equations. He arbitrarily added a so-called cosmological constant that changed the predictions. After astronomer Edwin Hubble showed that we do indeed live in an EXPANDING UNIVERSE, Einstein dubbed his cosmological constant "the biggest blunder of my life."

Today, nearly all astronomers agree that the universe is expanding. It began with a BIG BANG, a near infinitely hot and dense point some 10 to 20 billion years ago. Gravitation is the most important force acting on large scales in the universe. Unlike electromagnetic force, it is always attractive; its effects add up rather than canceling out. In consequence, gravitation *slows* the expansion of the universe. It is still not known whether expansion will continue indefinitely or whether, at some stage, the universe will recollapse to a single point—the so-called Big Crunch.

As the universe has expanded, it has cooled, resulting in several PHASE TRANSITIONS in which the stuff of the universe has taken different forms, as when steam condenses into water, and then water into ice. The evolutionary process to which this has led can be divided into seven eras, each with its distinctive features. (See Table One.)

Observational astronomy, with the aid of better telescopes, radio telescopes, satellite observations, and many other techniques, has greatly expanded what we know about the universe, which is, in many ways, dynamically changing and violent. Not only earthly life forms like ourselves but stars and the universe itself are born, live, and die. The development of particle physics in this century has provided good theoretical models of at least some of the universe's high-energy processes, like the nuclear reactions that power the STARS. The confluence of cosmology with observational astronomy and particle physics is leading to a unified branch of science. But the very earliest and highest-energy processes in the universe, which are perhaps the most philosophically interesting, still elude our full understanding.

Of the seven eras in the cosmic evolution, the brighter objects in the latest two—the galactic and chemical eras—are directly observable by telescope. In observing something like a quasar that is $10^{10}$ light-years distant, we are seeing it as it was $10^{10}$ years ago. Thus we can see that there used to be many more radio galaxies than there are today. But there is a barrier to data that can be gathered by any kind of telescope.

Before the universe was 300,000 years old, it consisted of PLASMA, a hot, opaque "soup" of interacting radiation and charged particles. It eventually became cool enough for electrons and nuclei to recombine into neutral, light atoms, and so become transparent. The COSMIC BACKGROUND RADIATION we see today was formed at this time when the universe was nearly, but not completely, uniform. (See WRINKLES IN THE MICROWAVE BACKGROUND.)

Today's high-energy physics gives us definite models of the energy processes between the times the universe was $10^{-34}$ second old and 300,000 years old. This was the period during which quarks and other fundamental particles and forces were formed. THE STANDARD MODEL of particle physics describes these particles and their relationships, and its predictions can be tested in particle accelerators back to energies that existed at about $10^{-10}$ second after the Big Bang. All experimental data collected so far are consistent with the Standard Model, and the

TABLE ONE

ERAS OF OUR UNIVERSE

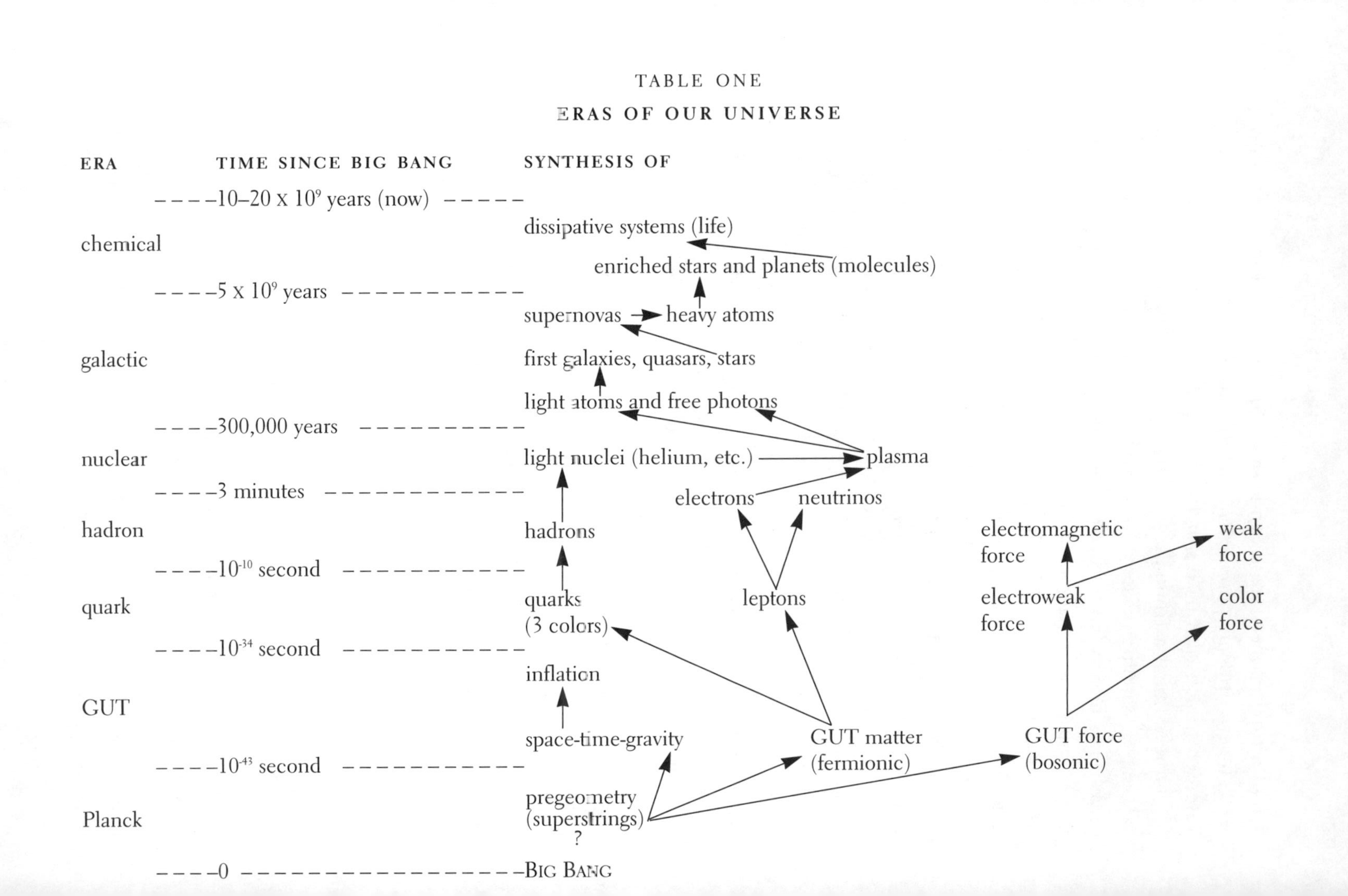

model's predictions about the chemical makeup of the early universe are consistent with astronomical observations. This gives cosmologists a positive cross-check between the Standard Model and experimental verification, back to only three minutes after the Big Bang. Nevertheless, this model leaves much still unexplained, and most cosmologists continue to look for a more unified, fundamental theory.

Most ideas about the nature of the universe during the first two eras of its evolution are speculative. Several versions of GRAND UNIFIED THEORIES (GUTs) exist. These allow for the possibility of inflation, a huge and sudden expansion in the very early universe that would resolve several problems in the Big Bang model. (See INFLATION THEORY.) But we have no possibility of testing these theories experimentally—to do so would require a particle accelerator bigger than the solar system! For the very first era of the universe, THE PLANCK ERA, we lack a fully developed theory, although various concepts raise fascinating philosophical questions. This one era may, for all we know, contain many further sublayers of evolution.

## Particle Physics

Cosmology and particle physics go hand in hand in our study of the universe. Physics looks at the behavior of matter in space-time when acted on by forces. The fundamental nature of each of these three dimensions of existence has been analyzed in stages. By the end of the nineteenth century, gravitational and electromagnetic forces were reasonably well understood. Matter was thought to be composed of about ninety kinds of atoms, held together to form molecules. Since then, particle physics has recognized further layers of fundamental structure found at increasingly high energies. (See Table Two.)

From our present state of knowledge, the standard model of particle physics, there are only two basic kinds of (fermionic) matter particles, quarks and leptons, and their associated ANTIMATTER particles. Aside from gravitation, which is negligible at the small scale of particle physics, there are two basic kinds of (bosonic) force, the color force (which produces the strong nuclear force), affecting only the quarks, and the electroweak force, which affects both quarks and leptons. (See BOSONS; FERMIONS.)

TABLE TWO

**THE ANALYSIS OF MATTER**

| ENTITIES | CONSIST OF | SHARED, EXCHANGED PARTICLES | FORCE |
|---|---|---|---|
| molecules | atoms | electrons | chemical force |
| atoms | nucleus and electrons | photons | electric forces |
| nuclei | hadrons (neutrons and protons) | pi-mesons | strong nuclear force |
| hadrons (protons, neutrons, mesons) | quarks | gluons | color force |

BOX TWO

**THE STANDARD MODEL**

matter (fermions)
- quarks (3 colors × 6 flavors)
- 6 leptons
  - 3 electrons
  - 3 neutrinos

forces (bosons)
- color force (8 gluons)
- electroweak force
  - electromagnetic force (photons)
  - weak force ($W^{\pm}$, $Z^0$ bosons)

The matter/force classification depicted in Box Two corresponds to the quark era and the hadron era in the evolution of the universe. It presents interesting symmetries between the fermionic and bosonic branches. (See SUPERSYMMETRY.) QUANTUM CHROMODYNAMICS describes the effect of the color force on quarks; electroweak theory describes all actions of THE ELECTROWEAK FORCE on fermions. The two theories together constitute the Standard Model, which can explain all experimental results so far collected in particle physics. To describe higher-energy phenomena, which we cannot experiment on, might require a Grand Unified Theory or a Theory of Everything.

Although the Standard Model of particle physics is complete for the energies it covers, we need to place it in the context of a wider framework to understand it. At the kind of energy level about which we are speaking, SPECIAL RELATIVITY has to be applied. Particles can be created out of, and dissolve back into, energy. Elementary quantum theory cannot incorporate such events, and it has been further extended with the development of quantum field theory, in which "particles" are not viewed as permanent objects, but rather as excited states of an underlying QUANTUM VACUUM, like vibrations on a string.

## Symmetry

In high-energy physics, concepts of symmetry are important. Space-time symmetries are abundant in the everyday world: the circular disk of the sun, the right and left sides of an animal body, a wallpaper design, or a repeated musical rhythm. More abstractly, the talk of justice as "even-handed," or the law of gravitation as the same everywhere, refers to a kind of symmetry. Symmetry has become a fundamental way of expressing the laws of physics.

Something has symmetry if the same theme is repeated two or more times. The degree and kind of symmetry are described by the group of ways in which examples of the theme play musical chairs with one another—the so-called symmetry group. Mirror symmetry exhibits the theme just twice, as object and its reflection, like the right and left halves of a person. Its (discrete) symmetry group has just two elements. Positive and negative electric charges have mirror symmetry. But all the rotations of a circle leave it looking just as it did. This is the infinite (continuous) symmetry group called U(1) by mathematicians. It is also important in electromagnetic theory.

If an object has symmetry, it has fewer features to describe, and it contains less INFORMATION. We need to describe only the theme and the number of ways in which it is repeated, like one house in a row of identical houses. Scientific laws can be put in these terms: Something is the same at all places and times. Conversely, if systems undergo phase transitions to a more differentiated state, they have more features to describe and now contain more information. Water looks the same in all directions (is isotropic), but when it changes and becomes a snowflake, it becomes more complex. A snowflake crystal looks the same in only six directions. This is an example of SYMMETRY BREAKING, an important process in both biological evolution and the evolution of the universe through its different eras. Wherever structure becomes more complex, symmetry—or at least the original symmetry—is lost. (See CHAOS AND SELF-ORGANIZATION.)

There is an important mathematical theorem, Noether's Theorem, stating that whenever physical laws have CONTINUOUS SYMMETRIES, some related physical quantity is conserved. It can be shown that, since the laws of physics are the same everywhere, in America and China, say, momentum must be conserved in physical processes. This approach to dynamics via symmetry is as powerful as older approaches, and perhaps easier to grasp intuitively. It applies to abstract (nonspatial) symmetries, in particular to the GAUGE FIELDS now used to represent all four forces of nature.

Philosophically, it is interesting that there seems to be a tension between symmetry and the conservation of physical properties like energy and momentum on one hand, and more dynamic evolution, self-organization, and symmetry breaking on the other. Nature can either preserve itself or take a great leap forward, which raises several interesting, but so far unanswered, questions: Is conservation or evolution more fundamental? Are there ultimate fixed laws that assure a balance of the two? Are there evolutionary tendencies at every level of the universe's existence?

Discrete symmetries are also important in physical laws. Three mirror symmetries are charge (positive versus negative), parity (right- versus left-handed), and time (forward versus backward). (See CPT SYMMETRY.) A supersymmetry between fermions (matter) and bosons (forces) has been proposed, but at this stage remains speculative. Finally, the LEPTONS and the QUARKS of the Standard Model and the many HADRONS fall into regular patterns, just as chemical elements fit into the periodic table a century before.

Cosmology and its companions, particle physics and quantum field theory, are among the most exciting areas of the new science in terms of the philosophical questions they raise. Perhaps no other branch of science goes so directly to the heart of the origin, nature, and context of human existence, or so urgently raises the wish that it might answer some of the basic "whys" of our existence. We still do not know whether our universe is finite or infinite, or how it came into existence. Nor do we know whether any other place in the universe harbors life or intelligence, or what role, if any, consciousness plays in the unfolding physical universe. It is no longer clear that the scientific study of cosmology will *ever* answer these questions. But cosmology may at least help to suggest a direction or model for attempting answers to these questions. (See THE ANTHROPIC PRINCIPLE; INTELLIGENCE IN THE UNIVERSE; THEORIES OF EVERYTHING.) One central motive for studying astronomy, by amateurs and professionals alike, is to seek some understanding of the world and our place within it.

## Units and Scales in Cosmology and Particle Physics

The purpose of this list is to give the less experienced reader more feel for the quantities involved.

**A.** Very large and very small numbers are conveniently written in exponential notation.

$10^n$ = 1 followed by $n$ zeros

e.g., $10^6$ = 1,000,000 = one million

$10^{-n} = 1/10^n$

e.g., $10^{-3}$ = one thousandth

**B.** The basic *units* used in science are:

Length: centimeters (cm) 1 centimeter = 2.54 inches

Mass: grams (g) 454 g = 1 pound

Time: seconds (sec) $3 \times 10^7$ sec = 1 year (approximately)

Temperature: degrees Kelvin (K; degrees absolute)
absolute zero = −273°C; $x$ K = $x$°C + 273

C. APPROXIMATE NUMBERS

Atoms in a glass of water: $10^{25}$

People on earth: $6 \times 10^9$

Neurons in the human brain: $10^{11}$

Stars in our galaxy: $10^{11}$

Galaxies in the observable universe: $10^{10}$

(N.B.: The whole universe may be much larger than this.)

Particles in the observable universe: $10^{80}$

D. APPROXIMATE MASSES

Proton or hydrogen atom: $10^{-23}$ g

Adult human being: $5 \times 10^4$ g = 120 pounds (approximately)

Sun ($M_\odot$): $10^{33}$ g

(N.B.: A human being is as many times heavier than a proton as the sun is heavier than a human being.)

E. APPROXIMATE LENGTHS

Planck length: $10^{-33}$ cm

Range of the weak nuclear force: $10^{-16}$ cm

Range of the strong nuclear force: $10^{-13}$ cm

Diameter of a typical atomic nucleus: $10^{-13}$ cm

Atomic diameter: $10^{-8}$ cm

Wavelength of visible light: $10^{-4}$ cm

Adult height: $1.7 \times 10^2$ cm

Sun's radius: $10^{11}$ cm

One light-year: $10^{18}$ cm

Our galaxy's radius: $10^{23}$ cm = $10^5$ light-years

Observable universe's radius: $10^{28}$ cm = $10^{10}$ light-years

**F.** APPROXIMATE TIMES

Planck time: $10^{-44}$ sec

Lifetime of a pi-zero meson: $10^{-16}$ sec

Lifetime of a neutron: $10^{3}$ sec

Lifetime of a human being: $2 \times 10^{9}$ sec = 70 years (approximately)

Age of the human race: $6 \times 10^{13}$ sec = $2 \times 10^{6}$ years (approximately)

(N.B.: Ten seconds is, in terms of ratio, about halfway between the lifetime of a pi-zero meson and the age of the universe.)

Age of the earth: $10^{17}$ sec = $4 \times 10^{9}$ years (approximately)

Age of the universe: $5 \times 10^{17}$ sec = $15 \times 10^{9}$ years (approximately)

# Absolute Zero

Is there a lowest possible temperature in the universe, a condition in which all molecular motion gives way to absolute stillness?

Heat a balloon, and it begins to expand. Place it in a refrigerator, and it sags. When the temperature of a gas is increased, its molecules move faster and faster. This increases the volume of the gas or, if it is confined within a fixed volume, the pressure it exerts. Increasing molecular motion also means an increase in ENTROPY. Is there a lowest possible temperature for a gas, in which molecules cease to move and entropy falls to zero?

A real gas liquefies or solidifies before this temperature is ever reached. Scientists therefore picture an "ideal gas" whose molecules have no internal structure or mutual attraction, a gas that will never liquefy. Theory dictates that the volume vanishes to zero at −273°C. Absolute zero is therefore considered the lowest possible temperature in the universe. At absolute zero all molecular motion ceases, and the entropy of an ideal gas falls to zero. (See STATISTICAL MECHANICS.)

Other thermodynamic processes also measure temperature. (See THERMODYNAMICS.) Heat a piece of iron, and it begins to glow dull red. Increase the temperature, and it turns orange. The color of the light it emits is an indication of its temperature—the higher the temperature, the shorter the wavelength of light, and the higher the frequency of the light. As the iron cools, its color fades until it is radiating infrared light of a lower frequency than visible light. Below infrared are other forms of electromagnetic radiation, each associated with a particular temperature. Microwave radiation has a wavelength measured in millimeters; radio waves can be many meters long and correspond to a still lower radiation temperature. What happens when the wavelength of light becomes infinite and its frequency and energy fall to zero? The answer again is absolute zero.

A variety of other temperature-indicating phenomena all agree that absolute zero is the lowest possible temperature. This suggests a new scale, the thermodynamic scale, in which temperature is registered in degrees Kelvin (K). Absolute zero is the starting point, 0 K corresponding to −273°C. Liquid helium boils at 4 K, ice melts at 273 K, and water boils at 373 K. Since increasing temperature implies an increase in entropy, the Kelvin or thermodynamic temperature scale can also be thought of as measuring entropy.

In the farthest reaches of deep space, beyond any stars, the temperature should be close to absolute zero. In fact, it is higher: 3 K. Radiation left over from the creation of the universe is still present. Stretched out by the space of an expanding universe, it has fallen to a temperature of 3 K.

Absolute zero is not quite attainable, although modern technology can get within one millionth of a degree. (One can never get absolutely clean in a bath for similar reasons.) This is the so-called Third Law of Thermodynamics.

# Actuality and Potentiality in Quantum Mechanics

In the quantum realm, things are always more than can be seen or measured. What can be seen and measured is actuality, the *what is* of a thing or situation. But any quantum entity or system has more—its potentiality. This is the reality waiting to be born as the system evolves, the *what might be* of the system, or its range of possible futures. In quantum physics, what might be has ontological status, a kind of reality of its own. A system's potentiality is one of its properties—the energy, creativity, and flexibility sleeping within it. When physicists make calculations on quantum entities, they take into account both actuality and potentiality.

We can say of an ordinary window that it is of a certain size and shape, made of glass or plastic, and so many millimeters thick. But the window also has the potentiality to let light through, or to break if struck by a stone. Similarly, many of the terms we use to describe people, such as name, sex, age, and eye color, are their actualities. Other terms for describing humans describe their potentialities: They are artistic, visionary, quick-witted, or likely to lose their tempers. What interests us here is how these two levels of description, actuality and potentiality, are related in physics.

In classical Newtonian physics, a thing's potentiality has no real existence. All potentialities are reducible to underlying actualities. The transparency or fragility of window glass is due to its underlying molecular structure. The abilities and personalities of people are reducible to their brain structures and earlier experiences. The properties of any larger whole are reducible to the properties possessed by the parts. The whole Newtonian approach is a hard-nosed "nuts-and-bolts" one, where facts—seeable, measurable, and quantifiable—reign supreme.

In ancient times, Aristotle, a biologist, wondered how simple things grew into vastly more complex structures. How, for instance, did an acorn become an oak tree? Aristotle believed that the acorn possessed an essence, a kind of inner blueprint of its potentialities, that guided its actual material growth. Today we think we know better. Molecular biologists tell us that the acorn's potentialities are carried by an underlying material structure, its DNA code. There is no need in either physics or biology for a concept of potentiality. Yet the new science of the twentieth century makes us question whether the reductionist view is the last word. (See REDUCTIONISM.)

One of the major interpretations of quantum physics, the propensity interpretation, suggests that quantum reality consists of two *kinds* of reality, actual and potential. The actual is what we get when we see or measure a quantum entity; the potential is the state in which the entity exists before it is measured. The Schrödinger wave function describes an infinite spread of things we *might* see if we measure the entity at any given time, or in any particular context. (See THE WAVE

FUNCTION AND SCHRÖDINGER'S EQUATION.) Though they are "only" possibilities, these states have an effect on each other and on the real world. They can evolve and interfere with each other; their interactions can give rise to actualities; they can initiate real processes. (See QUANTUM TUNNELING; SUPERPOSITIONS; VIRTUAL TRANSITIONS.)

Until an act of measurement (see THE MEASUREMENT PROBLEM) causes COLLAPSE OF THE WAVE FUNCTION and converts all these possibilities into a single actuality, like Schrödinger's cat becoming alive *or* dead, the possibilities extend through time and space. Experiments on NONLOCALITY show that potentialities can travel faster than the speed of light, giving rise to instantaneous correlations. Special Relativity demonstrates that actualities cannot do so, but potentialities, being a different kind of reality, can.

If we can grasp the fact that potentiality is a second domain of existence, and thus that possibilities are to some extent real entities, we can begin to understand the nature of THE QUANTUM VACUUM and its relationship to daily existence. The vacuum, described in QUANTUM FIELD THEORY, is the underlying, lowest-energy state of all, the source of everything that is. The vacuum does not, however, "ex-ist" in the strict Latin meaning of the word, which has the connotation "to stand out." We cannot see, touch, or measure the vacuum. It is a sea of pure potentiality, a kind of preexistence whose excitation gives rise to existence. Thus potentiality is the source of existence, while existence itself is a plethora of actualities or "manifestations." This kind of thinking is familiar to mystics, particularly Eastern ones, but it is alien to mainstream Western thought and illustrates one of the crucial ways in which quantum physics heralds a new paradigm.

Whenever a new paradigm emerges, those who are committed to the old paradigm fight back. Thus there have been two much more conservative interpretations of quantum mechanics, which try to explain away or avoid the ontological status of potentiality. The first, articulated by David Bohm and known as the hidden variables theory, holds that quantum entities are really just actualities underneath, but our techniques for looking at or measuring them are too crude, and thus our observations appear fuzzy or indeterminate. This view is philosophically appealing to those who like to believe that the world consists of actualities at all levels, but there is no evidence to support it, and its implications lead to much wilder implausibilities when applied to other problems in physics.

The second conservative interpretation is Niels Bohr's COMPLEMENTARITY, which holds that all we can say about quantum systems is that they give rise to certain measured results if we do certain experiments on them. Quantum systems themselves are regarded *only* as potentialities, and measured results, which are classical realities, only as actualities. Bohr's interpretation thus divides reality into a quantum realm and a classical realm and is not concerned with any subtle interplay between the two.

The most interesting view of how actuality and potentiality relate in mainstream quantum theory is HEISENBERG'S UNCERTAINTY PRINCIPLE, which shows that not every potentiality can be made actual at any given moment, or in any given context. Reality is a kind of evanescent dance between potentiality and actuality, facts and possibilities as yet unborn. If we actualize a particle's position, its momentum becomes fuzzy, and vice versa. Here there is scope for both observation (what is) and creative vision (what might be).

## The Anthropic Principle

The Anthropic Principle is a set of ideas that concerns how conscious creatures such as ourselves came to be in the universe. Do we owe our existence to a long chain of "cosmic coincidences," or was our eventual existence somehow "necessary"? Why are such physical constants as the four fundamental forces, the average density of matter in the universe, and the masses of the fundamental particles so delicately poised that it is *just possible* for us to exist?

There are two main versions of the Anthropic Principle, the weak and the strong. Both hold that the universe in which we live seems to favor life and consciousness. The Weak Anthropic Principle notes this as a *fact*. Out of all the possible values the physical constants of the universe might have had, they seem, improbably, to have arrived at the narrow range of constants that favor the existence of carbon-based life forms. Some critics accuse the Weak Anthropic Principle of being a tautology: Because we are here, it *must* be possible that we can be here.

The Strong Anthropic Principle goes further. It claims necessity for the presence of people like us: The universe *must* have those properties that allow life to develop at some stage in its history—and this requires explanation.

Those who follow the Strong Anthropic Principle usually give one of three explanations for why we exist. First, there is a God who *wanted* life and consciousness built into the universe. Second is the "many-worlds" hypothesis. (See THE MEASUREMENT PROBLEM.) Quantum many-worlds theory argues that *every* possible sort of universe exists, so in at least one of them we should find life and consciousness. Third, the evolutionary account of life argues that the universe is evolving in such a way that life and consciousness would be favored at some point.

In line with this third explanation, American astrophysicist Lee Smolin has proposed a theory that our own universe and others have been born out of BLACK HOLES, and have slightly altered physical constants. Black holes are necessary to make new universes, stars are necessary to make black holes (which are collapsed stars), and carbon atoms are necessary to make stars. Hence, the eventual preponderance of universes containing carbon atoms and stars. (See THE PLANCK ERA.)

But even Smolin's work, if verified, doesn't support the Strong Anthropic Principle; there is no necessary connection between carbon atoms and life and consciousness. Even given stars and planets and the right chemical elements, the formation of life still may seem against the odds. There is as yet no firm *scientific* explanation for why we are here.

# Antimatter

Are there worlds of antimatter, planets on which we can never set foot? The world of elementary particles has a myriad of mirror reflections or symmetries. (See CPT SYMMETRY.) One of these is symmetry between a particle and its antiparticle. Corresponding to the negatively charged electron stands its antiparticle, the positron. A negatively charged antiproton mirrors the proton. There are even antineutrons and antineutrinos.

If every particle has an antiparticle, do science fiction worlds of antimatter exist, antiplanets with antirocks and antirivers and even antilife? Because an antiproton and a proton are annihilated in a starburst of energy when brought together, it would be impossible for antirocks to exist on our planet, or even for antiplanets to exist for long within our solar system. Some astronomers have speculated that distant galaxies of antimatter could in fact exist. The problem is that we see pairs of colliding galaxies, but never the spectacular fireworks that would result from mutual annihilation.

It is difficult to create stable antimatter in the laboratory, since it must be constantly shielded from surrounding matter. Attempts have been made to bind antiprotons, antineutrons, and positrons together to form antiatoms. It may even be possible to create simple antimolecules. Since mass is converted into pure energy when matter and antimatter meet, it has been proposed that antimatter could fuel rocket ships that travel between the stars. Contained within magnetic bottles, antimatter fuel would release great amounts of energy when brought into contact with normal matter. But this is well beyond our present technology.

# The Arrow of Time

Why do clocks run in the same direction, from the present into the future? Why are our memories of the past, never of the future? Why, throughout the universe, do the various processes that measure the passage of time all point along the same arrow?

Take a video of cooking an omelet, a growing tree, or making coffee. If the video is played backward, it defies common sense. Events move in one direction only. Cracked eggs do not repair themselves; trees do not return to seedlings; poured milk does not leap back into the jug. Time has an arrow.

The mystery of time's arrow deepens when we realize that nearly all the fundamental laws of mechanics are time-reversible. Turn a video camera on two billiard balls as they collide on an idealized frictionless table. Now play the video in reverse. The collisions still make sense. In an ideal, frictionless world, any motion permitted by the laws of mechanics can go in either direction—from past to future, or future to past. Why then does time end up flowing in an unambiguous direction? This is one of the great mysteries of science, a mystery that, even after the revolutions of relativity and quantum theory, remains unsolved.

The universe displays a number of different arrows, all of which agree in their direction. Does one of these arrows act to drive the others? Or is time's arrow the cooperative result of the intertwining of many different processes? Let us look at some of these temporal arrows.

- **Cosmological time:** The universe is expanding. If we had a series of photographs of the universe over the past thousands of millions of years, we would see that the bigger the universe, the later the time. Most physicists believe that the entire cosmos began in THE BIG BANG. Did time exist before this primordial instant, or was time, and its arrow, created along with the universe itself? If it was, what happens to time's arrow if the universe eventually starts contracting again toward a Big Crunch?

- **Entropy:** A vast number of processes in the everyday world attest to time's arrow. We age; our cars rust; the house needs repainting; a vase shatters; hot coffee cools; the battery in a flashlight runs down; a clock needs rewinding. In these and many other ways, the passage of time from the present into the future can be observed and measured.

  Each of the above cases involves an increase in ENTROPY. Left to themselves, systems increase their entropy, moving toward greater disorder and less coherence. Another name for "the arrow of time" is "the increase of entropy."

  Is entropy the answer to the arrow of time? The problem is that when systems are broken down into their component parts, these parts are describable in terms of time-symmetric laws. Where does the inexorable increase in entropy come from? Does time drive entropy, or entropy drive time?

- **Light:** The fundamental laws of light are time-reversible. Switch on a lamp, and its light expands to fill the room. But Maxwell's equations, the laws that govern the movement of light, also permit the time-reversed situation in which light collapses inward from the night into the room, and arrives at the lamp at the instant its switch is pressed. Why do we never see this time-reversed situation? What principle of nature operates to select only one of these cases from its temporal mirror image?

- **Quantum collapse:** Unlike everyday objects, which are always well defined, quantum systems are inherently ambiguous, a combination of possibilities or potentialities. When a quantum measurement is made, the potentialities collapse into a single outcome. The direction of the process is always the same—as is, for example, the disintegration of a radioactive isotope. Time appears to have a direction at the quantum level. How does this connect to the cosmological arrow and the march of entropy? Do quantum processes drive the entire universe?

- **Psychological time:** Our memories are always of the past, never of the future, and our internal, psychological experience of time agrees with the temporal arrows exhibited by clocks and other material processes. Is this because the brain processes associated with consciousness have a time asymmetry?

- **Symmetry breaking:** Could there be no deep underlying explanation for time's arrow? Is the whole thing simply a matter of chance? The fundamental laws of nature have a very high degree of symmetry, which the material universe does not exhibit. Antimatter, for example, is permitted by quantum theory equally with ordinary matter, yet the universe is predominantly built out of matter. In many cases, there are two possible solutions of the laws of nature, each the mirror image of the other.

  Physicists believe that the particular solution chosen by nature is a matter of chance, but once the universe has made this particular "symmetry-breaking" choice, events lock together in a consistent way forever. Time's arrow may have similar origins; an initial chance break in time symmetry became fixed as the many temporal processes in the universe coupled together to move in the same direction. But if different regions of the universe broke this symmetry independently, there would be mismatches where regions having opposite senses of time met. This is hard to imagine without paradox.

- **Particle physics:** There is one rare process (the decay of $K^0$ mesons) that is not symmetric in time. (See CPT SYMMETRY.)

So far, there is no consensus among scientists about which of these aspects of time's arrow is fundamental.

# Artificial Intelligence

Artificial intelligence (AI) is the study of how to build computing machines that can behave intelligently. Basic AI philosophy rests on the assumption that the human mind works like a computer. Therefore, it argues, computers that will work as effectively as minds can be developed. These two deep assumptions underwrite a whole paradigm of human psychology as the "mind machine," generating such popular metaphors as "We switch on or off," "We blow our fuses," and "We are programmed for success or failure." The AI paradigm describes as valid or "scientific" only those aspects of human thinking that can be simulated by computers. Some critics argue that this model is too limited.

The idea of devising "mechanical men," such as chess-playing automata, is at least three hundred years old, but the practical means to bring it about had to wait until the development of high-speed electronic computers in the 1940s. Today AI has extensive scientific and commercial application. Scientific computers, word processors, and business machines now perform many tasks formerly requiring armies of clerks. How far future machines can perform more "intelligent" tasks, such as mechanical translation from one language to another, remains to be seen.

AI engineers employ two different kinds of data processing in machines. SERIAL PROCESSING, modeled on the brain's one-to-one neural tract connections, seems to perform all clearly defined logical and mathematical procedures, including rule-following ("algorithmic") activities such as mental arithmetic. (See THE CHURCH-TURING THESIS; TURING MACHINES.) Most ordinary personal computers are serial, and their step-by-step computing capacities are limited only by their "logical space" (the size of their memory stores) and their speed.

A second family of AI machines, known as parallel processors, NEURAL NETWORKS, or connectionist models (see CONNECTIONISM), are based on the brain's own neural networks, in which thousands of neurons are complexly interconnected. Parallel processors are quicker and more efficient at such tasks as voice and face recognition (pattern recognition) and are capable of simple "learning," but they are less useful than serial processors for rule-following tasks. It is possible to *simulate* parallel processing on a serial processor.

Some AI successes are the construction of EXPERT SYSTEMS (for medical diagnosis or legal precedents) and the modeling of vision, formal language, and memory. Some failures are in the modeling of informal but efficient problem solving, and of informal language use (employing context, metaphors, allusions, and so on).

Language translation machines have been largely unsuccessful, and it has been impossible to model mental domains involving visual imagery or the formation of new concepts on either serial or parallel processors. (See THINKING.)

Most typically, an AI expert concentrates on either serial or parallel processing and studies only one kind of problem-solving behavior. Unlike COMPUTATIONAL PSYCHOLOGY, a study of how human beings carry out information processing, AI is a purely mathematical, engineering discipline, concerned only with the behavior of its models. Like BEHAVIORISM in psychology, it has no model of experience. The most common philosophy among AI workers is FUNCTIONALISM, the belief that when we have described how something *behaves*, we have said everything important about it. Human beings, by contrast, do not regard their own experience or intentions as unimportant or irrelevant to their behavior. Thus there is a logical gap between AI functionalism and the awareness human beings have of their own behavior. (See THE MIND-BODY PROBLEM.)

# Artificial Life

Life as we know it is carbon-based, but could there be living forms based in the world of silicon chips? John von Neumann's machines, Stanislaw Ulam's cellular automata, and John Conway's GAME OF LIFE all have the ability to reproduce themselves. Computer viruses can spread through an entire network, attacking computer memories and using silicon chips for their own ends. Is there a sense in which all of these could be called alive?

Artificial life has become more feasible since science entered the realm of molecular engineering. Machines so small that they are visible only under the electron microscope can be built. Minute pumps and engines placed in blood vessels can drive through the human body to carry out repairs and do other medical interventions. One day such machines may be given the power of self-reproduction.

Robots that will function for years without any human intervention, repairing and even reproducing themselves, may be built under the sea, in outer space, or on the surfaces of other planets. International computer networks may develop to such a degree of complexity that they will evolve their own goals and personalities. Will machines, like animals today, have legal rights of their own? Is consciousness necessarily present in such systems? (See CONSCIOUSNESS, TOWARD A SCIENCE OF.)

Life involves the ability not only to reproduce but to modify its environment and enter into competition, or cooperation, with other organisms. What role will artificial life take? Will artificial systems enter into a mutually beneficial symbiosis with human society? Will they be our slaves? Or will they start to replace human beings on earth, or redefine our human situation?

The goals that ethics and artificial life present raise age-old questions about the meaning of life. They may sound like the speculations of science fiction, but they are already being taken seriously by some scientists and philosophers.

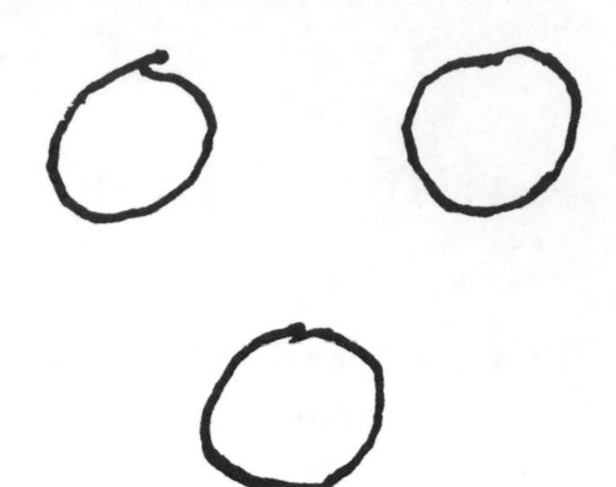

# Atomism

Atomism has left its indelible mark on the whole of Western culture. Basically, it is the view that the universe consists of tiny, separate, indivisible, and indestructible particles, each an individual player in the dynamic and sometimes violent drama of the material world. Atomism holds that all of nature's multiplicity can be reduced to these few simple units.

According to Democritus, in the fifth century B.C., "The atoms struggle and are carried about in a void because of their dissimilarities . . . and as they are carried about, they collide and are bound together." Democritus' atoms all shared the same essential qualities, but they differed from one another in shape and size, and these differences, together with the many ways in which different atoms could combine, accounted for the observed differences of objects in the material world. Changes in objects, like a block of ice melting into a puddle of water, were thought to be due to changes in the way the atoms were joined with others, while the individual atoms themselves never underwent change or inner transformation.

Another early Greek philosopher, Empedocles, believed that everything in the world was made up of four elements—earth, air, fire, and water—a view compatible with atomism, although here there were four essentially different kinds of atoms. Empedocles' atoms were held together by love and driven apart by strife. Later, the Roman philosopher

Lucretius suggested that there were also atoms of the soul and atoms of desire.

Early Greek views on atomism were philosophical ideas that originated from the attempt to explain how substances could change, and how there could be so many different kinds of substance in a world that the Greeks thought of as essentially one. This atomism was not supported by observation or wider physical theory. Scientific atomism came into vogue with the mechanistic scientists of the seventeenth century because it was easy to combine with DETERMINISM and REDUCTIONISM and compatible with Newton's mechanics of particles. Newtonian particles, like the early atoms, were separate, indestructible, and unchanging, driven by forces to collide with one another. They flew off in different directions or stuck together, depending upon whether these forces were attractive or repulsive. Newton's basic image for them was closely akin to a billiard ball. Everything in the material universe, he thought, could be reduced to a collection of these particles and the forces acting between them. Nineteenth-century STATISTICAL MECHANICS was in this spirit.

In the nineteenth century, physics and chemistry became sufficiently developed to test the atomic idea properly. By then it was known that chemical compounds like salt, sugar, and water could be broken down into a few basic elements. In his atomic theory of 1808, English physicist John Dalton proposed that all these compounds were made of combinations of ninety or so kinds of atoms, which then got into combinations called molecules to make the more complex compounds, but the original atoms never changed their internal nature. That was why a given compound, unlike a mixture, always contained the same fixed properties of its constituent elements. It was easy to test this hypothesis by experiment. In one such, a drop of oil was placed on water. The oil spread out, but only so far, at which point it was assumed that it formed a layer only one molecule thick. From this and other experiments, physicists could estimate that all atoms have a diameter of about one hundred-millionth of a centimeter.

Bolstered by scientific theory and experiment, atomism became a wider paradigm for modern political theory, psychology, and philosophy. The seventeenth-century political philosopher John Locke made atoms his model for individuals as the basic units of society, and all liberal political theory ever since has held that "society"—that is, the group—can always be reduced to a collection of these separate individuals. Adam Smith adopted the same model in his economic theory, reducing all large-scale economic activity to the economic behavior of individuals. For liberal political and economic theorists, individuals, like atoms in a chemical compound, remain unchanged by any social combinations or relationships they may enter. The individual is "prior" to all his or her associations.

In psychology, Sigmund Freud, too, turned to the atomic model when writing his theory of "object relations." For him, the individual is primary and never changes internally through relationships. Individuals never get "inside" each other. They relate externally as objects; like so many Newtonian billiard balls, they bump into and clash with one another and go their separate ways. The Western myth of the isolated individual who pursues his or her lonely path through life comes from the same source. The early experimental psychologists used atomism as their model for analyzing experience into isolated units of sensation, which then combined.

In philosophy, scientific atomism inspired the "logical atomism" of philosophers like Bertrand Russell and Ludwig Wittgenstein, who argued that truth is composed of "atomic facts" like "That is green" or "He is standing," which then get combined into more complex statements. Early efforts to translate one language into another by computers tried in the same way to treat individual words as the atoms of which sentences are composed. But these broke down when people realized that the possible combinations are infinite, and that the meaning of words often depends on context.

Even in the time of Plato and Aristotle, atomism seemed inadequate to explain some things. Aristotle in particular felt that it could not account for the change and variety in nature, particularly that associated with biological order; nor did it seem to provide an adequate explanation for human purpose. Toward the end of the nineteenth century and into the twentieth, science itself raised problems for the simple atomic hypothesis.

Although atoms behave like indivisible units as far as chemical reactions are concerned, we know now that they are themselves made up of smaller units. Electricity, for example, is a flow of electrons, which are one important component of atoms. Atoms give off particles and break down into simpler atoms through the process of radioactive decay. (See overview essay D, THE COSMIC CANOPY.) SPECIAL RELATIVITY shows that matter and energy are interchangeable, and one of the central pillars of QUANTUM THEORY is the finding that entities that behave like particles in some circumstances behave like waves in others (WAVE/PARTICLE DUALITY). The whole atomistic paradigm has been called into question by the emphasis on HOLISM in twentieth-century science.

# Attention

Our ability to pay attention, to focus our minds on a particular thought, or to concentrate on some part of a visual scene is one of the outstanding features of consciousness and an important survival technique. It is also one of the great mysteries. Human infants are born with the ability, and it is an important tool with which they learn about the world. Stress, sleepiness, or excitement can diminish our capacity for attention, as can brain damage. We know something about the mechanics the brain uses when paying attention, but we know very little about the process itself.

"Who" in the brain pays attention? How do we decide what is worth attending to? Why is it sometimes important to let our attention "wander"? Have the underlying mechanics of attention and its impact on consciousness something to do with our need for sleep? None of these questions has any definite answer within known mind science, and most of them defeat ARTIFICIAL INTELLIGENCE models that portray all-important brain activity as some form of computation.

At any moment, brain activity may be fully conscious, dimly conscious, or unconscious. The brain processes much of its incoming data in parallel (see NEURAL NETWORKS) and selects some for more detailed study, often via SERIAL PROCESSING. A person's attention may be caught by a strong stimulus (a loud sound), a novel or unexpected stimulus (a sudden movement), or a stimulus regarded as possibly important (a whisper in the ear). Sometimes we concentrate intentionally. But intentionality is another of the mysteries of consciousness.

Attention's information capacity is quite limited, making it difficult for us to attend to more than one thing at a time. This is borne out by tests in which a research subject wears a set of earphones, through which a different message is directed into each ear. The subject attends to one message and recalls the other as only a vague background noise. Despite this, the message that is not attended to has some subliminal effect on the way subsequent experimental tasks, such as word association, are performed. It is difficult to determine whether the more subliminal message was only dimly conscious in the first place and then quickly forgotten, or whether it was unconscious. (Similar problems of interpretation arise with people apparently affected by experience while under light anesthesia.)

PET (positron emission tomography) scans show an increase in blood flow to those parts of the cerebral cortex being used in the performance of a mental task. (See METHODS OF STUDYING THE BRAIN.) The greater detail brought to the fore seems to require a larger supply of energy. (The brain uses much more of the body's energy than its weight would suggest.) There are still many questions about how increased energy and blood flow are linked to attention. Parts of the cortex are involved in parallel processing, but overall control seems to reside in the thalamus, the more primitive forebrain. Nearly all sensory paths are relayed through the thalamus en route to the cortex. The thalamus is the highest brain sensory center in more primitive vertebrates, and even in human beings it is the background substrate for conscious sensations.

There is a two-way detailed map between the thalamus and the cortex. The thalamic reticular formation, a primitive neural network, is thought to turn the "volume" up or down on parts of the more specialized cortex. The reticular formation itself may be under voluntary control to a greater or lesser degree.

Such mechanics of attention, like those of MEMORY, can be studied in greater detail as the technology for brain scans develops. But this mechanistic approach, however detailed and sophisticated, seems to leave out the more holistic *experience* of paying attention, and of the self who attends. We are still very far from understanding the links among attention, memory, and consciousness.

## Attractors

Many nonlinear systems act as if compelled to repeat a certain type of behavior. These are systems in the grip of an attractor. Watch the way water spirals down the drain of your bathtub. Put your hand in the vortex, and it is temporarily disrupted, only to appear again. The motion of the water has created a so-called limit-cycle attractor that sustains the vortex.

The pendulum of a clock is also in the grip of a limit-cycle attractor. At each swing, the pendulum speeds up as it reaches its midpoint, slows down as it reaches the top of its swing. Over and over, the pendulum oscillates between maximum speed, zero displacement, and zero speed, maximum displacement. Blow on the pendulum and you disturb its swing. For a time it speeds up or slows down, but very soon it will have settled back into its regular steady beat. Like the vortex in the bathtub, the pendulum's motion is repetitive because of its limit-cycle attractor.

The nonlinearity of the pendulum that produces this limit-cycle attractor arises from the fact that, with each swing, it activates an escape mechanism, which then gives the pendulum a tiny push. (In many modern clocks, the pendulum is simply a piece of window dressing, since it is being driven by signals from a tiny quartz clock inside the mechanism!)

Yet another example of a limit-cycle attractor is the idealized PREDATOR-PREY system in which pike eat the trout in a lake. Season after season, the numbers of pike and trout oscillate in an exactly repeatable way. If there are too many pike, the trout population falls, and the pike lose their food supply. But as the pike begin to die out, the trout population flourishes, having no predators to bother it. In this way, pike and trout follow each other through an endlessly repeating limit cycle, unless some external factor intervenes.

Many systems exhibit limit-cycle behavior, including a variety of machines and electrical circuits. There are also more complicated systems, such as two pendulums coupled together, or a population of anglers added to the lake stocked with trout and pike. Now, instead of a single repeating behavior, the system shows an oscillation within an oscillation. Its attractor is no longer a circle but the surface of a doughnut, or torus. This may be a little difficult to imagine at first. The thing to remember is that the attractor is a shape that is not within our ordinary space but within what could be called "behavior space" or, more precisely, "phase space"—a graph of all the relevant variables of the system.

There are also point attractors that, in behavior space, are like valleys or pudding bowls. Switch off the escape mechanism in a clock, and the pendulum continues to beat out the seconds. But soon the effects of friction and air resistance begin to take over, and the pendulum slowly swings to rest—attracted to the fixed point of stillness.

In the heat of summer, a river flows slowly and smoothly. Each region of water is in the grip of the same point attractor. Throw two twigs into the water and see how they continue to sail along at the same speed and the same distance apart. The momentum of the water and the relative separation of neighboring regions are constant.

Then the rains come, and the speed of the river increases. The two twigs begin to drift apart as one of them is picked up by a faster current. Limit cycles are generated, and regions of water turn into vortices. Powered by the faster-flowing river, these vortices are quite stable and can persist over long periods. If the rainfall continues, the fast-flowing river will exhibit an even greater complexity of vortices within vortices. It is now in the grip of a torus or doughnut attractor.

At each stage of this process there is a bifurcation point, a region where behavior jumps from one type to another. A slow-flowing river is in the grip of a point attractor until, with sufficient rate of flow, a bifurcation point is reached where it jumps into limit-cycle behavior. The next bifurcation point involves a jump from a limit cycle to a torus attractor. In each case, it is the system itself that generates its own attractor. The limit cycle or point attractor is not something that exists externally to the system and influences its behavior; it is an expression of general nonlinear dynamics.

If a fast-flowing river moves from vortices to vortices within vortices, what is the next step? Vortices within vortices within vortices? Yes. But eventually these bifurcations lead to a fractal—a strange attractor. (See FRACTALS.) With an attractor of fractal shape and dimension, the system's behavior, the momentum and changing position of each element of water, is being attracted to a figure of infinite complexity. Every part of the water becomes engaged in a highly complex, chaotic motion. The result is turbulence. Generally the dynamics of a chaotic and random system are all in the grip of a strange attractor. Their motion has infinite complexity and is, in practice, unpredictable over a long period. It nevertheless remains deterministic. Like the weather or the stock market, tomorrow will probably be like today, but we cannot predict next year.

## Autopoietic Systems

Autopoietic systems are able to maintain their complex internal organization even under changing conditions. There is a wide variety of stable, far-from-equilibrium systems, like a vortex or the convection patterns in a pan of heated water, in which structures persist over time. Autopoietic systems go further: It is not so much a question of preserving a fixed internal structure as of the stability of a whole system of feedback loops, iterations, and other nonlinearities. Autopoietic systems are defined at a high level of organization in which the focus is not on material structure but on organization itself. (See SYSTEMS THEORY.)

Autopoietic systems are holistic; their existence is an expression of their overall meaning. Within the human body, the immune system is an example of an autopoietic system, a complex pattern-recognition system that maintains the body's integrity even when damaged or under attack. Entire ecologies, like cities and human societies, can be thought of as autopoietic systems.

# Becoming

The new sciences, according to the Nobel Prize–winning chemist Ilya Prigogine, are more concerned with Becoming than with Being. Focusing on the dynamic and creative aspects of time, physics, chemistry, and biology, they offer us a deeper understanding of spontaneous ordering, the emergence of novelty, and the creation of new structures.

Traditional science confined itself to systems in dynamic EQUILIBRIUM, or very close to equilibrium; to linear systems, closed systems, and systems involved in regular repetitive behavior. They are predictable, and nothing surprising will ever occur in them. Time within such systems is mechanical and does not lead to the emergence of the new. This was a science of regularly orbiting planets, swinging pendulums, oscillating circuits, smoothly running machines—in short, a science that focused on the state of Being. (See overview essay A, KINDS OF BEING.)

When systems are opened to the environment and forced far away from their comfortable state of equilibrium, self-ordering and the emergence of new structures take place. Systems in which positive FEEDBACK plays a role are capable of evolution and sudden change. In all such cases, as Prigogine points out, time takes an active, dynamic role, and Being is transformed into Becoming.

The sciences of Becoming are concerned with the development of chemical complexity, the appearance of life, the evolution of new species, the nature of consciousness, the discovery of language, the way learning takes place, the development of social organizations, the growth of cities, and the transformations of economic systems. Its theoretical tools are NONLINEARITY, bifurcation points, CHAOS AND SELF-ORGANIZATION, positive and negative feedback, and COMPLEXITY theory.

The philosophical roots of the notion of Becoming can be found in the "Logic" of G.W.F. Hegel. Many philosophical systems, such as Descartes's "I think, therefore I am," are based on the notion of Being as the most secure ground for any logical system. Hegel pointed out that when one attempts to think of pure Being, without any sense of definition or differentiation, the mind moves into the contemplation of Nonbeing, of that which is not, of that which lies outside distinction and definition. Yet the mind cannot rest in Nonbeing.

At first sight, Being and Nonbeing appear to be logical contradictions. On examination, they are closer to one of those paradoxes of visual perception—a duck that turns into a rabbit, or a vase that turns into a pair of faces. The mind is unable to rest in either of two positions (thesis and antithesis) but constantly moves between them. The resolution of the apparent dichotomy lies in the movement itself, where Being has given way to Becoming.

Hegelian dialectics is the elaboration of this constant movement of Becoming whereby a thought is investigated until it yields its contradiction and, in so doing, moves to a new and higher logical plane. Hegel believed that the world itself is a constant dialectical process, an evolution from Being to Becoming, and that the universe is the manifestation of the World Spirit thinking itself into existence.

The eclectic thinker Gregory Bateson suggests that Becoming is characteristic of the evolution of life and the development of consciousness. Within evolution, nature transcends logical levels and escapes the confines of limiting contexts. Bateson takes this as evidence of what he calls "Mind in Nature."

Computers, by contrast, live only in a world of Being and are easily trapped within the confines of fixed logical types. The Cretan declaims, "All Cretans are liars." If this statement is taken as true, then the Cretan is lying, and his statement is false. But if the statement is false, all Cretans cannot be liars, and his statement is true. But if it is true . . . Faced with this paradox, a computer performs an endless logical oscillation between the two possibilities. Human intelligence is, however, able to distinguish between declarative statements, statements that are about statements, and statements about statements that are themselves about statements. Because consciousness lives in a world of Becoming, it is free to escape from the confines of particular contexts and leapfrog over logical types. Becoming, for Prigogine and Bateson, is the essence of creativity, intelligence, and human consciousness.

# Behaviorism

The behaviorist movement was launched by J. B. Watson in 1913, and intentionally modeled closely on the physical sciences. Since the interpretation of subjective experience and the results of introspection are unclear and unrepeatable, Watson turned away from psychology as a study of the "inner life." He wanted to concentrate only on what a hard scientist could see and measure—the relation between a stimulus and the organism's response to it. The proper subject matter of psychology was therefore deemed to be observable behavior, the laws connecting stimuli to responses and conditioned responses. Both mind and brain were ignored. (See THE BLACK BOX.) After Darwin, Watson assumed that any differences between human beings and lower animals were differences of degree only. Since it was easier and cheaper to study something like the behavior of rats running through mazes, he decided to conduct experiments on these lower animals and apply the results to a theory of human learning.

The behaviorist view, which became very popular for a time, led to a much better understanding of simple learning processes. Pavlov

had earlier studied conditioned reflexes in dogs, where the simple repetition of linked stimuli forms an association. (When the ringing of a bell precedes the giving of food enough times, the dog learns to salivate at the sound of the bell, even if it is unaccompanied by food.) A second, somewhat similar process—operant conditioning, or trial-and-error learning—was also studied. Cats shut in boxes would do things at random until they discovered the mechanism that opened the box. Then they would gradually learn the rewarded action and escape more quickly each time.

There seems to be little doubt that such simple learning underlies many human habits, but Watson wanted to go further. He argued that simple conditioned training of infants and children was all-important. Through the right association of stimulus and response, anybody could be trained to do anything. But this enthusiastic dream (or nightmare?) has not been borne out in practice. The best brainwashing efforts of many dictatorships have failed to abolish dissent, and behavior therapy based on these forms of learning has not been entirely successful. Most psychologists now recognize that higher, more complex mental processes exist, too, and have an influence on behavior. (See GESTALT AND COGNITIVE PSYCHOLOGY.)

Behavior therapy assumes that psychological symptoms such as fear, phobias, and addictions can be seen as conditioned responses. Treatment consists of re- or deconditioning. This can take two forms—aversion therapy and desensitization. Aversion therapy is used in the treatment of alcoholism and sexual deviance. Alcoholics are given a drink while simultaneously receiving a drug that makes them sick. After some hours or days of this, they come to associate drinking with nausea and develop an aversion to alcohol. Unfortunately, the effect often wears off after a few months, and the method cannot be used directly on a condition like pedophilia (sexual attraction to children), where treatment has to be combined with the use of photographs that introduce the mental concepts the behaviorist is trying to avoid.

Desensitization therapy works well on something like an animal phobia. Some patients may be so terrified of birds that they fear to go out of doors. After relaxation exercises, they may be shown a single feather at the far end of the room. After learning to tolerate this for one or two more sessions, they progress to stuffed birds, then to caged birds, and ultimately to free birds in the park. The rationale behind this therapy is that the internal fear is itself a conditioned response to some unpleasant event involving birds, which has possibly been forgotten. But while this seems an adequate account of some phobias, Freud argued that other neurotic symptoms may have more deep-rooted, symbolic causes. Agoraphobia (the fear of open spaces), for instance, does not respond well to simple behavioral therapy.

Concentrating purely on behavior as a stimulus-response mechanism, behavioral psychologists have ignored consciousness. The more radical behaviorists deny that it exists; the less radical simply deny its importance in psychology. They regard consciousness as an epiphenomenon, an irrelevant "extra," not part of the chain of causal events. From a theoretical point of view, such denial raises problems. If all views are merely conditioned responses, why should we ever believe what anyone says, including behavioral psychologists? Gestalt psychologists, cognitive psychologists, the linguist Noam Chomsky, and others have shown that there are higher forms of thinking and learning. Behaviorist theory and method exclude much that is important to us as human beings, though they do focus successfully and with great scientific precision on *simple* mental processes. How far other, more "human" methods of study used by psychology can be classed as scientific remains a further question.

# Bell's Theorem

Bell's Theorem proposes that certain predictions made by quantum mechanics are *impossible* in any terms previously understood by physics. If these quantum predictions are *true*, we must use whole new categories of thinking about physical events to understand them. It is not good enough to "fix up" or readjust classical physics.

Specifically, Bell's Theorem is concerned with the special correlation relationships between distant particles predicted by quantum NONLOCALITY. Classical physics holds that any relationship existing between two or more particles must be mediated by a local force—attraction, repulsion, or at least some signal. Quantum mechanics predicts that two particles can be correlated instantaneously in the *absence* of any such force or signal, that somehow certain particles' properties can be linked nonlocally across space and time.

Initially Bell's Theorem, a variation on an earlier version conceived by Einstein, was related to a thought experiment devised by David Bohm. In this experiment, two protons are shot off from a common source toward opposite sides of a room. According to quantum mechanics, Bohm pointed out, the spins of the two protons, when measured, should be instantaneously correlated—that is, when the spin of proton A is found to be "up," the spin of proton B should be found to be "down," at exactly the same moment.

Einstein believed that this thought experiment proved the impossibility of quantum nonlocality. He saw it as a paradox that is known in the literature as EPR, or the Einstein, Podolsky, and Rosen Paradox. Bell's Theorem simply states that if the protons behave as predicted, this cannot be explained in any terms previously known to physics. Classical mechanics would have argued that a force or signal must travel from A to *B* to communicate information about A's spin, yet relativity shows that this is impossible, because no force or signal can travel from A to *B* instantaneously—that is, faster than the speed of light.

Bell's Theorem was tested in an experiment on photons carried out by the French physicist Alain Aspect in the 1970s. Two photons were emitted by a common source, and their polarization was measured when each had reached the opposite end of a large room. Quantum mechanics tells us that the polarization, the angle at which the electrical vibration takes place with respect to a plane, of both photons is indeterminate until they are measured, and predicts that when the polarization of A is fixed by measurement, the polarization of *B* will be fixed at an opposite angle at exactly the same time. In other words, the photons' polarization is instantaneously correlated. Aspect's experiment verified that this is true. When he measured the polarization of A, that of *B* was fixed (and found opposite) at the same moment. Einstein was proved wrong, and quantum mechanics was vindicated.

To understand the nonlocal correlation of the two photons' polar-

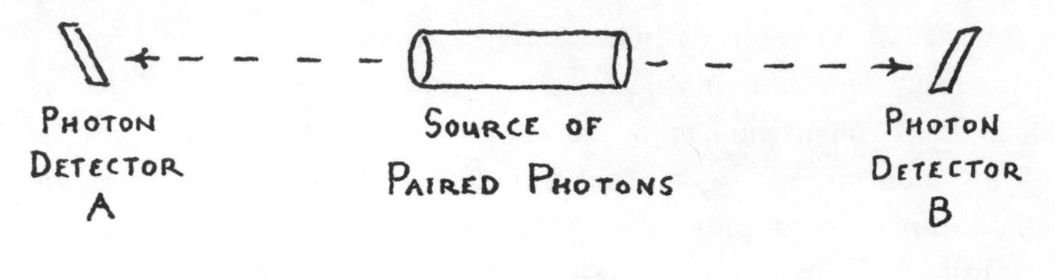


Aspect's Experiment

izations, one must take into account quantum holism and the entanglement of potentialities in the quantum domain. The two photons are correlated when measured because their potentialities—their indeterminate, wavelike aspects—were never separate in the first place. Born of the same source, they were interwoven, two parts of a larger whole. Thus it is not necessary for any force or signal to pass between them for one to "know" what the other is doing. Such holism is possible only with indeterminate potentialities, and Einstein never could accept the fact that these exist. He could not, in fact, accept quantum indeterminacy at all, believing that underneath were "hidden variables" that really determined the behavior of quantum systems. He could understand the photon correlations only in terms of some causal influence traveling from one to the other—which he called "telepathy"—and that was impossible. Einstein branded nonlocality "ghostly and absurd," but Aspect's experiment proved him wrong.

Experiments similar to Aspect's have been done "across time," demonstrating that nonlocal correlations can exist between the past and the present. In one of these experiments, two separate photons from two separate laser beams are fired through one of two slits in a barrier at different times. Though only one photon can strike a detection screen at any one time (our kind of time, that is), an interference pattern on the screen indicates that *both* are effective simultaneously. The wavelike pattern, or potentiality, of the earlier photon crisscrosses with the wavelike pattern of the later one. The two potentialities "reach across time," producing an interference pattern, as though both photons were present in the same moment.

In a second, perhaps even more striking experiment, a single photon is shot at a mirror covered with a very thin coating of silver. Be-

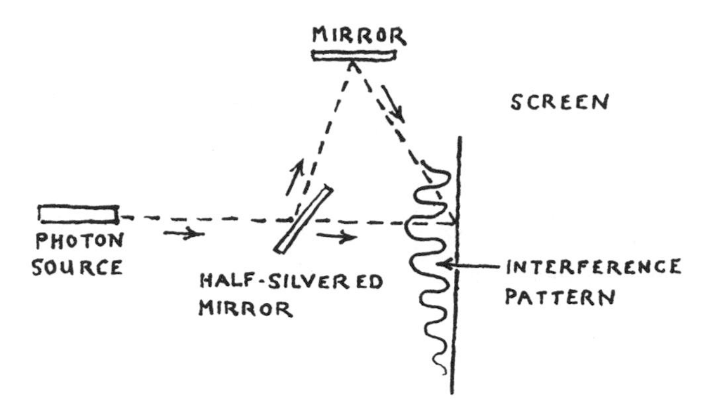


INTERFERENCE ACROSS TIME

cause the coating is so thin, the photon has a 50 percent chance of passing straight through the mirror as though it were plain glass, and a 50 percent chance of being reflected off it. If the photon goes straight through, it follows a short, direct path to a detection screen. If it is reflected off, it follows a longer, more indirect route, and arrives at the detection screen later. It should be either/or. Yet what an experimenter actually sees on the detection screen is an interference pattern indicating that the photon has interfered with—crisscrossed or become enmeshed with—its own *possibility*, or potentiality of arriving sooner or later. The mere possibility of the photon's arriving sooner or later has a kind of temporal reality that smears itself across time.

In both these experiments, each photon has potentialities to be both "now" and "then," which become entangled—that is, get into nonlocal correlation.

For some time physicists wondered whether a "Bell telephone" was possible—that is, whether it was possible to use nonlocal correlation effects to transmit information instantaneously across space and time. Quantum indeterminism makes this impossible. The potentiality of *B* "collapses" only when A is measured or observed, but the very outcome of A on measurement is itself indeterminate, so physicists have no control over the experiment. All they can ever know is that the photons will be correlated, not what value or direction either will possess upon any one measurement.

# The Big Bang

Big Bang theory, the bedrock of contemporary cosmology, suggests that the universe began as an almost inconceivably hot, dense point some 10 to 20 billion years ago, and that it has been expanding and cooling ever since. The first detailed version of the theory, referred to as "a hot Big Bang," was developed by George Gamow and his colleagues in the 1940s and 1950s.

The expansion of the universe since the Big Bang was not an explosion of matter into a preexisting space, but a coevolution of matter and space-time in the fashion described by GENERAL RELATIVITY. As it cooled, the stuff of the universe underwent several PHASE TRANSITIONS, from a hot PLASMA of particles and radiation to the formation of stars, heavy elements, planets, and ourselves. (See Table One in overview essay D, THE COSMIC CANOPY.)

Although Big Bang theory explains much about the early history of the universe and its evolution, outstanding questions remain. Why is the universe so isotropic (the same in every direction) and so "flat"? How did it come to be composed almost entirely of matter, rather than antimatter? These questions are addressed by INFLATION THEORY, which is based on a speculative Grand Unified Theory (see GRAND UNIFIED THEORIES) of high-energy physics and proposes a very sudden, inordinately huge expansion of the universe at one point in history. Another question is why the fundamental constants of the universe are so "finely tuned" toward the emergence of life and intelligence. This is addressed by THE ANTHROPIC PRINCIPLE, but here cosmology goes beyond the reach of observational science and borders on a more philosophical approach.

There are five main pieces of scientific evidence supporting Big Bang theory: THE EXPANDING UNIVERSE, COSMIC BACKGROUND RADIATION, CHEMICAL ABUNDANCES of the light elements in the universe, OLBERS'S PARADOX, and WRINKLES IN THE MICROWAVE BACKGROUND.

# The Binding Problem

The binding problem is neurobiology's current inability to explain how human beings unify, or "bind together," our perceptual experience. This problem seems to lie at the heart of the larger question of consciousness itself. If consciousness has a physical basis in the brain, and if the brain processes data from the outside world by way of its $10^{11}$ somewhat interconnected but ultimately separate neurons, why is there a unity to our conscious experience? Why do we perceive our perceptual field, our experience, as a whole, a unity? Why don't we see or experience it as fragmented into millions of separate bits?

In visual PERCEPTION, the image on the retina is analyzed into specific features in over twenty different areas of the brain, each of which deals separately with such things as shape, color, movement, or orientation. We do not know how these separate neural activities give rise to unified visual experience.

Not very long ago, it used to be supposed that there must be some special neurons or neural machinery whose job it was to recombine the output of the brain's separate visual feature detectors. If we saw something like our grandmother, there must be a special "grandmother neuron" that fires specifically in response to perceiving grandmothers. We know now that this is not so. Indeed, it *could* not be so; there are simply not enough individual neurons in the brain for one to be devoted to each possible experience that we might have.

The binding problem is complicated by the fact that we usually perceive several objects at a time and also group together different features of our experience from vision, hearing, smell, and other sensory brain areas. For example, we perceive a pink rose with green leaves. The color areas of the brain detect the pink and green, and the shape areas detect the rose and leaf shapes. Why is it that we don't see a scene consisting of green roses with pink leaves? How does the brain organize itself to get the right color with its corresponding shape?

A partial answer has resulted from the work of two groups of visual neurophysiologists since the late 1980s, the more prominent group led by Wolf Singer in the United States. In a case like that of the rose, it is now known that all the neurons in the brain that are receptive to pinkness or rosiness fire in unison at the sight of the rose. Similarly, neurons receptive to greenness and leafiness fire in unison. Thus coherent (synchronous) firing of all the neurons involved with the recognition of a given visual feature may be the code by means of which the features of various objects are linked to their neural substitutes in the brain.

The question of how this coherent firing comes about remains unanswered. What process or linkage brings one neuron into synchrony with others all over the brain that are sensitive to the same feature? And how do the *various* features of the object—the rose's pinkness, greenness, leafiness, and so on—get integrated through all *those* different groups of neurons firing in synchrony? Some linkages may be innate; others may be strengthened by repeated experience, such as recognition of familiar faces. But the speed with which the human brain can segregate an unfamiliar visual scene into its component objects and their different features deepens the mystery.

It takes the brain only about one tenth of a second to bring perhaps a dozen different areas of recognition to bear on a given visual experience. In this time, the brain integrates data and cues from areas having to do with color, movement, continuity, distance, past experience, and so on. Some of these cues involve "best fit" within the context of the whole scene, as when parts of a table are seen as one, though on different sides of an intervening person, and the parts belonging to the table are in turn distinguished from other pieces of wood that may be in the room. The amount of calculation required to arrive at such a best fit seems prohibitive for any known data processing in the brain. Parallel processing (see NEURAL NETWORKS) is not a sufficient answer, as one neural network can respond to only one familiar object or feature at a time.

The binding problem raises a philosophical stumbling block when we try to apply the paradigm of Newtonian physics to the mind. If all action and reaction is by way of contiguity or local forces, as maintained by Newtonian physics, how can the behaviors of such widely separated neural areas be part of one emergent reality—the unified perception of an object? It would seem that either we have to deny any necessary link between experience and its physical, neurological substrate, or we need to consider the possibility that the more nonlocal features of twentieth-century physics are some way implicated in brain function. (See CONSCIOUSNESS, TOWARD A SCIENCE OF; THE MIND-BODY PROBLEM; QUANTUM THEORIES OF MIND.)

# The Black Box

Many of nature's systems have a high degree of complexity, but when it comes to their interactions it is not always necessary for scientists to take into account the full details of their interiors. This is called the black box approach, a term that comes from electronics, where a piece of equipment can be treated as if it were a sealed box containing unknown components. By sending a variety of signals into the box and measuring the output, it is possible to make a model of what could be inside the box, or, to put it more precisely, what combination of electronic components would behave in such a way. The black box approach does not aim at providing an exhaustive description of an interior so much as giving a working model of behavior.

To a certain extent, the nucleus of an atom or the interior of an elementary particle is a black box. Elementary particles are shot into a nucleus; physicists measure the particles and the radiation emitted and attempt to build a mathematical model of the interior.

In psychology, the school known as BEHAVIORISM treats the brain as a black box. Behaviorists are uninterested in questions of subjectivity and interior mental states. As experimental scientists, they are more concerned with the correlation between external stimuli and behavior. Some, but by no means all, questions in psychology can be answered in this way.

The concept of a system of black boxes is fundamental to SYSTEMS THEORY. Complicated natural and social systems are treated as interconnected black boxes, the outputs of each being the inputs of others. Of particular importance are the various feedback loops between these subsystems. A representation of the total system takes the form of a flow chart in which information, decisions, or product flows link the various black boxes. Systems theory allows complicated forms of behavior to be modeled approximately, without needing to know details about the interiors of the various black boxes involved.

## Black Holes

A black hole is so dense that neither matter nor light can escape from the gravitational pull within its "event horizon," a critical surface limit. Such objects are one possible end result of stars that have collapsed, but they also have other causes.

Black holes were predicted by GENERAL RELATIVITY and by older theories of gravitation. They were thought to occur as the end point of massive stars, at least ten times the size of our sun, that have collapsed after exhausting their nuclear fuel. Large black holes were also predicted to be at the center of active galaxies, where falling matter from other stars would power QUASARS. There might also be scattered mini–black holes, formed just after THE BIG BANG. Our whole universe could conceivably be a black hole within a much larger universe.

Black holes cannot be seen directly, since light cannot escape from them. All that can be detected from the outside is their mass, electric charge, and rate of rotation. Some observational proof that they exist has resulted from studies of the orbit of Cygnus X1, a star that behaves as though it has a "black" binary companion; and from the extremely bright light emitted by quasars, the "nuclei" of active galaxies, which is thought to be caused by gas radiating as it falls into a black hole.

Black holes have many exotic properties. Ordinary space-time and our laws of physics do not apply at their central point (the "singularity"), because most physical quantities take on infinite values under that much gravitational pressure. Many physicists have felt uncomfortable with the total unpredictability this implies and have looked for ways around it. One idea is the cosmic censorship hypothesis, which suggests that any singularity is hidden behind an event horizon, so that we can never see it. There is no proof of this. Stephen Hawking has proposed a still incomplete theory of QUANTUM GRAVITY, in which there would be no pointlike singularities but instead rounded regions. This, too, may or may not be true.

Hawking has shown, however, that black holes are not *absolutely* black. According to quantum physics, particles can "tunnel" (see QUANTUM TUNNELING) out through the event horizon of a singularity, although this was not possible in classical physics. Even so, the rate of evaporation of a black hole would be negligible on the human scale.

# Blindsight

Blindsight is a condition in which some people with damage to the brain's main visual area seem able to respond to objects within their visual fields, yet remain unaware that they are "seeing" those objects. If asked, they will insist that they are blind, yet their behavior (locating objects, avoiding obstacles, finding their way in an unfamiliar place) indicates that they are not. It is simply that information resulting from the ability to process visual data does not reach consciousness.

A sound physiological understanding of blindsight was first reached by psychologist Larry Weiskrantz in the 1980s. The condition is usually the result of damage to the primary visual cortex. Philosophically, blindsight offers a powerful challenge to the behaviorist claim that there is no distinction between consciousness and the behavioral response to a stimulus. (See BEHAVIORISM.) Such a distinction is the condition's most marked feature.

In a test designed to verify the existence of blindsight, one of a set of small lights blinks on momentarily within the blind part of a patient's visual field. The patient insists he or she has seen nothing, but is asked to make a guess and point in the direction of the light that flashed. Scores on this test, carried out many times, run consistently above chance levels.

In normal vision, information from the retina passes to several places in the brain, including the primary visual cortex and a part of the primitive brain that controls some eye movement reflexes. In patients suffering blindsight, it is assumed that, while the visual pathway responsible for our *consciousness* of vision has been damaged, some visual information still reaches other parts of the brain that are responsible for behavioral response.

Phenomena similar to blindsight exist in other forms of perception. People whose brain damage prevents them from consciously recognizing faces may show all the physiological signs of recognition on a lie-detector test when shown photographs of familiar faces. People suffering from hysterical deafness will show an EEG brain response to sounds, even though they do not consciously hear them. Similarly, mothers often waken to the cry of a child (and not to much louder traffic noise), though they may not know what awakened them.

Some quite normal forms of "perception" are usually unconscious. Various physiological systems monitor and control things like blood pressure and blood sugar levels, and they do so without our conscious awareness. Some people have a magnetic sense that gives them a good sense of direction, but they are unconscious of any interaction between magnetic particles in their brains and the earth's magnetic field. We often do not notice the details of rough ground as we walk over it and adapt to its contours, and instances of drivers' getting their cars from one point to another without being aware of whole segments of the journey are numerous.

Such examples of unconscious perception are linked to what we call intuition, the ability to draw conclusions from subliminal or completely unconscious experience without being aware of how we have done so. Farmers, doctors, and stockbrokers sometimes seem to operate in this way. When challenged, they may be able to make the missing links in their chain of deduction explicit, but often they may not. They may instead say things like "I just sensed it," or "I had a feeling." Rational or scientific deduction is more publicly checkable, and therefore more reliable, but it is clearly not the only route to knowledge.

# Bose-Einstein Condensation

The quantum world is often thought of as applying only to the realm of the very small, to the level of atoms and elementary particles. But thanks to the curious phenomenon of Bose-Einstein condensation, quantum effects can be experienced in our large-scale world. Important examples of Bose-Einstein condensation are LASERS, SUPERCONDUCTORS, and SUPERFLUIDS.

The wave function, which characterizes a quantum system, generally makes its wave effects felt only over the dimensions of atoms or molecules. In our large-scale world, quantum interference effects are averaged out. This is the reason why classical mechanics works so well, and why the existence of the wave aspects of matter was not suspected until the twentieth century. When describing the motion of the moon or the fall of an apple, we have no reason to take into account the world of the atom. Yet there are unusual circumstances in which quantum phenomena intrude directly into our experience.

Quantum particles with whole-number spins, such as the photon (the quantum of light), are BOSONS and obey Bose-Einstein statistics. (See SPIN AND STATISTICS.) This means that any number of particles can occupy the same energy state. Under the exceptional circumstances known as Bose-Einstein condensation, an enormous number of such particles gather together in the same energy state and produce collective phenomena that spill over into our large-scale world.

In a laser, an astronomical number of photons congregate into a single quantum state. The result is a coherent wave function that extends over macroscopic dimensions. (See COHERENCE.) In an ordinary beam of light, the individual photons are all out of PHASE. No matter how tightly it is focused, the beam spreads out and weakens. In a laser, light travels in a coherent, cooperative way. It is all part of the same single quantum state with every photon in phase with every other. The light of a laser can be of great intensity, and a focused beam will travel to the moon without spreading appreciably.

Bose-Einstein condensation is also used to explain the strange properties of the superfluid helium II. Although a helium atom is made out of FERMIONS, not bosons, seen from a distance as a composite particle, it acts as if it had a whole-number spin. Furthermore, the molecules are light and not "sticky." In this way it becomes possible, at temperatures close to ABSOLUTE ZERO, for liquid helium to undergo Bose-Einstein condensation and behave in a coherent way, the entire fluid being described by a single wave function. In a normal fluid, individual atoms and molecules are constantly colliding and scattering off one another, producing viscosity and resistance to flow. Not so in a superfluid, in which every helium atom moves in the same coherent way. Superfluid helium is able to flow along a pipe without any resistance, the entire liquid being described by a single wave function.

Another phenomenon in which a wave function assumes macroscopic proportions is a superconductor. Since electrons have spin ½ and obey Fermi-Dirac statistics, they cannot associate in the same quantum state. But in a superconductor, weak attractive forces cause electrons (with opposite momentum) to pair up and form a "Cooper pair," which has whole-number spin. Many metals become superconductors near absolute zero. The result is a phenomenon somewhat analogous to the Bose-Einstein condensation, in which the flowing electrical current is described by a wave function that can extend for many meters. An astronomical number of electrons move through a wire in a way similar to a superfluid. But instead of scattering and producing electrical resistance, the entire river of electrons performs a coordinated ballet that passes smoothly around any obstacles in its path. A current circulating in a superconductor will continue to do so for thousands of years.

Physicist Herbert Fröhlich has speculated that Bose-Einstein condensation is characteristic of life itself. His controversial proposal suggests that organic FRÖHLICH SYSTEMS maintain their quantum coherence over the dimensions of a living cell.

# Bosons

Bosons are the fundamental class of particles that make up forces. All forces—gravity (we think), electromagnetism, the strong and weak nuclear forces—are composed of bosons.

All quantum particles are either bosons or FERMIONS, which are particles of matter. Bosons have an integral spin, which means any number of them can occupy the same quantum state. This allows BOSE-EINSTEIN CONDENSATION, the formation of substances so perfectly condensed that all their particles have coalesced into one giant particle. There is no equivalent in classical physics.

Bosons obey Bose-Einstein statistics. (See SPIN AND STATISTICS.) As well as being fundamental quanta of forces (particles of relationship), they can be composite particles consisting of even numbers of fermions.

# The Butterfly Effect

A butterfly's wing flapping in China affects tomorrow's weather in Kansas. This adaptation of an old Chinese maxim points to the interconnectedness of all things and nature's extreme sensitivity in certain regions of behavior. In modern times, its appropriateness was rediscovered by an MIT meteorologist, Edward Lorenz, as he attempted to solve equations that govern the world's weather.

The equations Lorenz developed were nonlinear. (See NONLINEARITY.) This means that, as weather develops, it feeds back on itself, weather in a particular region influencing that in neighboring regions. This developing weather in turn acts as an influence on further weather.

Lorenz supplied his equation with data on temperature, air pres-

sure, and wind directions, then computed a weather forecast. Because of his equations' extreme intractability, he was forced to solve them by using a process of ITERATION, in which the results of one stage of the calculation are fed back as data for the next stage. On one occasion, he happened to use a calculator that worked to fewer decimal places than usual, an error of only one tenth of 1 percent. To his great surprise, the weather forecast he obtained was radically different. A tiny change of one part in a thousand had resulted in completely different weather. Lorenz realized that the world's weather is so highly nonlinear that the effects of even the tiniest perturbation or error in initial data, constantly fed back and magnified, can accumulate until they swamp a particular form of weather. Thus, the world's weather is extraordinarily sensitive. It lies outside all possibility of prediction and control.

Weather, turbulent rivers, and insect populations are subject to the butterfly effect. So are economic and social systems such as the stock market. Because the behavior of such systems cannot be fully predicted or controlled, economists, sociologists, politicians, and managers are forced to develop new strategies that take into account the possibility of chaos—change, sensitivity, and openness. (See CHAOS AND SELF-ORGANIZATION.)

# Catastrophe Theory

The last straw, the melting of ice, the switching on of a light, a developing embryo, a breaking wave, a prison riot, anorexia, supersonic shock waves, heartbeats, the patterns of sunlight in a tea cup, the popping of a kernel of corn, are all examples of what the French mathematician René Thom calls catastrophe theory. Developed in the 1960s, the theory condenses the astronomical number of sudden natural, social, psychological, and economic changes in the universe into just seven basic "catastrophes."

The situations above are all examples of "singularities," regions in which the equations that govern a system break down. The stretching of a rubber sheet is described by a simple mathematical equation—blow a little air into a balloon and the equations describe how the rubber stretches, even the way a face drawn on the balloon expands. Blow in more air and the equations predict an additional expansion until, at a certain critical size, the rubber rips and the balloon bursts. It is at this point that a singularity occurs in the equations that govern the smooth expansion of the balloon. The equations fail.

The sudden, catastrophic breakdown of previously normal behavior—characteristic of a bursting balloon or a prison riot—would, at first, appear to lie beyond the grasp of science. Yet Thom was able to group together the "shapes" of these various catastrophes. Using a branch of mathematics known as topology, which deals with the way shapes and surfaces stretch and distort, he classified all reasonably simple singularities—jumps, discontinuities, and unexpected changes—into seven "elementary catastrophes," showing an unexpected unity in nature beneath the apparent diversity of so many different catastrophic changes.

Natural, social, and economic systems are driven by internal forces, and their behavior can be predicted by the appropriate equations. But these systems also exist in a surrounding environment that can exert its own influences. Thom calls the external elements that can push and pull a system "control variables." Small changes of the control variables usually result in small changes in a system's behavior, but in certain critical instances a radical, catastrophic change can occur.

Think of a roller-coaster car that reaches the top of a slope and plummets into a valley. A mild earthquake can slightly change the shape of the rails. This change in control variables produces a slight difference in the speed of the car as it approaches a curve. Distort the roller coaster a little more, and the car's speed also changes slightly. For a wide range of gradual changes in the shape of the rails (the control variables), there would be a corresponding change in the speed of the car.

Distortion of the rails continues until, at some critical point, the car suddenly leaves the tracks, flies off into space, and comes down in some different region. This accident corresponds to the simplest of Thom's elementary catastrophes, which he calls a fold catastrophe. The loading of a beam or spring until it fractures is similar.

On the next level of catastrophe, a "cusp catastrophe" encompasses a host of other sudden changes, like the way two dogs approach each other, growl, and show marks of aggression, until one of them turns tail and runs. The cusp catastrophe encompasses instabilities with two possible outcomes, such as manic depression or decision-making processes. Thom's catastrophe theory is a powerful way of classifying the various possible sudden changes in behavior and structure that can occur in natural and social systems.

Thom's theory was received with enthusiasm at first, but became less fashionable. It is a useful framework for classifying sudden changes, but, not being quantitative or predictive, it seems to lead no further. For other approaches to this area, see CHAOS AND SELF-ORGANIZATION; COMPLEXITY; NONLINEARITY; PHASE TRANSITIONS.

# Causality

Nature is in a perpetual state of flux and transformation. Nothing remains constant. Neither does anything appear to arise out of nowhere spontaneously, nor do things simply disappear into nowhere without a trace. Certain things follow others in patterns or relationships. Sunlight and rain bring about the ripening of crops; the collision between a nose and a door results in pain in the nose. We describe these relationships as cause and effect.

To ask the cause of an event or process is to ask its explanation. In everyday life, we constantly seek the cause of another's behavior if it is surprising or out of character. The notion that effects have causes is fundamental to common sense, science, and technology. Immanuel Kant described causality as one of the basic categories of human understanding. We can't imagine a world without it.

The early Greeks attributed change to various causes. Some felt it resulted from changes in the underlying substance—earth, air, water, and so on—of which the thing was made. Empedocles thought that love and strife were the causes of all change. Plato looked for explanation in the *form* of a thing. Aristotle, synthesizing the thinking of his predecessors, had the most fully worked-out theory. He distinguished four different kinds of causal explanation for any process—the material, the formal, the efficient, and the final. In his famous example of sculpting a statue, the marble from which it was sculpted represented the material cause, the shape it took on was its formal cause, the sculptor as agent of its making was the efficient cause, and the sculptor's vision or intention was the final cause.

We can understand the richness of Aristotle's four-part causal scheme by thinking of an everyday example: a game of cards or chess. What causal factors determine the progress of such a game? The dealt cards or placed chessmen are the raw material of the game, akin to Aristotle's material cause. The rules of the game determine the branching tree of all possible future moves, these are the formal cause. Usually a player has some choice; his or her actual move is the efficient cause of the next position. Each player takes into account his or her own purposes and strategies, and those that other players may have. These are the final cause. Mathematical game theory similarly analyzes the strategies open to intelligent players in some defined activity. It is used as a model for social psychology and economics, where each player seeks to gain an advantage. (See GAMES, THEORY OF; TELEOLOGY.)

Causality plays a prominent role in Newtonian mechanics and classical physics in general, but the scheme for understanding it is impoverished compared to Aristotelian thinking. There are no agents or intelligent players in Newtonian physics, no conscious beings whose purposes and intentions play an active role in the workings of the physical world. Thus it is impossible for there to be either efficient or final cause. (For Newton himself, a believing Christian, there were souls capable of efficient and final causation in the human and spiritual realms, but these played no role in his physics.)

The Newtonian scheme can accommodate only material and formal causes—the present state of a physical system (mass, position, momentum or energy, and so on) and the laws governing the behavior of the system. If we know both of these, and the forces acting upon the system, we can always predict the possible future states of the system. They are determined. (See DETERMINISM.) Thus cause and effect in classical physics is a little like those very simple children's card games where the players have no choice of strategy but must simply play the next card, whatever it is. In ancient physics, where the Greeks had the concepts of both the Fates and Tyche, the goddess of chance, there was indeterminism, as there is in interesting card games. Indeterminism comes back in QUANTUM PHYSICS.

In the wake of classical physics, philosophers have been confused about what kinds of causation it allows. This comes from a misunderstanding by Newton, who originally formulated his mechanics as if the forces acting on a system were efficient causes. A given particle would continue moving with the same velocity unless pushed or pulled by an outside force (Law of Inertia). It *seems* as if the forces cause a change, in much the same way as Aristotle's sculptor or the chess player. But this analysis breaks down if there is a system of *many* particles with forces acting among all of them. According to the "principle of least action," formulated by the great eighteenth-century mathematician Joseph-Louis Lagrange, the sequence of events in any dynamic system, including particles, fields, or, as we now know, relativity, is whatever sequence makes a defined mathematical quantity—the "action"—have the least possible value. Any forces acting among particles are *part* of the whole dynamic system, not something acting on it from outside. Thus the forces become absorbed within Lagrange's formal rule applying to the system, and any notion that force is an efficient cause falls away. It is, rather, part of a system's *formal* cause. Mathematicians and physicists understand this, but most philosophers still write as though there were efficient causality in classical physics.

The restricted concept of causality that classical physics recognizes describes the limited range of physical events analyzed by that physics, but as a model for wider thinking, it leaves out a great deal that both science and common sense consider important. Without efficient or final causation, how can we describe the role of the scientist himself in the design and execution of an experiment? How can experimental psychologists, who model their science on physiology and classical physics, discuss human behavior and leave out the purposes or goals (final causes) that motivate human beings? And how is it possible in artificial intelligence studies to describe the mind without the concept of a conscious agency?

The new twentieth-century physics casts even greater doubt on the narrow philosophical scheme of classical physics. Quantum physics reintroduces INDETERMINACY to physical law and dramatically demonstrates that there are kinds of physical relationships that are not describable in causal terms. (See NONLOCALITY.)

# Chaos and Self-organization

Chaos represents the overthrow of linearity, predictability, and ordered simplicity. It was the "clown" sleeping within Newton's tidy universal scheme. Since the middle of the twentieth century, it has become the focus of a new science in its own right.

After Newton's simple laws of motion and gravitation were shown to predict the tides and regular planetary motions, a clockwork universe became the popular image. If the exact state of the world were known at any moment, according to eighteenth-century physicist Pierre-Simon Laplace, all past and future states could be exactly calculated. Everything was smooth, determined, and predictable. The only exceptions were the actions of individual souls and of God. Later, a less religious age was left with only the clockwork.

Chaotic processes refute this view. A simple, predictable set of entities and laws may have a very complex, unpredictable outcome. Examples are weather, the stock market, insect populations from year to year, the timing of a dripping tap, and a turbulent river. Such systems are often balanced on a knife edge where the slightest deviation one way or the other will have large effects later on, as in THE BUTTERFLY EFFECT. The cumulative result of these small differences is that two initial conditions, almost exactly the same at their starting point, will diverge more and more. (See ITERATION.) Chaotic systems need not involve large numbers of components; three are enough. (See THE THREE-BODY PROBLEM.)

Chaos in this modern scientific sense is not the same as complete disorder, randomness, complexity, or unfathomability. It is a specific kind of process that has its own kind of loose order. (See ATTRACTORS.) Chaos is based on the fact that no level of accuracy is exact enough for long-term prediction. One image of chaos would be a pinball game in which the table continued to look rough and ragged, no matter what the degree of magnification. This kind of geometry, which produces roughness on every scale, is characteristic of FRACTALS and THE MANDELBROT SET. It is a good approximation of the shapes of things on our scale—coastlines, trees, cracks in stone, and so on—but when we get down to the atomic scale, things become much smoother. Quantum chaos has not been observed and may not exist. Nevertheless, chaos is a very useful model for many nonlinear processes in the macroscopic world. (See THE EDGE OF CHAOS.)

Self-organization is the reverse of chaos. In chaos, very simple patterns diverge into complex and unpredictable ones. In self-organization, a complex and unpredictable—in fact, chaotic—substrate develops a simple, larger-scale pattern. Examples of self-organization are whirlpools, crystallization, BOSE-EINSTEIN CONDENSATION, and living organisms. At first sight, self-organizing processes seem to decrease EN-

TROPY; in fact they always occur in OPEN SYSTEMS where matter or energy is flowing through. Entropy increases in the larger system as a whole, if that larger system is taken to include the environment of the self-organizing system. (See DISSIPATIVE STRUCTURES; NONLINEARITY; PHASE TRANSITIONS.) Self-organization occurs in biology as a fertilized egg develops. The mammalian brain contains far more information structure than the genes that encode it. The extent to which self-organization plays a role in evolution is still unknown, but certainly genes and cell bodies interact to form a nonlinear system.

Computer experiments on self-organization were done by the American complexity theorist Stuart Kauffman, beginning in the 1960s. It was known that the 100,000 human genes can be turned on and off by each other, or by the environment. Kauffman programmed a large array of computer components, each of which could be off or on at any instant and turned off or on at the next by inputs from others. When this computerized model of a gene network was started at random, if the connections were sparse it usually settled into a simple dynamic state when the states of most components (representing "genes") remained constant and a few were part of periodic cycles. This bears some resemblance to the behavior of actual cells. If there were more than two connections per "gene," Kauffman's network tended to become chaotic. With an average of two connections per gene, the system was at the edge of chaos, and interesting patterns appeared spontaneously. There is self-organization in the Kauffman network, but this, of course, is not proof of what happens in actual living systems. It does, however, offer a second way, in addition to DARWINIAN EVOLUTION, through which the order of living organisms might be created. Self-organization and evolution may operate in concert.

# Chaos Theories of Mind

Experimental research verifies the presence of chaos in some brain functions. This, together with the philosophical attraction of finding some new-paradigm physical theory of mind that would avoid determinism and reductionism, has led to chaos theories. These are less philosophically exciting than QUANTUM THEORIES OF MIND, but at present they are better supported by actual research.

Chaotic structures are in principle deterministic, but in practice unpredictable, and they will always remain so. They also display a kind of HOLISM, or weak EMERGENCE—their complexity cannot be reduced to simpler terms. Both features are reminiscent of human behavior and avoid the usual criticisms that more reductionist, computationalist models of mind cannot account for the richness of our experience. (See REDUCTIONISM.)

EEG tracings of brain activity show the presence of chaotic behavior. Walter Freeman's work on OLFACTORY PERCEPTION has revealed that there is chaotic electrical behavior in the olfactory bulb of the rabbit when no smell is present. This changes to coherent electrical activity when the rabbit processes some definite smell. However, this pattern indicates that chaotic activity does not underlie consciousness itself, since consciousness requires coherent brain activity as a background state.

Connectionist models of thinking (see CONNECTIONISM), based on NEURAL NETWORKS in the brain, often display chaotic modes of behavior. Their ATTRACTORS are often strange attractors, which has led some theorists to suggest that concentration, decision making, and commonly recognizable patterns or states of human behavior (such as personality types) may result from brain chaos. There have also been suggestions that Jungian archetypes may be strange attractors.

Both chaos and quantum theories of mind offer the possibility of more holistic, less predictable physical models for human behavior. As is usual in something so complex as the brain, both models may accurately describe different aspects of brain function.

# Chemical Abundances

The universal abundance of the chemical elements is one of five fundamental pieces of evidence supporting the cosmological model of the BIG BANG theory.

Each chemical element (see CHEMICAL ORGANIZATION) has a characteristic spectroscopic "signature," a complex pattern of wavelengths of light emitted when the element is excited. These patterns can be seen in light from the STARS, gas clouds, quasars, and other astronomical objects. By comparing their relative brightness, scientists can estimate how much of each element is present in each type of object, or in the universe as a whole.

The universe is nearly 75 percent hydrogen, the lightest element, and nearly 25 percent helium, the second lightest. Only 1 to 2 percent is made up of all the other elements together. The proportions are, of course, different on earth, where heavier elements predominate. It is assumed that the earth's gravity could not hold hydrogen or helium gas, most of which has blown away, although much hydrogen is retained as water.

How were the elements formed? Standard physics and Big Bang theory attribute them to two phases of development. The hydrogen, helium, and traces of other light elements were formed in the hot Big Bang. Nearly all the heavier elements were formed much later, by nuclear reactions inside stars. The more massive stars eventually explode as SUPERNOVAS, releasing some of the heavier elements into the interstellar medium, where they are incorporated into later generations of stars. Much of our world, even our own bodies, was once stardust.

Helium is too abundant to have been formed only in stars. But in the first three minutes after the Big Bang, both temperature and density were right to form the existing helium out of hydrogen. The hydrogen itself would have crystallized out of a previous "soup" of nuclear and subnuclear particles, at a temperature of $10^{10}$ K, when the universe was about one second old.

The formation of the heavier elements is a slower process. The temperature and density of the early universe were not sustained long enough to produce them. But conditions in the core of a star, where the temperature reaches millions of degrees Kelvin, are sufficient for the nuclear "cooking" of heavier elements.

# Chemical Organization

When, in the early years of the nineteenth century, the chemist John Dalton first proposed that chemical compounds are composed of atoms, he had in mind something in the nature of tiny building blocks that fit together to form molecules. In our own century, quantum mechanics has shown that, while molecules are indeed built out of atoms, the way they are assembled is far more subtle than mere mechanical connection.

An atom consists of a central core, or nucleus, surrounded by a cloud of electrons, which arrange themselves in a series of shells. Atoms in which the outer shell of electrons is filled, such as those of the inert gases helium, neon, and argon, are extremely stable and do not normally become involved in chemical reactions. Atoms that have an extra electron or a vacancy in the outer shell are chemically reactive. They try to borrow, lend, or share electrons in order to achieve stability. The result of this sharing or exchange is a chemical molecule.

Hydrogen (H) is a highly reactive gas. It contains only a single electron and seeks stability by either giving up that electron to another atom or finding a partner for it. The gas oxygen (O) has a gap for two electrons in its outer shell. Pairs of hydrogen atoms can team up with an oxygen atom and share their electrons to create $H_2O$, the stable molecule water. Quantum mechanically speaking, the wave function for the electron that originally centered on the hydrogen atom has now become delocalized over the entire molecule.

Most molecules found in nature have fairly simple constructions, since normally only a few atoms need to be brought together for chemical stability to be achieved. The exceptions are compounds containing carbon or silicon, which have the ability to join onto each other in long chains, rings, and other complicated structures. Organic chemistry is the study of molecules containing carbon atoms. Life on earth is based on such compounds.

Other atoms, including hydrogen, nitrogen, oxygen, potassium, and sulfur, can be arranged on a carbon backbone. The result is the seemingly endless diversity of molecules found in drugs, paints, explosives, pesticides, and living systems. Even the carrier of our genetic material, DNA, is an organic molecule, albeit a very large one.

The silicon atom, which, in the composition of its outer shell, is a chemical relation of carbon, is also able to form a wide variety of complex molecules. This has led some chemists to speculate that silicon-based life forms may have evolved on some distant planet.

## The Chinese Room

The Chinese Room argument, formulated by philosopher John Searle in 1980, has become a classic parable within the ongoing ARTIFICIAL INTELLIGENCE debate. It is central to the controversy surrounding whether machines can think like human beings, or whether FORMAL COMPUTATION is all that we mean by thinking. Searle developed his argument to illustrate that there is more to understanding (and hence to thinking) than just carrying out a set of logical procedures, as in a computer program. Since human beings understand what they are doing, or at least are *capable* of doing so, Searle concluded that their brains are not just computers, as the strong version of artificial intelligence claims.

The Chinese Room parable describes a man who knows no Chinese language sitting in a room. Messages written in Chinese keep being passed in to him through a mail slot. Though the man does not understand the messages, he responds by moving around various wooden counters with Chinese characters written on them, which he finds in his room. He manipulates the counters according to a rule

book he holds, which is written in English, and when all the step-by-step rules have been followed, the various counters add up to a new message. The man copies this message and passes it out through the mail slot.

From the vantage point of the Chinese who are sending and receiving the messages, the room is like a computer following a program, and it seems to "know" what it is doing. But the man inside the room has no idea what the messages he concocts mean. He is just following rules to manipulate abstract and, to him, meaningless symbols.

Responses to Searle's argument have been varied. Some people have suggested that the whole system—room, counters, rule book, and human being—"understands" the messages, even if the individual human being inside does not. Others have suggested that human beings can understand things because they are *more* than just computers—i.e., they do more than blindly follow rules. What more? Some philosophers think the composition of our biological system makes a difference—neurons are not silicon chips—an argument favored by Searle himself. Others argue that human beings behave in ways that no computer can simulate, that our understanding consists of more than formal computation. This argument is favored by Roger Penrose. (See

PENROSE ON NONCOMPUTABILITY.) Still others, largely nonscientists, feel that human beings understand things by virtue of their immaterial souls. This is, of course, the argument of dualists like Descartes, and of traditional Christianity.

## The Church-Turing Thesis

The Church-Turing thesis is a statement about the assumed nature of all mathematical computation and the possibility of this being carried out by information-processing machines. It is fundamental to the philosophy of ARTIFICIAL INTELLIGENCE and has been widely interpreted to imply that *all* thinking has the structure of FORMAL COMPUTATION.

In 1936–1937, the English mathematician Alan Turing described a type of theoretical computer known as a Turing Machine. (See TURING MACHINES.) At the same time, the American mathematical logician Alonzo Church described a scheme of computational logic known as lambda-calculus. The type of computing done by Turing's machines and the computation embedded in lambda-calculus were later proven to be equivalent—that is, any computation that could be carried out in one of the two systems could, after translation, be carried out in the other as well. Both lambda-calculus and the processing done by Turing machines are step-by-step, rule-bound computation, where each new procedure builds on the outcome of the preceding one and then leads in determinate fashion to the next.

Later, several further schemes of computation were described and also found to be equivalent to those of Church and Turing. One of these is the "Stone Age" computer described in the article on formal computation.

Today, an algorithm is defined as any rule-following process that can be carried out by a computational system of the Church-Turing kind. Adherents of "strong artificial intelligence" argue that all mental processes are algorithmic, while adherents of "weak AI" say that any mental process can be *simulated* by algorithmic computation. This does not seem to be true of intuitive or metaphorical thinking, but proponents of AI argue that these processes are not clearly defined.

Modern serial computers are equivalent to Turing machines, except that all existing computers have only a finite memory. This limitation is a bottleneck in principle, but not in practice. Even a simple, programmable pocket calculator would be capable of carrying out any algorithm, if its memory were unlimited and the algorithm were coded into its "language."

Since no computational processes that go beyond the theoretical Turing machines have been found, both Church and Turing assumed that there are none. Church's thesis was that all clearly defined mathematical processes are of this kind, whereas Turing's emphasis was that any *physical* mathematical processing device behaves in a way that can be simulated by a Turing machine. The two ideas are not the same; Church's thesis leaves open the possibility that biological computers (i.e., brains) might compute by some different process, which Turing machines cannot match. Roger Penrose (see PENROSE ON NONCOMPUTABILITY) suggests that the two ideas should be named separately as "Church's thesis" and "Turing's thesis." So far, no undisputed exceptions have been found to either thesis, although opponents of AI, like Penrose, believe there are some.

There are clearly features of human psychology other than thinking that lie outside the scope of computation—for example, experience, motivation, and intention. But since the Church-Turing thesis applies only to processes of thinking, these things offer no direct challenge to it.

# Coevolution

DARWINIAN EVOLUTION is based on the concept of competitive struggle between various animals and plants for food and space. Coevolution stresses the cooperative and symbiotic aspects of the evolutionary process.

Within the Darwinian approach, impacts on the environment act to promote evolutionary change, selecting certain varieties and hybrids, which then become dominant. But this modification in the balance of plant and animal populations is itself a new environmental change that requires fresh adjustments. Seen in this way, each species is locked with its entire environment in a number of FEEDBACK loops. Negative feedback creates stability; positive feedback emphasizes difference and accelerates change.

Coevolution stresses the advantages of cooperation in nature and points to the interdependence of plants and animals. Certain species of plants use insects for their propagation and so develop colorful flowers, perfumes, and other attractants. Bees, for example, become dependent upon a flower's pollen for their food and to construct their hives. Cooperation is clearly of advantage to both parties.

In Central America, grass hybrids were once selected and cultivated by humans. The result was a dramatic increase in grain size that produced the maize plant. A ready supply of protein allowed for a denser population and the birth of the Olmec civilization. By then the maize plant had become totally dependent on human farmers for its cultivation.

To borrow from the theory of games (see GAMES, THEORY OF), Darwinian evolution considers nature a zero-sum game in which any advantage will be at the expense of other species. By contrast, coevolution is a non-zero-sum game in which mutual advantages may exist through cooperative behavior. A wide variety of examples can be found in the natural world. It has even been suggested, by James Lovelock in his GAIA HYPOTHESIS, that the entire earth resembles a single living organism, operating in part on cooperative principles.

The notion of coevolution has also been applied at the level of molecular biology. Conventionally, DNA is considered to be the blueprint of life, a molecular message that determines the development of plants and animals. Changes in DNA are believed to come about only through random mutations. But suppose DNA can respond in more adaptive ways to environmental stresses. Barbara McClintock's discovery of "jumping genes" suggests that portions of DNA's message can be actively transposed within the context of a changing environment. It has even been suggested that, by using such mechanisms, coevolution could occur at the molecular level, with feedback signals being exchanged between the DNA of cooperative species in order to bring about mutually advantageous change.

In the real world, both cooperation and competition exist, both positive and negative feedback. The dynamics of such systems are not fully understood, but they are under active investigation. (See COMPLEXITY; SYSTEMS THEORY.)

## Cognitive Psychology

See GESTALT AND COGNITIVE PSYCHOLOGY.

## Cognitive Science

Cognitive science is devoted to the study of thinking or problem solving in general, whether this is done by human beings, animals, or machines. It is one of the most exciting areas of scientific research today, with enormous technological implications for the information revolution. There is extensive cross-fertilization with ARTIFICIAL INTELLIGENCE, which seeks to develop intelligent machines that can emulate or go beyond human thinking capacities. Together, cognitive science and AI try to *extend* human thinking by understanding its processes and finding faster and more efficient ways to supplement it with machines that do mathematical calculations, word processing, or information retrieval. Cognitive science is therefore much broader than COMPUTATIONAL PSYCHOLOGY, which limits itself to studying how human beings *actually* think.

Cognitive science has both theoretical and practical aspects. Theoretically, it is, although a new science, deeply steeped in the old mechanistic paradigm. Its two most prominent models for problem solving are SERIAL PROCESSING, as in personal computers, and parallel processing, as in NEURAL NETWORKS. Both can be replicated by machines, and this has led to controversy. Heated philosophical debate surrounds the question of whether computer models of mind can ever provide a complete approach to human psychology. See, for instance, FUNCTIONALISM, THE MIND-BODY PROBLEM, PENROSE ON NONCOMPUTABILITY, and THE TURING TEST.

At the moment, most research projects in cognitive science are devoted to testing the theoretical powers and limitations of the two computer-processing models. What can each do or not do? How do machines built according to such principles compare with what is known about human neural and psychological function? Can artificial systems solve problems as well as or more efficiently than we do? Can the way machines "think" guide us in understanding human thinking?

Cognitive science as presently conceived may have theoretical limitations. It may not be asking the best questions. But its research has led to practical devices that greatly increase the world's capacity for information processing and problem solving.

# Coherence

*Coherence* refers to the cooperative behavior of an astronomical number of particles in, for example, LASERS. Within a laser, individual light waves are all in PHASE.

The illumination of a light bulb or candle flame is produced when many atoms each emit a photon of light. These events are totally uncoordinated; neighboring atoms emit photons at slightly different times. This results in a beam of light whose component waves are all out of phase. Even when focused with a lens, such beams have a tendency to spread.

In a laser, large numbers of atoms are triggered to emit photons simultaneously. The resulting light waves are all exactly in phase, marching in step, and this coherent light can be focused into a tight beam that does not spread out.

The term *coherence* is also used to refer to other systems, such as SUPERCONDUCTORS and SUPERFLUIDS, in which large numbers of particles behave in an orchestrated way. (See also BOSE-EINSTEIN CONDENSATION; PLASMA.)

# Cold Fusion

Cold fusion, its supporters claim, produces limitless energy from the cheapest substance on earth—water. The idea is pure nonsense, say its detractors—a combination of bad science and wishful thinking.

For more than fifty years, scientists dreamed of harnessing the energy of the stars in what was called thermonuclear fusion. Conventional "atomic energy" is based on nuclear fission in which the nucleus of a heavy radioactive isotope like uranium-235 splits in two, emitting energy in the process. Thermonuclear fusion is the reverse. Two light nuclei fuse together and produce even larger amounts of energy.

The nucleus of heavy hydrogen (deuterium) contains a single proton and a single neutron. When two of these nuclei fuse, they create a nucleus of helium and give off energy in the process. A barrier to this reaction lies in the strong electrical repulsion between the two nuclei, which can be overcome through fast and energetic collisions, which means extremely high temperatures. Apparently thermonuclear fusion on earth is feasible only with expensive large-scale apparatus capable of sustaining temperatures as hot as the center of the sun. (See PLASMA.)

Fusion reactions in the laboratory are technically possible, but, despite decades of work by teams all over the world and an enormous expenditure of money, they have never been maintained for long enough periods to make fusion energy production a commercial possibility. Imagine the shock, in March 1989, when scientists in Utah announced that they had achieved nuclear fusion at room temperature, with an apparatus that cost only a few dollars!

The claim was made simultaneously by two teams working independently and using somewhat different approaches—Stanley Pons and Martin Fleischmann at the University of Utah, and a team headed by Steven Jones at Brigham Young University. Both groups maintained that they had solved the problem of bringing deuterium nuclei close enough together for a nuclear reaction to take place, not by colliding them at high speeds, but through gentler means.

Platinum metal has the ability to absorb large quantities of deuterium. Deuterium nuclei enter the metal lattice and occupy gaps between the platinum atoms. Many deuterium nuclei can be forced into the lattice, and the researchers argued that if the nuclei could be packed tightly enough, a fusion reaction should begin. The theory behind this claim was that the platinum lattice would act to shield the electrical repulsion between nuclei, thus allowing them to approach close enough for fusion to occur.

The apparatus used by Pons and Fleischmann was simple—a test tube and a pair of electrodes. An electrical current was to be passed through a cell containing heavy water (deuterium oxide), which forced deuterium nuclei to pack even more tightly into the platinum electrode. When the nuclei were sufficiently close, a fusion reaction was supposed to begin, and the cell would heat up. The result, researchers claimed, would be limitless energy generated in relative safety and without the need for expensive equipment.

Is this only a pipe dream? Did cold fusion ever happen? It is certainly true that heat was given off by the cells, but did it have a nuclear origin, or was it nothing more than the release of energy that had already been pumped into the cell? For the reaction to begin, a cell must be initiated for several hours, by using an electrical current to pump up the platinum electrode with deuterium. Critics of cold fusion suggest that what happened was that, at some point, all the stored energy was released, producing heat. Cold fusion cells were nothing more than electrical batteries. In the initiation phase, the cell was charged until it gave off this energy as heat. Heat was not being created; stored heat was simply released.

The crucial test is to discover if the total energy output of the cell is greater than the energy input. If it is, the cell is truly operating as a power source. Such measurements are, however, very difficult to perform in an accurate, controlled way. Several teams have carried out such measurements, with conflicting results. Most physicists are skeptical.

Additional evidence for a nuclear reaction would come from the presence of radioactive by-products in the cells, or from emission of radiation and elementary particles during the reaction. Claims and counterclaims have been made to the point that there has been considerable animosity between the supporters and detractors of cold fusion. The entire debate is an interesting example of how science actually works; when the stakes are high, scientific objectivity gives way to passion and strongly held beliefs. Like other people, scientists sometimes see only what they believe they will see.

## Collapse of the Wave Function

Quantum reality is described by Schrödinger's wave function as a spread-out of superposed, sometimes contradictory possibilities. When the quantum world is measured or observed, these many possibilities become one single actuality—the "quantum hussy" marries and settles down; Schrödinger's cat is found either dead or alive. This transition from many to one, from possibility to actuality, is known as the collapse of the wave function. What happens at, or what causes, the moment of collapse is still a mystery. (See THE MEASUREMENT PROBLEM; A QUANTUM HUSSY.)

## Color—What Is It?

Why is it that our world is a rich tapestry of many colors instead of just black and white, or many shades of gray? What are the origin and actual mechanics of color? Why can quantum mechanics explain color when classical physics never could?

The story of color is essentially the story of the atom's structure and how electron energies are balanced and distributed within the atom. In the early 1900s, the atom was known to consist of a heavy, positively charged nucleus and negatively charged electrons held to it by electric forces. The number of electrons, which determines the kind of chemical element the atom represents, varies from one for hydrogen to ninety-two for uranium. The atoms of some artificial elements have even more. Chemical bonds depend upon interactions between the electrons of different atoms.

When an atom is stimulated, by heat or by passing electric sparks through the vaporized element, it gives out visible or ultraviolet light—the yellow glow of sodium-vapor streetlights or the brilliant and varied colors of fireworks. This emitted light is always of a definite wavelength, and patterns of various wavelengths, known as spectral lines, are associated with each chemical element. Physicists imagined that these characteristic wavelengths of light were given off because electrons in the atoms were in some sense being "kicked" onto higher energy "shelves," which they then "fell off" when they returned to a settled state. It was assumed that they gave off photons of light in the course of falling from a higher shelf to a lower one.

In 1913 Niels Bohr presented his famous solar-system model of the atom, in which the nucleus was like the sun and the electrons revolved around it like planets. Physicists attempting to explain color associated the energy shelves onto which electrons climbed with the various electron orbits in the Bohr atom. When a photon was emitted, it must be because the electron was falling from one orbit to another closer to the nucleus.

But this left many questions unanswered. Why, for instance, were there only certain possible orbits in the atom? If there were any number, there would be no distinct energies (colors) emitted; hence everything would be black-and-white or gray, because the colors would all mix together and cancel each other. But why didn't all the electrons of any atom fall into the lowest-energy orbit? Why were they spread around on different orbits?

It took quantum mechanics to answer these questions. Bohr had shown that electrons are restricted to certain orbits because they are "quantized"—that is, each orbit is associated with a definite energy level. In the mid-1920s, Wolfgang Pauli's exclusion principle demonstrated that each electron orbit would allow only a specific number of electrons (two) into any one orbit. And the mathematics of Schrödinger's equation answered the mystery of why only certain electron orbits are possible, rather than an infinite range. (See THE WAVE FUNCTION AND SCHRÖDINGER'S EQUATION.)

We can get an intuitive picture of why electron orbits are quantized by remembering that electrons, like all matter, have both wavelike and particlelike aspects. Considered as a particle, an electron can be knocked from one orbit to another. But when an electron is considered as a wave, any one electron orbit is a wave pattern circulating in a ring around the nucleus, like a snake biting its own tail. A stable wave pattern must join onto itself seamlessly; its "head" and its "tail" must be in the same phase of motion. Hence, there must be a whole number of waves in each circuit, complete cycles with no fractions.

When numerical calculations about the atom's internal structure became possible, the orbits were found to be exactly those that would account for the observed spectral lines (lines of color) associated with each element. This was a triumph for quantum mechanics and its wave/particle duality. (See BOSONS; FERMIONS.)

# Complementarity

Waves and particles have radically different behaviors and properties. How can we possibly understand that light is *both* a wave *and* a particle? In what terms can we describe it? Niels Bohr's Principle of Complementarity, which he first proposed publicly in 1927, states that each description excludes the other, but both are necessary—they complement each other. We will, like William Bragg, teach about waves on Mondays, Wednesdays, and Fridays, about particles on Tuesdays, Thursdays, and Saturdays. The course as a whole will add up to a complete picture. Other complementary pairs are position and momentum, and energy and time.

Complementarity quickly became a magic wand that Bohr, and later his followers, waved at all apparent contradictions and conflicts between the emerging world view of quantum mechanics and the older picture of reality offered by classical mechanics. Bohr accepted the fact that quantum experiments require quantum categories, like WAVE/PARTICLE DUALITY and quantum mathematical description, but argued that when we *talk* about reality we can use only classical terms. The quantum and classical worlds complement each other, but we can never unite the two in a single act of understanding.

Bohr used his Principle of Complementarity to argue that there was no point in *trying* to describe the quantum world, or to understand its apparently bizarre picture of reality. Understanding of the way things really are, as opposed to what we see when we measure them, was, he claimed, not the business of physicists. (See THE MEASUREMENT PROBLEM.) His colleague Erwin Schrödinger accused him of trying "to complement away all difficulties" by refusing to discuss them.

If we were to apply the Principle of Complementarity to a particularly complex person, who behaved pleasantly in some circumstances and unpleasantly in others, we would have to say that his or her two sides complement each other. Any psychologist or novelist who tried to get beneath the apparent contradictions to describe "the real person underneath" would be told that there is no such person, or that we were wasting our time trying to describe one.

Bohr himself applied his Principle of Complementarity widely in fields outside physics. In a series of papers and public talks, he argued that many things were complementary or mutually exclusive: thought and action, subjectivity and objectivity, feeling and reasoning, male and female, the truths and values of one culture and those of another. Physicists and philosophers of Bohr's generation liked this way of thinking because it rested within the dualist either/or paradigm of the old world view and required no revolution in thinking. But younger physicists, particularly many of today's philosophers of physics, feel that complementarity is just an excuse for avoiding the kind of both/and thinking that quantum physics makes both possible and necessary.

To accept that light is *both* a wave *and* a particle, and to learn to live conceptually with that kind of ambiguity, is one of the creative leaps quantum physics calls upon us to make. Applied in other fields, both/and thinking requires us to see that there may be two or more mutually contradictory ways of doing something, or of looking at something, *all* of which are valid. Seeing the truth of *all* tells us something more profound about the situation. Some people *may* have both pleasant and unpleasant sides, and learning to see both at once may give us a deeper understanding of the kind of people they are.

# Complexity

*Complexity* may refer to anything that is not simple. If we define the complexity of a system as its shortest description in some standard computer language, then such things as a heap of rubbish or a string of meaningless numbers are more complex than the works of Shakespeare. On the other hand, complexity might refer to the information content of a structure, in which case Shakespeare's works would be more complex.

In twentieth-century science, complexity has often meant something very specific—order at THE EDGE OF CHAOS. In this sense, it may herald a new science that offers as significant a challenge to the old paradigm of Newtonian science as do QUANTUM PHYSICS and chaos. (See CHAOS AND SELF-ORGANIZATION.)

Newtonian science was about organized *simplicity*. In physics, biology, and chemistry, and in associated thinking to do with economics and society, those who followed Newton's example sought to reduce the observed world to a few simple laws and components. Scientific method itself was concerned with viewing systems in isolation from their environment, breaking them down into their simplest component parts, and then using these parts to predict the unfolding future of the system. Simplicity—including smooth linear development, determinism, and predictability—was the cornerstone of this approach.

But there are naturally complex structures that defy so reductive an approach, things like the brain, a cell, a city, a rain forest, or a beehive. If we are to study these at all, they seem more suited to the new scientific thinking of the twentieth century, which is concerned with organized *complexity*. Quantum theory, chaos, and now complexity theory all concentrate on the emergent whole that cannot be reduced to the sum of its parts, on discontinuous, nonlinear change that leads abruptly to surprising new states or forms, on indeterminacy and unpredictability. Chaos and complexity, opposite sides of the same coin, concentrate on the turbulent disorder lurking at the edge of many supposedly predictable systems, and then, in turn, on the surprising new order discovered at the edge of chaos.

This new order acts as an attractor, or "strange attractor," that pulls energy or matter into a complex pattern. (See ATTRACTORS.) The whirlpool is one such self-organizing pattern. It pulls water molecules surging around it into a tight funnel, and the shape persists although the water flowing through it is different at every moment. The human body also is such a complex self-organizing system. In its case, it takes seven years for all the material content to flow through, but the pattern persists.

Complexity studies are the central focus at the Santa Fe Institute. One of its leading scientists, Stuart Kauffman, believes that the model of chaos and strange attractors can be applied to the problem of how order arises in evolution. According to him, living communities are the most striking examples available of organized complexity. The genome system of any higher metazoan cell encodes 10,000 to 100,000 genes that orchestrate development in the embryo from a single cell. The human immune system deploys about 100 million different antibody molecules in complex, harmonized patterns. Each of these systems is the result of evolution. Kauffman challenges the orthodox Darwinian idea that such wonderful order and complexity could result from the random selection and mutation of "an accidental machine."

Kauffman's theory is that the same dynamics that allow complex polymers to reproduce collectively in the absence of any genome or template are available to an evolutionary process working in partnership with natural selection. There is an order, he suggests, *there*, in nature, waiting to unfold—a natural direction to evolution pulling evolved forms toward complexity. Morphogenesis, or the growth of form, may be at least partially a consequence of inherent self-organization.

Why should complexity be of value? Why should nature embody it as a possibility? Simple living creatures like bacteria or insects are quite successful in their own ways. But if intelligence is the ability to discriminate, predict, and respond appropriately to different circumstances, it is of survival value. In a complex environment, equally complex computations would be required from thriving organisms.

There is no doubt that nature contains complex systems that cannot usefully be reduced to simple rules or components. But there is controversy surrounding the viability of the "science of complexity." It is, of necessity, a computer science. Nonlinear equations that apply to complex systems are too difficult for human beings to solve, but if they are run through computers, amazing trends and patterns begin to emerge. This gives rise to what critics argue may be an illusion that there are, after all, quite simple laws at work within complexity. The computer simulation of one complex system, a beehive, looks quite similar to that of another, the human brain. Complexity scientists believe they have found a unifying principle. Critics believe all they have found is the unifying principle behind the way computer-generated simulations work.

Complexity is real, and examples of it in nature are manifold. "Complexity science" may be no more than a disguised effort to apply reductionist science to what cannot be reduced. It may be that there are no simple, general principles of complexity.

Before the Santa Fe Institute and the emergence of complexity science as a study of systems at the edge of chaos, there were other attempts to describe systems as opposed to separate working parts, and to find unifying principles for them. Each has produced interesting, though limited, applications. See, for example, CATASTROPHE THEORY; CHAOS AND SELF-ORGANIZATION; CYBERNETICS; DISSIPATIVE STRUCTURES; INFORMATION; SYSTEMS THEORY.

# Computational Psychology

Suppose the premise of ARTIFICIAL INTELLIGENCE, that human thinking processes are essentially like those of computers, is correct. What insight into human behavior might such a model give us? Computational psychology is a companion field to AI that tries to understand human psychology as though it were a series of computational processes. With the acceleration of computer theory and technology since the 1940s, this branch of psychology has developed into a broad, controversial discipline.

Computational psychology is an outgrowth of cognitive psychology (see GESTALT AND COGNITIVE PSYCHOLOGY), which is itself an attempt to grow beyond simple BEHAVIORISM. Where the behaviorists structure human psychology in terms of stimulus and response, cognitive psychologists focus on the interaction between the two. They ask questions such as what knowledge, beliefs, perceptions, memories, thinking, imagery, or language we bring to bear when responding to a stimulus. They do not, however, regard emotion, intention, or consciousness as cognitive (thinking) phenomena, and thus do not attempt to investigate their role in behavior.

Computational psychology tries to bring cognitive abilities into relation with the knowledge derived from NEUROSCIENCE (biological brain structure and function) and also with computational models developed by AI. This has been very successful in fields like PERCEPTION and MEMORY, where both neuroscience and computer science are sufficiently advanced to account for known facts. Some computational psychologists question whether neuroscience or computation is *sufficient* to explain higher mental processes such as problem solving and understanding natural language. Because of such doubts, this minority restricts itself to building detailed models of limited areas of cognitive ability, such as perception. Other computational psychologists concern themselves with general theory, including problem solving. All are united in concentrating on one aspect or another of human information processing. (See FORMAL COMPUTATION; NEURAL NETWORKS.)

Computational psychology goes beyond artificial intelligence; it is concerned with how *human beings* actually do things. AI workers are more concerned with simply developing intelligent machines. Unlike AI, computational psychology must harmonize its theory with facts and observations derived from wider psychology and from neuroscience. An AI theorist working on memory knows that a (serial) machine can remember large numbers of facts simply because it has large enough memory storage, comprising individual elements for each fact. But human memory is more complicated. Each person may, in his or her lifetime, store $10^{14}$ or more items in memory. That is one thousand times more than the number of individual neurons in the brain. Thus we cannot assume that each memory is stored on a separate neuron. We do not remember our grandmothers because we have a "grandmother neuron." Some more sophisticated theory of human memory is required by computational psychology.

Computational psychology restricts itself to studying those human cognitive capacities that can be replicated by, or at least *simulated* by, information-processing machines. But it recognizes that human capacities may arise in a different way from similar machine capacities. Like the cognitive psychologist, the computational psychologist does not investigate any role in behavior played by intention, emotion, or consciousness.

## Connectionism

Connectionism is a school of thought within philosophy and computational psychology that argues that most important human cognitive functions can be explained in terms of a neural network model. That is, of the brain's two known sorts of neural connections—one-to-one neural tracts, which do a kind of SERIAL PROCESSING, and the more complex NEURAL NETWORKS in which thousands of neurons are interconnected to enable a kind of parallel processing—it is the neural networks that predominate in human thought and perception. Connectionism became a popular view in the 1950s, fell out of favor, and then began to flourish again in the 1980s with the successful development of neural network theory and processing machines.

Connectionist models seem too one-sided in the light of other knowledge from psychology and the neurosciences, but they represent an important part of the truth. Neural networks excel at pattern recognition, something at which human beings are very good and serial processors are very weak. The microarchitecture of the brain's cortex is networklike, and neurons are richly interconnected, apparently at random, within a 1- to 2-millimeter radius. Larger-scale architecture in the cortex contains precisely wired neural tracts as well, like the optic nerve. The cerebellum is also precision-wired.

Some types of memory and problem-solving ability are not localized, but distributed throughout the cortex. This supports a connectionist case. Localized damage, or the gradual loss of neurons with age, does not affect any specific memories or intellectual abilities, but causes a general and gradual falling off. Such nonlocalization and "graceful degradation" are typical of neural networks, but not of serial computers. Other mental functions, however, such as sensory, motor, and language abilities, are known to be localized. Damage to some specific part of the brain can cause severe and specific damage to these abilities.

The extent of the human memory also tends to support a connectionist case. We have too many memories or possible experiences for each one to be represented by a single neuron. It is more likely that each such representation is distributed over a group of neurons.

Finally, the speed with which human beings carry out certain mental functions supports the connectionist model. It takes us about one tenth of a second to recognize a face or a visual scene. In that time, individual neurons can fire at most 100 times, meaning that our "recognition program" could have at most 100 steps. But 100 is far too few steps to be adequate within a one-by-one serial processing program. In a parallel or neural network program, fewer steps are necessary, because each step can carry out several processes at once in a cooperative fashion, due to the large number of neural interconnections.

All these arguments in favor of the connectionist model suggest that neural networks must play some very important role in human computation. Serial processing could not be the whole story. But parallel processing also has its weaknesses. It is poor, for example, at rule-following tasks like mental arithmetic and grammar. It seems most likely that the brain uses neural networks embedded in its microstructure for concrete perceptual, motor, and associative processes, but uses serial processing for higher-level cognitive tasks. It may also use other forms of computation, or some third emergent form of computation that integrates the serial and parallel. (See CHAOS THEORIES OF MIND; QUANTUM THEORIES OF MIND.)

# Consciousness, Toward a Science of

One of the first lessons we learn as children is to divide the world into persons and things. Both have an objective, material side, but persons also have a subjective, mental side. That which is objective is, in principle, observable to all. It can be seen, touched, weighed, and measured. But those who observe the objective qualities of *things* are themselves *subjects*. Their experience is private. Matter is unconscious and inert and follows the laws of physics. Mind is a conscious agent that may or may not have some relation to the laws of physics. It is a process rather than a thing; it cannot be pinned down; it does not look the same from every point of view. How, then, do we study it?

The problem of consciousness—how to describe it and how to make it fit into our scientific paradigm—is at the cutting edge of late-twentieth-century science. It is the "hard question" that every ambitious neurologist and many physicists want to answer. Today's philosophers struggle to deal with the reality and properties of consciousness in a "scientific" way. The process of doing so may push us to broaden our whole understanding of science and the scientific method.

Science as we have known it since the seventeenth century is about the detached, repeatable study of objective reality. Its whole credo rests on the exclusion of subjective phenomena, which follows from Descartes's clear distinction between the properties of mind and those of matter, and from the later Newtonian preoccupation with matter as the proper object of scientific inquiry. Consciousness plays no role in Newton's physics.

With the rise of psychology and the brain sciences at the beginning of the twentieth century, mind became an acceptable target for scientific study. But the "new" sciences of mind have employed the "old" scientific paradigm: They have applied the assumptions, standards, and procedures of reductive objectivity to a phenomenon that is, in large part, emergent, holistic, and subjective. The result is a large body of information about neurons, the effect of drugs on brain capabilities, and the mechanics of perception. But the *process* of consciousness itself eludes such study. It is as though these scientists were attempting to watch a movie in a still frame or, to use an expression coined by the Californian polymath Alan Watts, as though they were "eating the menu instead of the dinner."

In the twentieth century, relativity theory, QUANTUM PHYSICS, and chaos theory have all contributed to a greatly changed understanding of matter. (See CHAOS AND SELF-ORGANIZATION; GENERAL RELATIVITY; SPECIAL RELATIVITY.) Matter is now widely conceived as process itself, as "patterns of dynamic energy." The extent to which its properties can be measured "objectively" is often in question. The new sciences require us to come to terms with uncertainty, ambiguity, and the importance of context and relationship. They focus on emergent, creative, and holistic physical processes. In the mind sciences, there are now fledgling attempts to apply the paradigm of the new sciences to a study of consciousness. There are QUANTUM THEORIES OF MIND and CHAOS THEORIES OF MIND. So far, such theories remain at the cutting edge, promising but speculative. Their greatest contribution may be a move toward a conceptual revolution in science itself, toward reintegrating the study of mind with the study of body, or toward reintegrating subjective and objective paradigms.

# Construction Copier Machines

IS ARTIFICIAL LIFE possible? Can robots that repair and reproduce themselves be constructed? If machines are ever brought to life, they will be based on the principles of Construction Description.

In the 1940s, mathematician John von Neumann investigated the mathematics of self-reproducing machines. Imagine an elaborate robot called the Universal Constructor that wanders around the earth, collecting spare parts and building other machines. Given a description of any machine, it immediately sets about building it. With a description of itself, the machine will build its exact replica.

Is this machine alive? No, said von Neumann. The newly created copy cannot reproduce itself because it does not contain a description of its own being. The clone can construct only machines for which it has been given descriptions. For life to exist, von Neumann reasoned, an original machine must not only reproduce itself, but also make a copy of its own description and attach this to its offspring. Now the offspring can reproduce and in so doing will copy their description and attach it to the next generation. In this way a Construction Copier or Construction Description machine will continue to propagate itself without limit.

When DNA was discovered, the cell was found to have exactly the properties von Neumann had predicted. The DNA molecule is an active description of a cell that also contains the machinery for its own reproduction. When a cell divides, it makes a copy of its own DNA.

Von Neumann's colleague S. Ulam showed that self-reproducing systems could be simulated on a computer in the form of what he termed a cellular automaton. Some scientists have begun to question if, in a certain sense, cellular automata could be called living systems. The science-fiction writer Arthur C. Clarke, in a sequel to the film *2001*, suggested that advanced intelligences might use von Neumann machines to bring about planetary engineering. A single machine is placed on Jupiter and uses elements in the planet's atmosphere to build and reproduce itself. One machine becomes two, two become four, four become eight, and soon the surface of the planet is covered with machines that will totally transform the planet's atmosphere.

Science-fiction or engineering of the future? It is now becoming possible to build machines and engines of molecular dimensions, powered devices so small they can be seen only under the electron microscope. If they are given the power to reproduce, they will become miniature von Neumann machines, multiplying and spreading across an oil spill, for example, to consume its pollutants.

# Contextualism

The realization that in quantum physics the very existence and identity of a particle are bound up with its overall environment or context is known as contextualism. Like homonyms, words that look the same but have different senses depending upon the context in which they are used, quantum reality shifts its nature according to its surroundings.

In classical physics, things are what they are. In quantum physics, there is more of what a philosopher might call an "existential" dialogue among the particle, its surroundings, and the person studying it. This is one upshot of the WAVE/PARTICLE DUALITY of light and matter.

Light sometimes behaves like a wave, and sometimes like a particle. Which way depends upon the circumstances or, more strongly, how we *want* it to behave, or how we *look* at it. The somewhat eerie reality of this is borne out by one of the most famous experiments in quantum physics, the two-slit experiment.

In this experiment, a stream of photons is emitted from a light source. Just in front of the photon source, the experimenter erects a barrier with two open slits, which allow the photons to pass through. On the other side of the barrier are placed *either* two particle detectors (two photomultiplier tubes near the slits) *or* a wave detector (a screen), with which to observe the photons after they have passed through the slits and met again. If the experimenter selects the particle detectors, thus measuring the photons separately, they travel through *one* of the two slits and cause a click in one of the detectors. If, on the other hand, he or she chooses a screen and thus measures the photons collectively, they travel through *both* slits and leave a wave interference pattern on the screen. If the physicist looks for a particle, a particle is found. If the physicist looks for a wave, that is what's found.

In the two-slit experiment, it is not possible to say that light (the photon) is *really* a wave that sometimes acts like a particle, or vice versa. Light is deeper or richer than either of these partial realities, and which side of its dual *potential* nature it decides to show depends

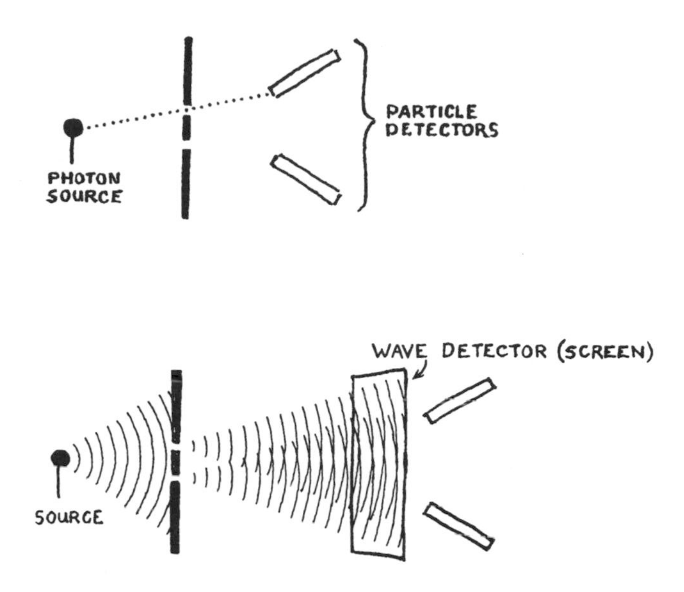


entirely upon the experimental context in which it finds itself. We can never observe light outside of some context.

The wave/particle duality is one of many complementary pairs of variables. (See COMPLEMENTARITY.) For each pair we can devise, there is a context in which one or the other of the pair can be seen or measured.

Though quantum contextualism sounds eerie when we are speaking about elementary particles that change their identity at the drop of a slit, we are very familiar with similar behavior in everyday life. We have all experienced feeling different when at home and when on holiday in some exotic environment. We know that in some relationships, or some jobs, we feel more alive and more creative than in others. Quantum contextualism simply shows us that the adage "Nothing really is as it seems" applies at the most basic level of physical reality. As the French philosopher Maurice Merleau-Ponty put it, when we speak of truth, we can only "define truth within a situation." The same kind of thinking is borne out by Einstein's SPECIAL RELATIVITY.

# Continuous Symmetries

In Newton's world, the laws of nature were fixed equations. Everything was clear and unambiguous; plot a graph of the variable in an equation, and you have a picture of the path of an actual body moving through absolute space and time. Then along came Einstein, who showed that it was possible to transform these equations in a wide variety of different ways. In each case, while the descriptions of what is taking place look different, there is something unvaried about the underlying physics. All possible transformations of the laws of nature are permitted, providing that the basic form of the equations, the mathematical pattern, is preserved. The formulations express the same meaning in different languages.

The laws of nature are written in terms of space-time coordinates, and Einstein's insight means that we can move freely from one space-time background to another, provided that a basic mathematical form is unchanged. What was taken, in Newton's world, as the force of gravity becomes, in Einstein's GENERAL RELATIVITY, a change of background geometry. At the heart of Einstein's theory, like all twentieth-century physics, lies the concept of symmetry. Even matter itself, Einstein dreamed, will one day be reduced to the twists and curves of geometry.

Quantum theory is a radically different approach to nature, yet both it and relativity have the same reliance on symmetries and the conservation of mathematical forms. Earlier in the century, elementary particles were thought of as "the fundamental building blocks of matter." Physicists pictured a world in which atoms were built out of these stable units, and molecules, in turn, out of atoms. Now it appears that elementary particles are themselves in a constant state of flux and transformation. Two particles collide, and a starburst of new particles is created. Wait long enough, and a particle will spontaneously decay into several others. At every instant, elementary particles are creating a cloud of VIRTUAL PARTICLES around themselves, constantly borrowing and paying back energy from the vacuum field of the universe. If building blocks have given way to ceaseless transformations and fluxes of energy, then what remains the same? The answer again is symmetry.

The new physics is based on symmetries of all kinds. One of its most fundamental insights is that wherever a continuous symmetry is discovered, there must be a corresponding conservation law (Noether's Theorem). A conservation law can be thought of as the preservation of a particular mathematical form under all possible transformations and processes. This is an elegant, fundamental, and intuitively appealing type of explanation in physics.

Among the first symmetries to be studied were those that operate in space and time. Both relativity and quantum theory assume that space is everywhere the same. (The reader may object that, when you look into the night sky and see stars, galaxies, and collections of galaxies, it does not look as if this were true. Astronomers assume that this is a secondary effect, SYMMETRY BREAKING, and that the underlying space is uniform.) Move all the matter from one location to another, and everything should be the same—space is homogeneous. Without anything to break the symmetry, it does not make sense to talk about one location or another. Similarly, look in any direction and the properties of space are identical—space is isotropic.

These fundamental symmetries give rise to two fundamental conservation laws of nature—the conservation of (linear) momentum and the conservation of angular momentum. (Momentum is mass multiplied by velocity.) Conversely, if a force is acting, the properties of space lose their symmetry, and momentum ceases to be conserved.

How does conservation of linear momentum follow from the homogeneity of space? If space is everywhere the same, there are no marks of reference to tell you in what direction you are moving or how fast. You can't even say that a lump of rock in empty space is rushing past you, rather than that it is you who are rushing past the rock.

In a homogeneous space, every location is equivalent, and the motion between these locations is everywhere the same. The quality of the motion, the pattern of the motion, is always the same. Physicists call this quality momentum and prove that momentum must be conserved. If the properties of space were different from point to point, as when you try to run through a crowd, the pattern of the movement from one point to the next would always be changing, and momentum would not be conserved.

Conservation of momentum is enormously important in quantum theory. Whenever an elementary particle breaks apart into other particles, the total momentum of all the products must be equal to the momentum of the initial particles. Only these transformations of particles, which preserve linear momentum, are possible. Other imaginable processes are strictly forbidden.

Space is isotropic, the same in all directions. This presents another important conservation law—the conservation of angular momentum. Angular momentum is strictly conserved in all elementary particle reactions.

Yet another law deals with the conservation of energy, or, more precisely, with the conservation of matter-energy. This comes about through the symmetry of time. The fact that the times and dates on our watches all read the same is simply a human convention that allows us to arrive at appointments together. There is nothing special about calling midday twelve o'clock. Physicists believe that time is totally symmetric, like space, so that the laws of the universe should be the same at one instant as the next. (Again, the reader may object that if the universe is expanding and had its origin in THE BIG BANG, this cannot be exactly true. But the rate of expansion on the human time scale is negligible.) Time symmetry means that whenever transformations take place, no matter how radical they are, something remains unchanged in time. This conserved quality is energy. In any elementary particle process, total mass-energy must be exactly the same before and after a collision.

These three conservation laws are equally valid for quantum theory and classical physics. They refer to what are called continuous symmetries. There are discrete symmetries, too. (See overview essay D, THE COSMIC CANOPY; CPT SYMMETRY; SUPERSYMMETRY.)

The conservation of energy, conservation of momentum, and conservation of angular momentum all have to do with symmetries in our everyday space and time. Other conservation laws can be expressed as continuous symmetries in more abstract spaces. For example, electric charge is preserved in all processes. This can be represented as a symmetry, not in space-time but rather in an abstract "space"—the phase of the wave function representing the charged particle. (See GAUGE FIELDS.) Again, the strong nuclear force, the color force, acts the same way on a neutron or a proton. This can be represented as an internal symmetry of these particles and a corresponding conservation law. (See QUANTUM CHROMODYNAMICS.)

# Cosmic Background Radiation

The cosmic microwave background is a nearly uniform field of thermal radiation that fills the entire universe. It is a remnant of THE BIG BANG, and its experimental detection in the mid-1960s gave a huge boost to proof that the universe had actually begun with the Big Bang. Today, the cosmic microwave background is one of five fundamental observations on the large-scale structure of the universe.

As early as the 1940s, American theoretical astronomer George Gamow and his colleagues proposed that there might be microwave radiation as a relic of the early universe, but there were no instruments to test for this. In 1965, the discovery was made quite accidentally by two American Bell Laboratory scientists, Arno Penzias and Robert Wilson, who were calibrating a large radio receiver shaped like a horn, to be used for satellite communication. Their radio picked up background noise from the earth, the sun, and our galaxy, as expected. But there was also an extra microwave signal that seemed to be coming from all directions in the sky equally. The scientists wondered if the unexpected noise was the result of pigeon droppings. But cleaning out their radio antenna made no difference to its reception. They had, in fact, detected Gamow's predicted cosmic microwave background radiation.

After the universe began with the Big Bang, it was originally filled with thermal radiation reacting with opaque matter. After 300,000 years, it had expanded and cooled enough (to about 4000 K) for electrons, protons, and neutrons to recombine into hydrogen and helium atoms. Suddenly, space became transparent to radiation, which has been traveling undisturbed through it ever since. The universe has continued to expand, and this has stretched the wavelength of the background radiation—that is, redshifted or cooled it—until today it has reached its observed temperature of just under 3 K. All this is in line with Big Bang theory predictions and difficult to explain in any other terms.

The cosmic background radiation is isotropic, the same in all directions, to 1 part in 10,000. But it is not perfectly isotropic, nor should it be. (See WRINKLES IN THE MICROWAVE BACKGROUND.) Our whole galaxy is moving relative to the microwave background, and is thought to be gravitationally pulled by a very massive, very distant concentration of matter. (See THE GREAT ATTRACTOR.)

# The Cosmological Principle

The Cosmological Principle holds that the universe is much the same everywhere. What we see from our vantage point is a fair sampling of the whole. This contradicts much of ancient belief. In earlier times, many people believed that the earth was somehow unique, and that the sun, moon, planets, and stars moved around it in circular orbits. The Ptolemaic model was accepted in the West for 1,500 years, and Judaism, Christianity, and Islam all held that God had created a unique earth, inhabited by a unique human species.

The Cosmological Principle seems attractive, in that, with no evidence to the contrary, we assume that other people are like ourselves, other places like ours. How far does this actually apply? The universe has evolved and had a different makeup at different times. (See THE BIG BANG; THE PERFECT COSMOLOGICAL PRINCIPLE.) We know that ours is not the only planet. Whether there is intelligence elsewhere in the universe remains an open question. We constantly search for messages from elsewhere, so far without success. (See INTELLIGENCE IN THE UNIVERSE.)

The most direct way to test the Cosmological Principle is to investigate the large-scale distribution of matter in the universe. On small scales, the universe is clearly "lumpy." We have planets, stars, galaxies, and clusters of galaxies, up to a scale of about $10^8$ light-years. Beyond this scale, there is still evidence of lumpiness. (See THE GREAT ATTRACTOR; WRINKLES IN THE MICROWAVE BACKGROUND.) This means that on any scale so far testable by us, the universe is not uniform, thus not as the Cosmological Principle would lead us to expect. The total universe may be much greater, perhaps infinitely greater, than those parts we can observe, and only very rare regions of it may be suitable for life. (See THE ANTHROPIC PRINCIPLE.)

# Cosmology

Cosmology is the study of the universe as a whole. (See overview essay D, THE COSMIC CANOPY.) Modern cosmology is a meeting point of observational astronomy, philosophy, and particle physics. The current dominant view in the field is a development of BIG BANG theory.

# CPT Symmetry

There are three important mirror-image symmetries in the laws of elementary particle physics. *C*, *P*, and *T* stand for symmetries of charge, parity, and time.

C symmetry holds that for every particle there is an antiparticle, possessing the same mass but the opposite electric or color charge—like the electron and the positron, or the up red quark and the up antired antiquark. P symmetry says that for every possible "left-handed" process, there is an equal and opposite "right-handed" one. T symmetry holds that, for every possible physical process, there is another possible process running in the opposite time direction.

The CPT symmetries were long thought to hold exactly. In fact, all three of them are violated, at least by weak nuclear interaction (see THE ELECTROWEAK FORCE), as though they were being reflected in a slightly distorting mirror; this violation was proposed theoretically by Lee Tsung-Dao and Yang Chen Ning, two young Chinese physicists, in 1956 and then proven experimentally shortly afterward. Most physicists were greatly surprised. Wolfgang Pauli complained that "God is a weak left-hander." But in retrospect, we now realize that none of these three symmetries holds exactly, even at our macroscopic level. The universe as a whole contains far more matter than ANTIMATTER. (See GRAND UNIFIED THEORIES.) Our bodies contain *only* left-handed amino acids in their proteins. And, through the increase in ENTROPY, we experience an ARROW OF TIME. It is perhaps not surprising that these macroscopic asymmetries are related to microscopic asymmetries.

As a final and fascinating twist to the symmetry story, the overall *combination* of CPT symmetry is conserved exactly, even though the individual C, P, or T may not be in a given situation. For example, a video of a left-handed neutrino, played backward, would look *exactly* like a right-handed antineutrino.

# Crick's Hypothesis

In 1962, Francis Crick shared the Nobel Prize with James Watson and Maurice Wilkins for the joint discovery of DNA. Since that time, Crick has devoted himself to neuroscience, specifically to the problem of consciousness. The brunt of his work is the claim that the nature and existence of consciousness can be fully explained by the nature and function of neural circuitry in the brain. In his 1994 book on consciousness, *The Astonishing Hypothesis: The Scientific Search for the Soul*, Crick states his hypothesis boldly: " 'You,' your joys and your sorrows, your memories and your ambitions, your sense of personal identity and free will, are in fact no more than the behavior of a vast assembly of nerve cells and their associated molecules." Explain these nerves cells, he claims, and we will explain away the mystery of consciousness.

Accepting the fact that the total functioning of the human brain is too complex for study, Crick has focused his research on the visual system, about which a great deal is known and on which further research is possible. His view is that current knowledge of vision, combined with appropriate research in psychology and cognitive science, will reveal the basic mechanics of consciousness.

His critics argue that his whole approach is too reductionist (see REDUCTIONISM) and lacks philosophical rigor. Few people seeking a scientific understanding of the mind would deny that neural activity is the *necessary* basis of our conscious life, but fewer still would agree that, on its own, it provides a *sufficient* explanation of conscious experience. Vision itself is not yet fully understood, particularly THE BINDING PROBLEM, yet we know a great deal about what neural activity accompanies our perception of, for example, red. The thrust of the criticism against Crick's hypothesis is that a bridge has not yet been built between the neural firing that *accompanies* our seeing red and our actual *experience* of redness. Where does that experience come from? What links the objective observation of certain neural activity with a subjective experience like "I am seeing red" and all its associated feelings? How do we get from electrochemistry to feeling?

Crick's hypothesis is a vast improvement over necessarily unscientific dualist theories that argue that mind and body are different substances, loosely attached by God or by mystery. (See THE MIND-BODY PROBLEM.) But Crick's confident claims that neurology on its own, or at least as it is understood today, can explain mind are, at the very least, misleading. Many of his critics argue that consciousness is an "emergent" property of neural activity—something that comes into being as a result of neural activity, or in association with it or its complexity, but cannot be simply reduced to it. In that notion of EMERGENCE lies a vast area of things still unknown about subjectivity.

# Cybernetics

Cybernetics is the science of systems that regulate themselves without the need of human intervention. Cybernetic systems appear intelligent; they are able to steer a course throughout the fluctuations and exigencies of the outside world. (See SYSTEMS THEORY.)

An elementary example of a cybernetic system is the thermostat connected to a home heating system. This device senses when the temperature in the room is too cool and switches on the heat. When the temperature rises above a pre-set level, the thermostat instructs the heating system to switch off. Thermostat and furnace are locked in a negative FEEDBACK loop. Negative feedback is the key to cybernetics, the method whereby a system detects that it is deviating from its preplanned path and acts to correct itself.

Cybernetic systems can be biological. Examples include the way our bodies regulate their internal temperature and chemistry (homeostasis), and the feedback among organisms on earth, which acts to maintain a constant salinity in the oceans and the composition of atmospheric gases. Cybernetic systems can also be artificial, such as the many feedback loops employed in industrial robots and for the steering of spacecraft. Entire factories and manufacturing processes can be regulated by cybernetic devices. The checks and balances in an economic and political system also cause it to act as a cybernetic system.

Cybernetic systems can also be discussed in terms of INFORMATION flows. Self-regulating systems use information about their relationship to the environment in order to steer their behavior along a predetermined path. Their various feedback loops are essentially flows of information. Today, complex self-directing systems are all controlled by computer chips. A variety of sensors supply information about, for example, the system's orientation in space and the various tasks it is performing. This information is processed by the chips and used to send out corrective signals.

The biologist C. H. Waddington noted that living systems go beyond the simple cybernetic behavior that arises through interlocking feedback loops; when a growing plant or animal is interrupted in its development, it has the ability to self-correct. Unlike a mechanical cybernetic system, however, it does not return to its original state but jumps forward *to the point it would have developed to had it not been disturbed*. Waddington called this developing path a chreod.

## Dark Matter

Dark matter is the mysterious "missing" mass of the cosmos that we know must be there but we cannot see.

The stars and planets that we can see, by optical, radio, or other telescopes, amount to only a small fraction of the mass of the universe. They total about 1 percent of the critical density (about one hydrogen atom per cubic meter) required for a "flat" universe. (See RELATIVISTIC COSMOLOGY.) But stars in spiral galaxies orbit too quickly to be held in orbit by the mere gravitational attraction of the other stars we can see. The actual mass of the galaxies must be nearer 10 percent of the critical density, and they must contain a halo of dark matter much wider than the visible galaxy. On the larger scales of clusters and superclusters of galaxies, their relative motions imply even more dark matter. The average density of the universe is known to be delicately balanced between so little that it would cease to hold things in place and so much that the whole thing would collapse. The measurements arrived at require the presence of a great deal of dark matter.

Today's universe is very lumpy, with huge voids and large concentrations of matter hundreds of millions of light-years across. (See THE COSMOLOGICAL PRINCIPLE; THE GREAT ATTRACTOR.) This foamlike structure is what we would expect if the average density was the critical density. In some regions, a larger density would lead to crowded areas; in others, a smaller density would lead to voids. (See WRINKLES IN THE MICROWAVE BACKGROUND.)

There are many variations on three ideas about what dark matter consists of. Some believe it may be ordinary (baryonic) matter that does not emit light—burned-out stars (white dwarfs), neutron stars, extremely faint stars, planets, or BLACK HOLES. But based on the CHEMICAL ABUNDANCES left over from the epoch of nucleosynthesis, no more than 10 percent of the critical density of the universe could be accounted for by this kind of matter—protons, neutrons, and electrons. A second theory suggests that the dark matter may consist of neutrinos, but we don't yet know whether neutrinos have *any* mass. A third range of speculations suggests that the dark matter may be some as-yet-undiscovered kind of particle or particles.

# Darwinian Evolution

Charles Darwin held that evolution is a chance process. The appearance of new plants and animals is motivated not by teleological (goal-directed) ends but through competition for food and territory. (See TELEOLOGY.) Our modern understanding of NONLINEAR systems, with its notions of bifurcation points, self-organization, and negative and positive FEEDBACK, places Darwinianism on a more scientific footing.

It was once thought that all nature's richness was the result of God's plan, the earth having come into existence in 4004 B.C. Clearly, the seas, rivers, and mountains—the plants, insects, birds, and animals—must all have been created at the same time, since six thousand years was not a sufficient period of time for new forms to emerge. In the nineteenth century, however, geological studies showed that the earth was far older and had passed through a variety of phases, including the building and eroding of mountains and the shifting of geological strata. The discovery of fossil plants and animals suggested that life itself was not static, and that there had been eras in which quite different plants and animals had existed.

It was within the revolutionary atmosphere of these new geological discoveries that the young Charles Darwin made his famous voyage on HMS *Beagle* and observed the patterns of plants and animals in different parts of the world. He made a study, for example, of the different forms taken by the beaks of finches that lived on neighboring islands, each with its own ecology. He reasoned that plants and animals are always in competition for a limited territory and food supply. Under such conditions, only those best fitted for a particular environment will survive. On a remote island, the plant and animal population will be singularly stable, unchanged over hundreds of thousands of years. But when an environmental change takes place, the entire ecosystem becomes stressed. Under such conditions, a new variety of plant or animal will have the edge over its competitors and will quickly become a dominant form.

It is in this fashion that the wide variety of nature's forms comes about, each one surviving by adapting to a particular ecological niche. Even behavior comes about in an evolutionary way. Grazing animals herd together for mutual protection and, when under attack, form a protective circle to shield their young. The evolutionary goal is not the survival of an individual, but that of the group or species. In this way what at first sight appears to be altruistic behavior may have been evolutionarily selected for the advantages it brings in a world "red in tooth and claw."

Darwinian evolution can now be expressed in terms of modern genetics. New varieties emerge not by design but through random mutations of genetic materials. Most mutations are harmful, or do not offer any overwhelming advantage for survival. Nevertheless, through crossbreeding, some enter the "gene pool" of that particular species. This pool could be thought of as containing a reservoir of information about individual variations within a particular genus or species. When an environment is stressed, a currently dominant form may cease to have the edge. But the potential for new varieties is already present within the gene pool, and in this way a change in a plant's or animal's abilities to survive will surface and come to dominate that particular evolutionary niche. In this sense, Darwinian evolution is not powered by the survival of the individual, or even of a particular genus, but by the survival of the gene pool itself.

Since evolution is a slow process, it is difficult to discover examples in action. One famous instance is that of the peppered moth, which exists in two varieties—one light-colored, the other dark. In earlier centuries, naturalists noticed the predominance of light moths, which camouflaged themselves on the bark of trees. Yet as the industrial revolution proceeded, trees in the north of England became covered with soot and other pollutants. As a result, the dark variety of moth had an advantage in hiding from predators and it predominated in that particular region.

Critics of Darwinian evolution argue that there was simply not enough time for a process of chance mutations to have produced anything so specialized as an eye, or so diverse as the wide variety of plants and animals on earth. There is also the question why, if human beings and apes evolved from a common ancestor, they do not continue to change. Why is the human brain not increasing in size; why do we not develop new abilities and lose others?

An understanding of nonlinear systems supports the Darwinian position. (See NONLINEARITY.) Just as negative feedback in a stable marketplace means that companies remain relatively stable over a long time period, so, too, plants and animals cease to change within a stable ecology. Yet once the environment is stressed, nonlinear effects allow self-organization to come into play as new orders emerge out of the chaos produced. (See CHAOS AND SELF-ORGANIZATION.) In a competitive market, for example, through positive feedback a minimal advantage can be rapidly amplified until it overpowers its competitors. In a similar way, when the environment is stressed, small fluctuations within the gene pool will be rapidly selected to produce a new variety or hybrid in a short space of time. Neo-Darwinians picture evolution as involving periods of relative stability interspersed with intervals of rapid change. (See PUNCTUATED EQUILIBRIUM.)

Darwinian evolution is based upon competition, and, in this sense, it is a zero-sum game (see GAMES, THEORY OF) in which any evolutionary advantage is always gained at the expense of some other variety or species. By contrast, COEVOLUTION suggests that mutual advantages can be gained through cooperation. Seeing nature as a non-zero-sum game suggests that naked competition may not, in the long run, be the best strategy.

# Determinism

Must things always happen the way they do? Is there a pattern or set of laws that determines the outcome of every event, or are there "escape clauses" in life and in nature? Determinism is the philosophy that holds that there are not. Whatever happens, it argues, there are conditions that indicate things could never have gone any other way.

Determinism has taken many forms. When people have predicted a phenomenon successfully, they often wonder whether *everything* is predictable. The ancient Babylonians mastered the art of predicting eclipses and phases of the moon, but they were unable to predict the weather or foresee epidemics of disease. More at home with the heavens and their apparent orderliness, they invented a system of celestial deities and tried to predict earthly occurrences from the movements of the sun, moon, and five visible planets. Our seven-day week derives from this.

From earliest times, the Greeks believed Fate was an impersonal power that corrected imbalances in the behavior of matter, human beings, and even the gods. Everything had its appointed role, every feeling or action its appropriate expression. If these boundaries were overstepped, through pride or hubris, Fate reestablished the balance. Fire eventually consumed itself and became smoke and ashes. A human fault or excess was punished in the end. But like the Eastern doctrine of karma, the Greek view of justice could coexist with chance and human freedom. Some Greek philosophers were determinists. Many were not. Similarly, the doctrines of original sin and predestination are determinist, although many Christian thinkers believe in at least some form of free will.

Classical physics is rigidly determinist, both in the predictions of its equations and in its broader philosophical foundations. There is no scope for chance, surprise, or even, in any meaningful sense, creativity. Everything is as it has to be. Newtonian mechanics is determinist as far as it goes—Newton's three laws of motion accurately describe the necessary behavior of any particle acted upon by a force—but Newton himself was both a Christian and an alchemist, and he never fully articulated, indeed may never have fully appreciated, the wider deterministic implications of his own work. It was left to the great mathematician Pierre-Simon Laplace a century later to work out the determinist philosophy underpinning the classical physicists' picture of the universe. Laplace saw that the iron laws of physics exactly determine any future event. If we know the initial state of any system and the forces acting upon it, we can always predict exactly (in principle) how *B* will follow A.

Although human consciousness, goals, and purposes play no role in classical physics, this scientific philosophy exerted an enormous influence on nineteenth- and twentieth-century psychology and sociology. Freud articulated a determinist view of the self, with the psyche driven by the tempestuous forces of the id and adult behavior determined by early childhood experiences. Sociologists argued that environment and social conditions determine social behavior, and this view crept into legal procedure with the familiar defense tactic that a criminal couldn't help committing his crime because his circumstances or early experiences "led" him to do it. Experimental psychology extended the model of Pavlov's dog to human behavior and supposed human beings were conditioned to have a given response to any particular stimulus. These views threw into question how much responsibility human beings bear for their actions.

Twentieth-century physics challenges the old physics' determinism (see INDETERMINACY), but people still cling to the old paradigm. Steadfastly maintaining that "God does not play dice with the universe," Einstein himself, a founder of quantum physics, could never accept its lack of determinism or "rules of the gaming house" applied to physical reality. Some physicists supported David Bohm's attempt to formulate a causal theory for the evolution of quantum systems (see COLLAPSE OF THE WAVE FUNCTION), and some interpreters of chaos theory emphasize that, even though chaotic systems may be too complex to be predictable, underneath they are fully determinist. (See CHAOS AND SELF-ORGANIZATION.) But these views are not supported by scientific evidence.

# Dissipative Structures

The Nobel Prize–winning scientist Ilya Prigogine coined the term *dissipative structures* to describe systems that spontaneously come to order. Closed, isolated systems and systems that are in equilibrium with their surroundings maximize their ENTROPY and move toward a featureless state. By contrast, in a dissipative system matter and/or energy constantly flows through the system from its environment. This enables the system to remain in a far-from-equilibrium state, in which the development and maintenance of internal structure become possible. Thanks to this flow of matter or energy, the system's internal entropy is reduced and order emerges out of chaos.

Examples of dissipative structures range from the vortex formed as water runs rapidly past an obstruction, stable convection patterns formed in a pan of heated water, and certain autocatalytic chemical reactions, to living cells, cities, and patterns within traffic flows.

# Distance Measurements in Astronomy

The crucial measurement of a star's or galaxy's distance from us cannot be made directly. A light source of a given brightness may be close and intrinsically dim, or distant and intrinsically bright. Consequently, astronomers rely on a dozen or more indirect methods to measure distance, and the results are uncertain within a range of up to double the minimum distance.

One method of distance measurement is the "parallax" method. The stars nearest us can be observed to move their apparent positions against the more distant background, as the earth moves around the sun. This allows a calculation of their distance. Modern methods involving radar ranging of the planets and accurate measurements on earth of the speed of light have given us an accurate model of the solar system.

Another method utilizes a class of main-sequence stars, the Cepheid variables, which pulsate. Their average intrinsic brightness is accurately related to the period of pulsation (from one to one hundred days), as can be observed in any globular cluster where stars are all at about the same distance. Hence the intrinsic brightness of these stars in nearby galaxies can be calculated, and thus the distance of each galaxy. This method works for stars at up to about 10 million light-years away, beyond which Cepheid variables are too faint to be seen individually. It was this method of distance measurement that enabled Edwin Hubble to do his pioneering work on the distances of the nearby galaxies and THE EXPANDING UNIVERSE.

These are but two of many methods for measuring distance.

# DNA

DNA (deoxyribonucleic acid) is the "miracle" macromolecule that carries the instructions for life on this planet. It is like a tape whose parts, when "played" on the appropriate cell machinery, are translated into the various protein components of organic cells. It is organized into genes—long sequences of molecules—each of which is an instruction to assemble a particular kind of protein out of the twenty-odd protein components (amino acids) floating in the cellular "soup." (See THE HUMAN GENOME PROJECT.)

In addition to the message itself, DNA contains reading instructions and active punctuation marks that indicate where a particular message begins and ends. The molecule also makes use of error-checking strategies that ensure, for example, that the cell does not begin to read DNA in the middle of a sentence or skip part of an instruction.

DNA is the main memory store of each living cell. The cell reproduces templates of DNA—worker molecules called RNA (ribonucleic acid). These travel to the ribosomes, tiny machines that are responsible for building the various protein components required. Each component plays a specific role in cell structure or the chemical reactions that power and repair cells. Proteins are synthesized in exactly the right sequence and amount for the cell's proper functioning. Some proteins are enzymes, chemical catalysts that in turn regulate other metabolic processes. The control mechanisms are partly in the DNA and partly in the rest of the cell, as with computer software and hardware. During an organism's growth phase from embryo to adult, some of the DNA's instructions remain inactive until the requisite level of maturity has been reached, at which point they are stimulated into action.

DNA's special structure (a double helix) also allows a cell to reproduce itself. When the cell replicates, the twin strands of the DNA helix begin to unzip and separate; DNA even gives instructions for the manufacture of enzymes that will unzip it. When one strand is separated from its twin, its chemically active nucleotides—the letters of the genetic code—are no longer stabilized by their partners on the other strand. As a result, they attract the corresponding nucleotides from the chemical soup within the cell nucleus. Like a series of children's building blocks, an ordered sequence of nucleotides begins to align itself and, in the process, creates an identical copy of the original DNA strand. Following cell division, the DNA copy is incorporated into the nucleus of the new cell. (See CONSTRUCTION COPIER MACHINES.)

Despite DNA's ability to detect and correct errors in the copies it makes of itself, there are occasions when a DNA message is reproduced incorrectly, or where the molecule itself is damaged through chemical contamination, radiation, or viral attack. In most cases, this is so disruptive that the cell malfunctions and dies. There are also instances in which an incorrect DNA message will cause a cell to multiply in an uncontrolled way, becoming cancer. Errors at the earliest stage of development can result in birth abnormalities and mutations.

Yet, despite the fact that DNA is constantly under attack from natural levels of radiation, chemical contaminants, and viruses in the environment, such abnormalities are the exception. Those rare mutations able to survive into adulthood are generally not fertile. It is only in exceptional cases that DNA's message becomes transformed in such a way that a new variety or even species appears on the planet.

Despite several decades of international research, DNA holds many secrets. For example, only about 10 percent of DNA's genes appear to be read and used by the cell. What is the function of the silent 90 percent? Are they built-in redundancy to guard against the loss of part of a message? Are they ancient messages, no longer used, that had importance earlier in evolution? There is also a question of "jumping genes," portions of the DNA message—entire paragraphs, if you like—that are capable of leaving their position in the molecular sequence and jumping to some new location in DNA.

It is sometimes said that life exists only to serve DNA, but this is very one-sided. Life and DNA react to benefit each other, and evolutionary diversity is the real winner. A tadpole and a frog, for example, have exactly the same DNA, as do a caterpillar and a butterfly. But the ultimate way in which the DNA is expressed depends upon both the cellular machinery and the environment in which the organism finds itself at the time. Some DNA genes become active only at certain temperatures, or at some point in the organism's life cycle.

Not all genetic material is confined to the DNA in the cell's nucleus. Tiny bodies called mitochondria, found within the cell's cytoplasm, also contain portions of DNA. They play an important role in the cell's metabolism but are relatively autonomous of the nucleus's central program. Mitochondria are thought to be survivors of the first forms of life, protobacteria, which survived through a symbiotic relationship within the cell.

Viruses are also in large part a form of DNA. A virus cannot reproduce by itself because it does not contain the infrastructure of a living cell. It is like a tape (DNA) without a tape player (ribosomes). Instead, like a cuckoo in a nest, the virus enters a host and takes control of the metabolic functions of its cells in order to reproduce itself. In the process, it often destroys the cell, and sometimes the entire organism. At one level a virus is a parasite whose presence does not bode well for the host. But there is also speculation that viruses can inject sections of foreign DNA into a cell's genetic message. In this way, DNA may perhaps be able to communicate with its environment, generate new mutations, and accelerate evolution.

While much of DNA still remains veiled from biologists, the Human Genome Project is an attempt to read the entire molecular message in human genes. Through international collaboration, different research teams will each work on a selected segment of DNA and decode its molecular message. Biologists hope to identify various human characteristics and dysfunctions with specific instructions on the DNA molecule. Many scientists hail the project for the new knowledge and medical breakthroughs it will bring. Others point to the many ethical issues that must be debated.

# The Edge of Chaos

The edge of chaos is that critical band where physical systems display the greatest COMPLEXITY. It is delicately poised between order and chaos.

Crystals and gases represent the two extremes of order and chaos. Crystals are very ordered, gases entirely random. In a perfect crystal, each atom occurs in its place in a regular, repetitive lattice, like wallpaper. The system can be described very simply. In gases, the motion of each molecule is random. At the molecular level, the system is too complex and chaotic to be described at all. (At the macroscopic level, all the complexities average out to give an overall pressure and temperature.) Neither crystals nor gases have much interesting structure. Neither contains much INFORMATION. To convey information with either medium, we have to inscribe messages on the crystal or send sound waves through the gas.

The extreme examples of order and chaos represented by crystals and gases suggest that the most complex and interesting structures are to be found at *neither* extreme, but rather at the *boundary between* order and chaos—"the edge of chaos." This is indeed where we find complex structures like living organisms, ecologies, whirlpools, and perhaps conscious brains. All have both regular and unpredictable features. Computer models of ARTIFICIAL LIFE suggest that the most complex behavior occurs in a narrow band between repetitive and chaotic behavior.

Work at the Santa Fe Institute suggests that edge-of-chaos systems evolve naturally. Physicist Per Bak has built a model of "self-organized criticality" illustrated by a pile of sand. More sand is slowly added to the top of the pile, which spontaneously evolves into a conical shape. The cone remains stable until it reaches a critical slope at which the addition of even a *single* grain of sand will cause an avalanche, large or small. At that point, the system shows maximum responsiveness and maximum unpredictability. These characteristics—adaptability and fragility—seem to characterize many ecosystems and also such human phenomena as fashion trends and the stock market.

Theories of self-organized criticality suggest that evolution drives these systems to the edge of chaos, where they can be maximally creative but are also unstable. This may have very deep implications for the human desire to maximize security, to arrange nature or our own affairs in some static Utopia. Edge-of-chaos theories suggest that we might be wiser to aim at dynamic balance, to accept that given forms of order are impermanent and that nature itself evolves through a constant strategy of risk taking. Such thinking now emerges regularly as a challenge to business and financial circles, but at this stage it may be no more than a metaphorical extension of chaos science. (See CHAOS AND SELF-ORGANIZATION.)

# The Electroweak Force

The unification of electromagnetism and the weak nuclear force has proved to be one of the triumphs of modern theoretical physics. In addition to the electromagnetic force between electrons and the strong nuclear force (now reduced to gluons) binding protons and neutrons within the nucleus, physicists recognize a third force, intermediately strong between the other two: the weak nuclear force responsible for various forms of radioactive decay (beta decay), including the way a neutron decays into a proton, electron, and antineutrino.

Physicists naturally wanted to explain the weak nuclear force by means of a quantized field, in the same way as they had done with the electromagnetic force. As the quantized vibrations of the electromagnetic field, photons, carry the electromagnetic force, other BOSONS, intermediate vector bosons, carry the weak nuclear force.

The problem is that, unlike the electromagnetic force that acts over long distances, the weak nuclear force is virtually a contact force. Its range is much smaller than the size of a proton. Because the range of a force is related to the mass of the force-carrying particle, the hypothetical carriers of the weak nuclear force would have to be extremely heavy—much more massive than the proton or neutron.

There are still other differences between the two forces. The pho-

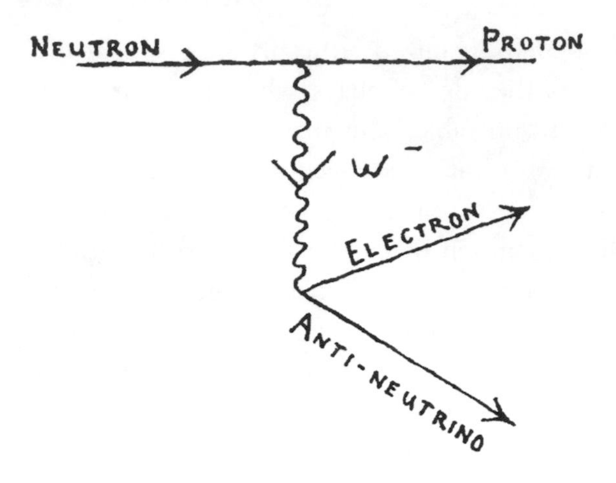


DECAY OF THE NEUTRON

ton is electrically neutral, because when an electron borrows and later pays back photons from the electromagnetic field, its charge does not change. This is not the case for the weak nuclear force, in which an electric charge can change during a wide variety of interactions. This means that a total of three massive bosons, one neutral ($Z^0$) and two electrically charged ($W^+$ or $W^-$), are needed to carry the weak nuclear force. Moreover, the force is "left-handed," i.e., unlike its mirror reflection. (See CPT SYMMETRY.)

Despite these differences, there are striking analogies between the weak and electromagnetic forces, which can both be expressed as GAUGE FIELDS. Inevitably, physicists proposed a single, unified gauge theory in which the two forces are aspects of the one field. But there was a difficulty: The carriers of a gauge field are always massless. This would mean that both forces were long range, a total contradiction to the contactlike nature of the weak force. It was left to theoretical physicists to demonstrate how spontaneous SYMMETRY BREAKING could save the day (the Glashow-Weinberg-Salam model, completed by 1968). Using what is called the Higgs mechanism, when symmetries are broken, some of the massless bosons acquire mass, and the electroweak theory thus contains two neutral particles, the massless photon and the massive $Z^0$, plus two massive charged bosons, $W^+$ and $W^-$. Experiments carried out at CERN from 1973 onward successfully identified traces of these particles, exactly as they had been predicted by the electroweak theory. The electroweak theory, a triumph of unification, has led physicists to hope that a single Grand Unified Theory (see GRAND UNIFIED THEORIES) will eventually be built out of the electroweak force and the color force between QUARKS.

The weak interaction converts one lepton into another, or one quark (within a hadron) into another. (See HADRONS; LEPTONS.) The decay of the neutron is represented by the Feynman diagram on page 136. The total quark number (quarks minus antiquarks) and the total lepton number are saved. These conservation laws can be expressed as CONTINUOUS SYMMETRIES.

# Emergence

Does anything really new ever come about in physics? Are properties or characteristics of things ever created or evoked "from nothing"? Emergence is a view that holds the answer to these questions to be, in some very important sense, yes.

A property of an entity or complex system is said to be emergent if it cannot be defined or explained in terms of the properties of its parts, or if it is not reducible to these properties and their relations. Many people have asked if life and consciousness are reducible to physiochemical processes in the body or brain, or if they are emergent phenomena—truly new phenomena displaying properties without a classical, physical antecedent. Philosophers have wondered whether things like conceptual thinking, mathematics, and "society" are emergent phenomena. Can the properties and behavior of any group be wholly defined in terms of the properties and behavior of its individual members?

Classical physics is rigidly reductionist. It holds that the properties of any whole can be broken down into the properties of simpler parts—ultimately of atoms—and the forces acting between them. (See REDUCTIONISM.) Taking its lead from this, a reductionist paradigm leads to explanations of biological processes in terms of simpler physics and chemistry, and to an explanation of human behavior in terms of animal instincts or, more recently, computer processes. Complex human experiences like love or spirituality are similarly "explained away"—although, so far, all attempts to reduce consciousness itself to Newtonian physical or chemical categories have proved unsuccessful.

By contrast, all properties of quantum entities or systems are emergent properties. Things like position, momentum, energy, spin, and so on simply do not exist (are indeterminate) until they are measured or observed—or until the Schrödinger wave function collapses. They exist as potentialities, but they cannot be used to make predictions or to explain the properties of actualities. Similarly, when two quantum systems meet, their wave aspects (potentialities) overlap to form a new, combined system that has properties not possessed by either of its constituents. The whole is greater than the sum of its parts. Thus, in the photon experiments to test BELL'S THEOREM, neither of the paired photons has a polarization until it is measured, even though the relation of their eventual polarizations ("oppositeness") is already fixed. The two photons are "entangled" together—that is, they acquire a new property that cannot be reduced to properties possessed by either of them singly. (See HOLISM.)

In quantum physics, the potentialities of most or all systems are entangled to some extent, so the emergent properties of their coming together are all-pervasive. In an important sense, *all* properties in the world are emergent to some extent, since all have an ultimate quantum foundation, but the properties of most very large things, like boulders or desks, can, as a good *approximation,* be reduced to the properties of their atoms and the forces acting between them. (Not all, however; see BOSE-EINSTEIN CONDENSATION.) This has enormous metaphysical implications and presents the possibility of looking at the material world in a whole new light. It tells us that nothing can ever be *wholly* reduced to the sum of its constituent parts. There is a surprising, creative edge to all existence.

Does emergence imply that something comes from nothing? Not strictly. In quantum physics, actualities emerge from an indeterminate sea of potentiality, but potentiality itself has a kind of ontological status. (See ACTUALITY AND POTENTIALITY IN QUANTUM MECHANICS).

It is tempting to wonder whether the properties of the world that have so far resisted attempts to reduce them may be based on emergent phenomena associated with quantum holism or self-organization. These include consciousness, conceptual thinking, and perhaps life itself. This would require biological systems (including brains) to contain large-scale emergent phenomena (quantum processes and/or COMPLEXITY), a possibility that is discussed in several places in this book.

The emergence/reductionism debate mirrors the holism/atomism debate. But the former is about the *properties* of things; the latter, about things themselves.

# Entropy

Automobiles rust; dead trees decay; old buildings fall down. These are examples of the inevitable march of entropy, the spontaneous movement from order to disorder. THE SECOND LAW OF THERMODYNAMICS dictates that in every one of nature's changes entropy, which means disorder, decay, dissipation, and the breaking down of patterns and structures, must either increase or remain the same.

In physics, entropy is discussed at two levels. Classical THERMODYNAMICS treats entropy as a parameter, along with the parameters of heat, work, internal energy, and temperature. Its definition lies in its relationship to these other parameters. Studies of the efficiency of machines have indicated that heat can never be totally converted into work, and that all machines therefore have less than 100 percent efficiency. The reason for this loss of usable energy is the inevitable friction and consequent increase in entropy.

The Second Law of Thermodynamics indicates that isolated systems move spontaneously toward maximum entropy. As energy circulates around a system, entropy is produced, and some of the usable energy is dissipated as heat, rendering it unavailable for work. When any machine—a steam engine, electric dynamo, or human body—is running, energy is lost through friction in the form of heat. No matter how well an engine is designed, nature extracts a payment in terms of entropy increase. Only for machines that approach what are called reversible processes (processes that take place very gently and are always close to equilibrium) is the entropy increase a minimum. Although these machines are the most efficient possible, they still cannot convert all the heat supply into useful work.

Entropy can also be explained at the molecular level by STATISTICAL MECHANICS. Just as temperature is interpreted as the degree of molecular agitation, so entropy is defined in terms of a breakdown in molecular order. A substance may be created in an ordered way, but at any finite temperature its molecules are in a state of agitation. The result is like shuffling a deck of cards that have been arranged by value and suit. In the first few shuffles, some of the initial order may still be preserved—several cards of the same suit may remain in sequence—but eventually even this will be lost. While, in a shuffle, the original sequence has as much chance of recurring as any other given arrangement, the number of possible arrangements is astronomical, and it will probably be many centuries before that particular sequence pops up again.

Molecular shuffling acts in the same way to break down order in a system. Whenever the circulation of heat is concerned, there will be a degree of molecular shuffling that destroys the system's original order, making it less able to perform useful work. Imagine two cylinders, one containing gas at high pressure, that are connected by a narrow tube. Since all the gas is collected in one half of the system, the entropy is lower than if the gas were equally distributed. The pressure difference between the cylinders gives them the capacity to do work, but very quickly the random motions of molecules cause a drift of the gas between the cylinders until the pressure is equalized. This increases the entropy of the system, and its ability to do work falls.

Molecules in ice form an ordered pattern. Place an ice cube in a glass of Scotch. As it warms, its molecules move faster and lose their initial pattern. Entropy increases, and the ice melts. In a sheet of iron, atoms are arranged in a lattice pattern. When the iron rusts, atoms of iron combine with oxygen in the air to form iron oxide. Not only does the regular metal lattice break down, but also the boundary between air and metal. Rusting is associated with an increase in entropy.

If entropy always increases and systems always move toward increasing disorder, how is it possible to make ice cubes? Ice has a more ordered structure than water—pound for pound, its entropy is much lower. Does this mean that the freezer trays of a refrigerator defy the march of entropy? The reason that the laws of statistical mechanics are still respected is that a refrigerator is an open system and the total entropy of ice plus environment does, in fact, increase. (See OPEN SYSTEMS.)

The concept of entropy and its increase is of enormous significance. It applies not only to steam engines and refrigerators, but also to the human body. We consume food and retain our internal order at the expense of increasing the entropy in the environment around us, in the form of heat and the body's waste products. Even a city produces entropy. It is an open system into which materials and energy flow, and out of which garbage, sewage, and heat issue.

The concept of entropy also plays an important role in INFORMATION theory, the study of the way messages are sent along transmission lines. No transmission link is perfect, and telephone lines and radio transmissions generate a certain degree of "noise"—random sounds that degrade the quality of a signal. Information theory provides a measure of the amount of information passing along the line and the amount that is lost through noise. Since a breakdown in the order of a signal is an increase in disorder, engineers use the twin concepts of information and entropy. The amount by which entropy increases is directly related to the degree to which information in a signal decreases. Information could therefore be thought of as negative entropy or "negentropy."

Since the entropy of the whole universe always increases and is still very far from its maximum, entropy must have been very low at the time of THE BIG BANG. We do not know how to explain this.

# Equilibrium

The idea of equilibrium has dominated physics for more than two hundred years. Only relatively recently have physicists discovered the interesting sorts of things that can take place when a system is forced away from equilibrium.

Equilibrium can be either static or dynamic. The scales used in an old-style pharmacy are an example of static equilibrium, which can be totally stable, like an olive lying at the bottom of a highball glass. Give the olive a push, and it rolls back to the bottom of the glass. Static equilibrium can also be unstable. An olive perched on the rim of the glass is perfectly balanced and will not move until it is given a tiny push. In this eventuality, however, it falls off the rim.

There is another, physically more interesting, type of balance called dynamic equilibrium. This occurs when competing processes match exactly. Turn on the bath tap, and the level of water rises. Pull out the plug, and the level falls. Dynamic equilibrium occurs when the rate of water entering the bath exactly matches the rate at which water disappears down the drain. The water level in the bath is static, but the overall process is a dynamic one—water constantly entering and leaving.

Ozone in the earth's upper atmosphere is produced by the action of sunlight on oxygen. It is destroyed by other natural processes. Dynamic equilibrium occurs when the destruction of ozone is exactly balanced by its creation and the concentration of ozone in the upper atmosphere remains constant. The "ozone hole" is a disruption of this dynamic equilibrium, produced by contaminants in the upper atmosphere that break up the ozone molecules faster than they are formed.

The temperature of the human body is another form of dynamic equilibrium, in which heat loss through cooling is balanced by the burning of sugars. Homeostasis—maintaining an internal equilibrium—is aided by sweating when the body gets too hot and shivering when it gets too cool.

Until well into this century, physics generally dealt with systems in dynamic equilibrium. Where deviations had to be considered, they were always assumed to be small, so that the system could adjust and come back into dynamic equilibrium. It was only relatively recently that scientists like Ilya Prigogine began to study systems that are far from equilibrium. Systems close to equilibrium also behave in a linear way. Prigogine was able to show that nonlinear systems (see NONLINEARITY), OPEN SYSTEMS, and far-from-equilibrium systems have a much richer behavior. They are able to form spontaneous structures and evolve novel forms of behavior.

# The Expanding Universe

According to the BIG BANG theory, our universe is steadily expanding from an almost infinitely dense point singularity. The expansion is being slowed by gravitational attraction, but whether it will continue indefinitely or one day reverse toward a "Big Crunch" remains a matter of conjecture.

In a Newtonian model of space, the Big Bang would have resembled an explosion, sending bits of the cosmos in all directions into a preexisting, absolute space. We know now that this is not correct. In GENERAL RELATIVITY, matter, energy, space, and time are inextricably bound together, and all of them were created at the Big Bang. A more accurate model of what happens is an expanding balloon. If we think of space as the surface of the balloon, and the galaxies in the universe as so many insects crawling slowly about on the surface, this means the recession of the galaxies from one another is due to the expansion *of* space, not the distribution of debris *through* space.

It is tempting, but scientifically impossible, to ask what happened before the Big Bang. Since time itself as we know it was created at that moment, there was no "before." Some religious thinkers argue that God created the universe at the moment of the Big Bang. A few scientists, including Stephen Hawking, have argued that the universe was a quantum fluctuation from a preexisting "vacuum." (See QUANTUM VACUUM.) But all that is certain is that our universe is not static, but evolving from an unimaginably violent beginning.

# Expert Systems

Expert systems are serial computer programs that encapsulate some body of human expertise. They are an application of ARTIFICIAL INTELLIGENCE that can save a great deal of time and money in cases where the knowledge and techniques of a human expert can be summed up as a simple set of rules or search procedures. There are expert systems to help chemists trace the structure of organic molecules, others to help doctors with diagnoses, and still others to assist lawyers in gathering relevant arguments based on past case citations. ELIZA, the computer program meant to simulate a psychiatrist's interview with a patient, is a famous spoof of expert systems.

Creating an expert system is costly and difficult. Human experts must be interviewed extensively about the knowledge they would call upon and the procedures they would use for analyzing or classifying something. Sometimes these interviews lead to clear-cut procedures and thus to successful expert systems, but as often as not they flounder on the main stumbling block to all artificial intelligence software: Human experts often don't *know* how they make their decisions, or at least can't formulate a method in explicit, logical sequence. Great doctors and lawyers often act out of intuition, or moments of insight, or simply with tacit skill. They have "hunches" or decide on the basis of probabilities. Asking them to outline their thinking step by step can be as difficult as asking a native speaker of a language to articulate the grammatical rules he or she uses unconsciously while speaking. On the other hand, if extensive questioning of human experts *can* eventually lead to constructing an expert system, the rules of that system may be much more clearly formulated than those used by the experts from whom it was learned.

There are expert systems for serial computers that function as TURING MACHINES and can function only according to clear, logical, step-by-step rules. They are not useful if a system is governed by many partial rules that have exceptions or interact (as in the translation of natural languages), or where pattern recognition is required (as in distinguishing smells or fingerprints). In some of these cases, NEURAL NETWORKS can act as "experts," but there are many human skills for which there are as yet no equivalent AI tools.

# Feedback

Why does a new product sweep away competitors and dominate the market? How does order emerge out of chaos? The answer is positive feedback. Cybernetic systems make use of negative feedback, small corrective signals, to keep themselves on track. Positive feedback acts in the opposite way, amplifying chance fluctuations.

The governor on a nineteenth-century steam engine is an example of negative feedback. If the engine runs too fast, the governor cuts down the supply of steam. When the engine slows down, the governor allows more steam to flow. Positive feedback works in the opposite way, magnifying fluctuations and forcing a system to enter new domains of behavior. Positive feedback is the reason behind the screech sometimes heard on a public-address system. The slightest sound picked up by the microphone is amplified by the system until it can be heard through the speakers. The sound from the speakers is, in turn, picked up by the microphone and amplified again, until it blares out of the speakers as a louder noise. Within a fraction of a second, this positive feedback loop creates an ear-splitting screech.

Positive feedback is the mechanism whereby new orders are able to emerge in open systems. Take, for example, the competition between Betamax and VHS for the video recorder market. Both systems were produced about the same time, and some experts argued that Betamax was technically superior. The two systems competed for the same market until VHS gained a very slight edge. At this point, positive feedback kicked in to amplify what was initially only a tiny fluctuation.

People often purchase on their friends' recommendations. Thus, after slightly more VHS machines had been purchased, new buyers tended to ask for them. Suppliers noticed the trend, and more films were made available in the VHS format. Prospective buyers, realizing that a wider choice of their favorite movies was available, chose the VHS system. At each stage of this positive feedback loop, more VHS machines were sold, more models placed on display, and a higher percentage of films made available. Very rapidly VHS dominated the market. The reason did not lie in the technical superiority of one system over the other, or even relative skills in advertising, but the operation of positive feedback, "the law of increasing returns" within a brand-new market. By contrast, well-established commodities and products tend to settle down into economic equilibrium, with any fluctuations in the market being stabilized by the forces of negative feedback.

The stock market itself is a complex system of negative and positive feedbacks. Economic competition tends to iron out fluctuations through NEGATIVE FEEDBACK. On the other hand, when the market is filled with uncertainty and speculation, investors try to second-guess what other investors will do. As a result, the slightest dip in a stock price causes others to sell, and a wave of positive feedback sweeps the market, rapidly depressing prices below their true values. At this point, those who previously sold short invest their profits and prices rise again. A combination of several such feedback loops makes the market highly nonlinear and subject to wide varieties of behavior, from stability to unexpected crashes and even chaos.

Positive feedback is an evolutionary driving force, in social and environmental as well as economic systems. A truck stop at a remote highway intersection can grow into a collection of shops and services, or even into a small community. The competing forces of negative and positive feedback could amplify this community into an entire town, or equally well cause it to wither away. In a particular environment, a new plant hybrid or animal may exhibit a slight advantage over its competitors and, thanks to positive feedback, dominate that region in a few generations. (See DARWINIAN EVOLUTION; PUNCTUATED EQUILIBRIUM.)

In certain chemical reactions, the presence of a particular substance catalyzes its own production. The results are autocatalytic reactions, which create self-sustaining structures in space and time. Something analogous occurs when water is heated in a pan. At first, the water behaves chaotically, warm water attempting to rise and competing in the same space with cooler water trying to fall to the bottom of the pan. (Hot water is less dense than cold.) Through positive feedback, these initial fluctuations become amplified until, in one region, a volume of warm water forms a rising column, while nearby a stable column of cool water falls to the bottom of the pan. Seen from above, water in the heated pan settles down into a stable convection pattern of cells of rising and falling water. Positive feedback is the mechanism whereby many of nature's systems evolve internal structure out of initial chaos.

# Fermions

Fermions are the elementary particles that embrace all matter. Their opposites are BOSONS, the particles of force or relationship. (See Box Two in overview essay D, THE COSMIC CANOPY.)

Elementary fermions, QUARKS and LEPTONS, have a quantum spin of ½, a mathematical description of their angular momentum. Mesons, which are combinations of two quarks, have an integral spin of 1 or 0, depending upon whether or not the spins are parallel. They behave like bosons from a distance wide enough that their composite structure cannot be seen. Similarly, atomic nuclei or atoms themselves may be composite bosons or fermions.

Fermions obey Fermi-Dirac statistics. (See SPIN AND STATISTICS.) Only one fermion can occupy any one state; thus they resist being pushed too close together. This aversion to bunching explains the solidity of matter. In the speculative GRAND UNIFIED THEORIES there is basically only one kind of fermion and one kind of boson. The diversity of these particles that we see at our level arises from SYMMETRY BREAKING, the process by which a symmetrical set of quantum possibilities becomes one asymmetrical actuality—like a lottery anyone might win that finally produces one definite winner.

# The First Law of Thermodynamics

The First Law of Thermodynamics is an expression of the conservation of energy in the universe. It describes the relationship between heat and work.

Scientists in the early eighteenth century believed heat was a fluid called caloric. Just as water flowed downhill, caloric flowed from hot to cold objects. According to this theory, the caloric in a cup of hot coffee leaked away into the environment, and as it flowed out of the coffee, the liquid cooled. At the same time, the surrounding air and the table under the cup became filled with caloric and heated up. The total amount of caloric in the universe remained constant; heat could flow from one place to another but was never created.

While watching a cannon being bored, Count Rumford (Benjamin Thompson) discovered a flaw in the caloric theory. He noticed that the cannon, the metal chips, and the boring instrument itself were all heating up. From where was the caloric flowing to heat up the metal? There were no hot objects nearby. The only answer was that some of the work being done in boring the cannon must be generating heat. This meant that heat could indeed be created, and at one blow the caloric theory was demolished. Some years later James Joule demonstrated that the work done as a paddle rotates in water acts to heat the water, and scientists were able to show that electrical and chemical energy can be converted into heat.

The First Law of Thermodynamics is a statement of total energy conservation—energy is not created or destroyed; it only changes its form. When a system performs work, or cools, it loses a little of its internal energy. Likewise, when a system is heated, or work is done on it, it gains internal energy. The First Law of Thermodynamics tells us that heat or work does not simply vanish but is interconverted in such a way that the total energy of the system and its environment remains constant. In a steam engine, for example, the First Law dictates that the total energy output is *equal to the amount of heat supplied*.

The First Law of Thermodynamics indicates that there is no free lunch in the universe. Work cannot be done for nothing; it has to be paid for through a change in the system's internal energy or by the application of heat. The First Law likewise shows that systems do not spontaneously heat up unless work is being done on them. (The Second Law of Thermodynamics places restrictions on exactly how heat can flow and work be done.)

In his theory of SPECIAL RELATIVITY, Albert Einstein transcended the First Law of Thermodynamics by showing that objects can heat up without the input of work. Marie Curie had noticed that the radium she purified was always warm to the touch. This heat, Einstein argued, is generated by nuclear disintegrations. A small amount of matter ($m$) in the nucleus is converted into pure energy ($E$). Einstein related the energy created to the amount of matter destroyed, in his famous equation $E = mc^2$, where $c$ is the velocity of light. Since the speed of light is so enormously large—186,000 miles per second—a great amount of energy can be released at the expense of a tiny amount of matter. This is why, for example, the sun and stars are able to "burn" for millions upon millions of years. The law of conservation of energy is thus not strictly true. It has been replaced by a more general law: The sum of mass and energy is conserved.

# Formal Computation

Formal computation is rule-bound (algorithmic), step-by-step information processing of the sort all serial computers use. It contrasts with the parallel processing done by NEURAL NETWORKS, in which many computer elements are interconnected and many stages of computation are carried out simultaneously. Formal computation is as old as recorded human thinking, and the first known serial processing "computer" was the abacus.

The mathematical theory of formal computation is abstract, but its implementation in a serial processing system can be illustrated with a very simple example: We might imagine a Stone Age "computer" consisting of four large pots labeled *A*, *B*, *C*, and *D*. Each pot is capable of holding zero, one, two, or any finite number of stones. The four pots stand in a cave that has a plentiful supply of stones on the ground. The "computing" system is operated by a man who has a list of instructions to work through—the "program." This is a program for altering the number of stones in each pot initially—the "data." When the man has finished shifting the stones about according to his instructions, the number of stones left in one or more pots is the "result" of the calculation.

The stone-shifting program's instructions are of a few simple types. The man begins with the first instruction, then follows through each of the others in order of presentation, unless instructed otherwise.

X+ means: Add one stone to pot X (where X = A, *B*, *C*, or *D*).

X− means: Take one stone out of pot X, if there is a stone in X, and throw it on the ground. If pot X is empty, follow the arrow to the next indicated instruction.

*X!* means: Display contents of pot X to observers.

*Go* means: Go next to the indicated instruction. (Follow the arrow.)

A high-speed electronic computer is no more complicated in principle than this system for changing stones in pots. The modern serial computer uses electric charges in transistors or magnetic areas on disks, instead of stones in pots, and it has more sophisticated input, output, and memory facilities, but that is all. The man operating the Stone Age computer is easily replaced by a mechanism, the central processing unit. But out of such simple elements, all forms of human logic and mathematics can be built. (See THE CHURCH-TURING THESIS.) A few examples illustrate how this can be done with the simple processes of the Stone Age computer.

1. To empty pot A, call the required operation $A^0$. So,

$A^0$:

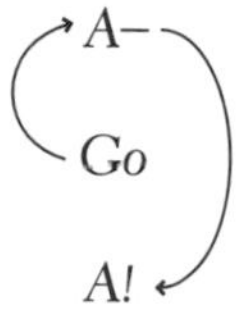

If A is already empty, the arrow on A− goes to *A!*, which ends the program. Otherwise one stone is removed from A. Then the *Go* instruction returns to the beginning and repeats.

This small program can appear as part of a larger program, in which case it is a "subroutine."

2. To add the contents of pots A and *B*, leaving the result in *B*, call the required operation $B \longrightarrow (A + B)$. So,

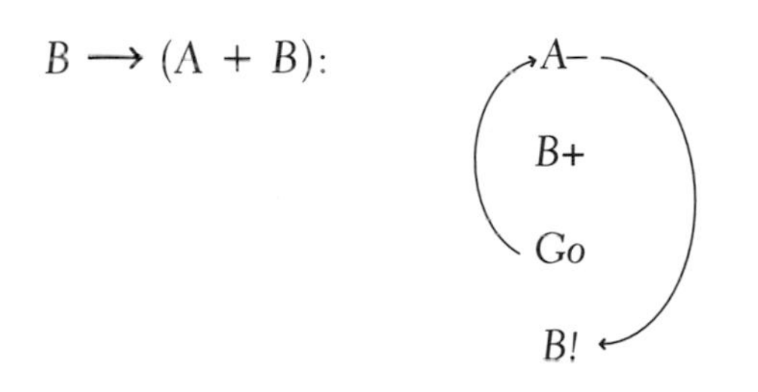


3. To copy the contents of pot A into pot *B*, leaving the contents of A unchanged, call the operation $B \longrightarrow A$. So,

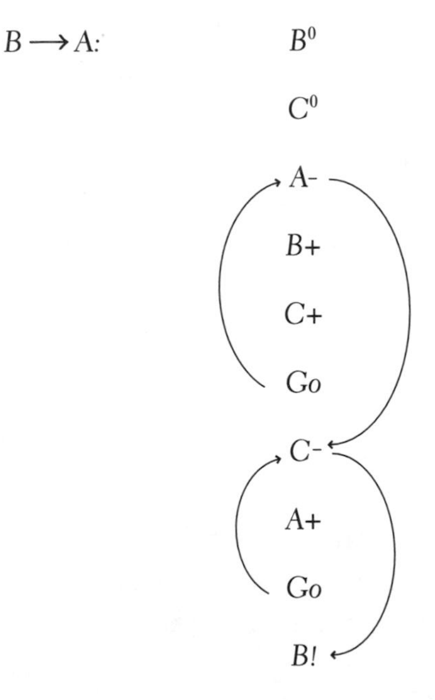


First, *B* and C are emptied, using the first subroutine. Then one stone is removed from A, one is added to *B*, and one to

> C. This operation is repeated until A is emptied, at which point both *B* and *C* contain replicas of the original A. Then the contents of C are transferred back to A. At this point, A is as it originally was, and *B* duplicates it.

We could go on to write programs for multiplication and all the other mathematical operations, but the above examples are sufficient to illustrate the basic procedure.

Any computing machine has states (for instance, the number of stones in each pot at any moment), which are called the "symbols" of the "machine language." The simple language illustrated by our examples contains four "instructions," which can form parts of more complex programs. Combinations of instructions into more useful subroutines become a higher-level language, like those familiar to users of personal computers. The central processing unit (man or machine) is used only to recognize and execute the four kinds of simple instructions. In more advanced, stored-program machines, the program as well as the initial data is stored as "the number of stones in a pot" and then translated and executed as required.

Although it is common to speak of a computer's "language" and the symbols it uses to represent things as "information," these are metaphors. The computer is a tool. Its states represent things from the user's point of view. The man (or machine) inside, like the man in John Searle's CHINESE ROOM, merely moves stones around according to instructions. He does not know what they represent. Indeed, the same states of a machine can be interpreted by an outside observer in at least two different ways.

Suppose, for instance, that [A] represents the number of stones in pot A, and the operations [A] + [*B*] and [A] × [*B*] have been defined. To an outsider, [X] may represent an item in the warehouse or business accounts. Or it may represent a statement in some logical relationship. Then we could define a language where the statement *X* is true if [X] is nonzero, and false if [X] is zero. In this case, it is easy to verify that:

> ([A] + [*B*]) represents "At least one of *A* or *B* is true."
> ([A] × [*B*]) represents "Both A and *B* are true."

So the same state of the computer can "mean" different things to an intelligent observer or user. The computer itself does not "know" what it "means" if its program results in an empty or nonempty pot at the end. It does not even "know" how many stones are in a given pot. It merely *is* in one state or another.

There is almost certainly a physical feature or process by which our brains are associated with consciousness. Consciousness may have some impact on the way human beings carry out computation or it may not. (See CONSCIOUSNESS, TOWARD A SCIENCE OF.) The argument above, showing that a computer's state can mean different things and that it means nothing whatever to the computer itself, simply illustrates the fact that there is no *necessary* justification for attributing personality or consciousness to a computer, any more than to a refrigerator or a television set. A computer does not *need* to understand the symbols of a formal operation in order to carry out that operation.

# Fractals

How long is the coastline of Britain? What shape is a cloud? What does the human lung have in common with a river delta and a stick of broccoli? The answer is fractals, infinitely complex shapes that fall between the cracks of dimensionality and serve as the generating principles of chaos.

Ever since the ancient Greeks, scientists have pictured the world in terms of simple idealized shapes—circles, triangles, cubes, prisms, and spheres. Within this vocabulary, a point has zero dimension; a line, one; a surface, two; and a volume, three. But how useful is a circle or a triangle when it comes to describing the pattern of frost on a windowpane, cracks in metal, or the gradient of a bumpy mountain path? MIT mathematician Benoit Mandelbrot realized that an entirely new mathematical description had to be developed.

What exactly is the length of the coastline of Britain? The answer,

Mandelbrot discovered, depends on how you make the measurement. A first approach would be to take a motorist's map and a length of thread, arrange the thread so that it follows the coast, and measure it against the scale at the bottom of the map.

But suppose you use a map with a larger scale. Greater detail is shown, and the thread must follow all the twists and turns of estuaries and promontories. The distance is now found to be much greater than before. If you take a yardstick and walk all around the coast, the answer will be surprisingly large. But is this the true length? Not really, because the yardstick averages out all the small bumps and misses the indentations.

What is the real length of the coastline? The answer, Mandelbrot pointed out, is that the coastline is infinite, precisely because of the infinite complexity of its shape. (We are ignoring the limit set by atomic theory.) Is it useless then to ask questions about coastlines? Not so, said Mandelbrot. The important thing is not the length but the shape, which can be given a clear and unambiguous measure.

Mandelbrot called the infinitely complex shape of a coastline a fractal figure. Each fractal has a dimensionality that provides an accurate measure of its inner complexity.

Mandelbrot showed that the same sort of answer is given when you ask about the gradient of a mountain. The value of the gradient keeps changing as you walk up the mountain and depends on how finely you carry out the measurement. On the other hand, the fractal dimension of a mountain is clear-cut. A wide variety of natural forms such as clouds, river deltas, lungs, broccoli, weather patterns, cracks in metals, electronic noise in a receiver, interstellar dust, gusts of wind, and tree branches require the use of fractals for their description, in place of the circles, triangles, and pyramids of the ancient Greeks.

How can something have a fractal dimension and lie between a plane and a volume, or a point and a line? The precise answer is mathematical, but simple analogies help. What is the dimension of a ball of string? Seen from a great distance, it appears as a dot, a zero-dimensional figure. As we approach, we realize that this dot is a three-dimensional sphere. Closer still, we notice that this ball is composed of a single thread—a twisted line, a one-dimensional figure. Look even closer and the thread becomes a long cylinder, a three-dimensional figure. Under a magnifying glass the cylinder is seen to be made up of individual fibers, tiny twisting lines. So what is the dimensionality of string?

Take another image of dimensionality. Scatter 10,000 grains of rice across a chessboard. A straight line on the board passes through only 200 grains. Now allow that line to become ever more complex, so that in its twists and turns it passes through more and more grains. Eventually, this complex line passes through every grain. Can it still be called one-dimensional? Of course, grains of rice are of a finite size, so now draw a curve that is so incredibly complex that it passes through every dimensionless point of the board. Such a curve exists, in theory at least. It is called a Peano curve and is so complex that it touches every point in the plane; clearly its dimensionality must be two and not one!

It turns out that different fractal lines whose dimensionality ranges between one and two can be drawn. Likewise, shapes can have a fractal dimensionality that lies between two and three (a plane and a solid). A complex dust of points can have dimensionality between zero and one. In each case, the "fractal dimension" is a unique way of classifying a natural form and comparing it with the shapes of other forms. In place of a world of Platonic solids, there now appears one of fractal dimensions.

Fractals have remarkable properties. They are infinitely complex, in the sense that any particular detail can be magnified to show still further detail, and so on *ad infinitum*. In some cases they also exhibit self-similarity, a particular shape being repeated under greater and greater magnification. Mandelbrot has referred to one particular fractal he discovered—THE MANDELBROT SET—as "the most complex shape in the universe."

Fractals are generated by a process of ITERATION and can easily be produced on a home computer using a simple algorithm. To take an example, draw a triangle and replace each of its three lines with the simple shape (called a generator) on page 156 (*a*).

The result is a star. Next replace each straight line in the star with a scaled-down version of the generator (*b*).

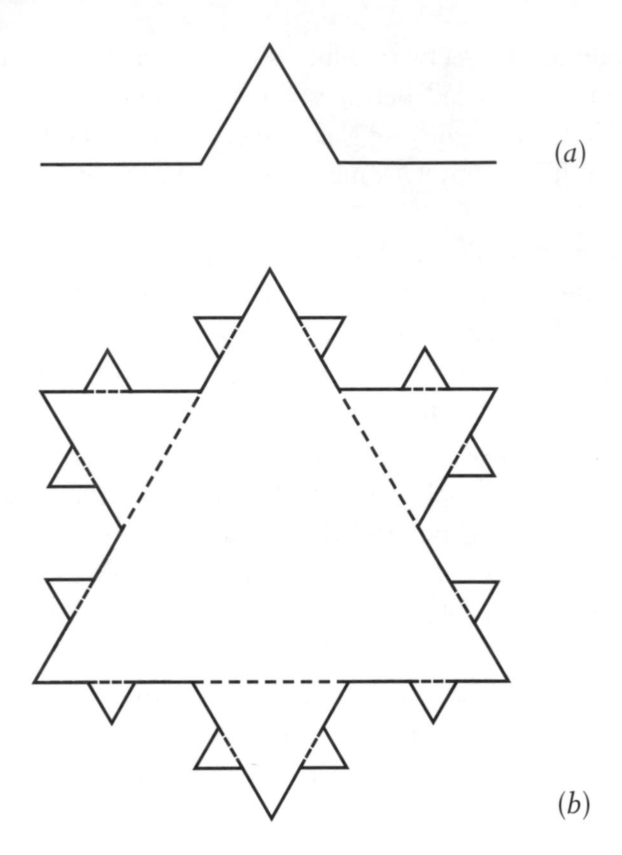


The result looks a little like a snowflake. Repeat the algorithm again and continue without limit. Of course, in the world of drawings and computer displays, this process cannot go on for ever, but at least you can take any part of the figure and magnify its details endlessly. The final shape has infinite detail and a fractal dimension. Because of the way it is created, it is also self-similar, the particular shape of its generator recurring again and again on ever-decreasing scales.

This triangle shape is one of the simplest fractals to construct. Other types can be generated by a process of branching at progressively different scales to produce figures that resemble river deltas, trees, lungs, frost, or the human circulation system. What is truly remarkable is the way nature employs branching fractal structures in so many of its forms.

Take, for example, the biological problem of getting blood to all parts of the body. Blood is the stuff of life but, metabolically speaking, it is such an expensive commodity that it makes up only 3 percent of the volume of the human body. How can nature design a system that will pump this blood close to every cell and organ in the body? The answer is fractal branching, in which major arteries branch between eight and thirty times before they reach individual parts of the body. The resulting fractal dimension of the circulatory system is close to three.

Lungs present another design problem: how to generate the maximum exposure of blood to air within the lung. In this case, iterative branching takes place ten times and then suddenly changes to a new form of branching iteration. The lung has a more complicated fractal structure than the circulatory system. It is not exactly self-similar, since the algorithm for its generation changes at a given scale.

Brains also have a fractal structure with a dimension between 2.79 and 2.73. Even the pattern of heartbeats is a fractal. When that complex fractal order changes into a regular, repetitive beat, it suggests that congestive heart failure is around the corner. On the other hand, if the fractal becomes too chaotic, fibrillation occurs. This fine line between healthy complexity and chaos is also present in the electrical activity of the human brain.

Fractals can be used to describe social and economic systems. The stock market is a complicated nonlinear system of positive and negative feedbacks and self-corrections. It exhibits all the features of nonlinearity—periods of stability, oscillations, and sudden jumps and crashes. Some economists have noted the way fluctuations of the stock market exhibit fractal self-similarity. Fluctuations that on casual examination appear random reveal self-similar repeating patterns at intervals of years, weeks, days, and even hours.

What appears to be absolute lawless chaos in the fluctuation of stock prices may be evidence of deterministic chaos, in which fluctuations occur only between certain ranges of possibilities. Some speculators are using the mathematics of fractals to buy and sell within a matter of hours, and so exploit the tiny differences between deterministic fluctuations and pure randomness. (See CHAOS AND SELF-ORGANIZATION.)

Fractals are generated by means of iterative or algorithmic procedures, whereby the result of one stage of a calculation is FEEDBACK into the next. Such iterative feedback can also be the cause of deterministic chaos. It is therefore no coincidence that the strange ATTRACTORS, which govern the dynamics of chaotic systems, all have fractal dimension. Normal systems tend to oscillate or otherwise behave in highly repetitive ways in the grip of regular attractors. But what happens when an attractor has a fractal dimension? In this case, the corresponding motion is chaotic and the attractor is called a strange attractor. Just as a fractal displays its infinite complexity, so, too, motion in the grip of a strange attractor is infinitely complex. At first sight, such motion could be called totally random. But a better description is that it has vast complexity in which certain patterns of behavior can be repeated on ever-diminishing scales.

# Fröhlich Systems

Living systems are characterized by their coherence and ability to coordinate complicated processes over the dimensions of a cell or organ. The physicist Herbert Fröhlich proposed that this involves macroscopic quantum states and is a manifestation of something analogous to BOSE-EINSTEIN CONDENSATION.

It is difficult to account for the particular properties of life in terms of mechanical interactions and short-range interactions. One of the most fundamental features of living systems is the way that enzymes assist the processes of a cell. Their activity (enzyme recognition) extends over fairly long distances on the atomic scale of things. Processes within the brain also involve the coordination of complex activity over different regions. What is the source of these long-range interactions and coordination?

Molecular interactions are relatively short range, yet SUPERCONDUCTORS and SUPERFLUIDS involve the coordinated behavior of astronomical numbers of quantum particles over centimeters or even meters. Could something analogous be operating in living systems? Is consciousness, for example, a phenomenon involving single quantum states across whole regions of the brain? (See QUANTUM THEORIES OF MIND.)

In a superconductor or superfluid, the temperature (kinetic energy within the system) must normally be reduced to close to absolute zero, so that subtle attractive forces are no longer swamped. (High pressure will do the same thing. See NEUTRON STARS.) Once the attractive forces are felt, they act to coordinate the motion of an astronomical number of particles. The result is a coherent system described by a single wave function of macroscopic dimensions. Fröhlich proposed that something analogous can occur when living systems are "pumped" by metabolic energy. Fröhlich systems are states of self-organization that extend over macroscopic distances. (See CHAOS AND SELF-ORGANIZATION.) They are modeled as molecules in cell membranes vibrating in PHASE. The result is analogous to Bose-Einstein condensation, but does not demand extremely low temperatures. Rather, it takes place at the normal temperatures found on the surface of the earth.

The notion of Fröhlich systems certainly answers questions about the coordination of life processes, but in spite of a few suggestive experiments it remains a hypothesis that has yet to be accepted by the general scientific population.

# Functionalism

Is a thing or a person ever more than its use? Once we have described how things and persons influence each other, or causally interact with their environments, have we said all there is to say about them? A functionalist would answer, "Yes."

Functionalism is the most common philosophical position of both COMPUTATIONAL PSYCHOLOGY and ARTIFICIAL INTELLIGENCE. It argues that every mental state, like hunger or anger, insofar as it *can* be described scientifically, is fully described in terms of its causal role—how it interacts with other mental states and with outer stimulus and response. Hunger can be described as a mental state that causes mental images and food-seeking behavior and might arise in response to the stimulus of a strong roast-beef aroma. Other mental states, such as anger or panic, would be described as states associated with the hunger's inability to cause *successful* food-seeking behavior.

Functionalism is a first cousin of BEHAVIORISM, but it goes further in trying to describe processes and relationships that *underlie* behavior. Behaviorism describes only behavior that we can see—how the dog actually moves when the bell rings. Functionalism, by contrast, describes how various segments of the brain, or various mental states and their interactions, function to *produce* visible behavior. Thus it is more detailed in its attempt to analyze the underlying causal factors of behavior. Functionalism takes account of various internal mental states, corresponding to an organism's different perceptions, beliefs, and desires, whereas behaviorism treats the organism like a "black box." (See THE BLACK BOX.)

Functionalism can accommodate to the fact that we may have many conflicting or competing motivations within one state of consciousness—I might feel hunger and an associated desire to seek food at the same time that I resist seeking food because I feel I am too fat—and psychology's common assumption that each personality contains many "subpersonalities." We can divide computer programs into "subroutines," elements that carry out a limited procedure (like "spell check") within a broader program's capacity. But functionalism, like behaviorism, seems to omit any attempt to describe our *experience* itself—the hungriness or desperation of hunger, the painfulness of pain, or our intentions and emotions. (See CONSCIOUSNESS, TOWARD A SCIENCE OF.)

Soft functionalists argue that, while the subjective phenomena of experience may well exist, they are outside the scope of objective science. Hard functionalists and many behaviorists are committed to the even more extreme view that subjective phenomena either do not exist or, at any rate, are unimportant and have no effect on observable behavior. This seems contrary to our common sense. (See THE MIND-BODY PROBLEM.) Hard functionalism is the philosophy underlying THE TURING TEST—if a system *behaves* as if it is conscious, there is no more that we can say about whether it indeed is or is not.

# Fuzzy Logic

Fuzzy logic presents a new model of thinking that challenges the traditional logic of Western culture. It is designed to cope with ambiguity and "gray areas," and is particularly suited to the conceptual and technological challenges of late-twentieth-century science.

Western logic began with Aristotle and is modeled on the precise thinking and categories of mathematics. In mathematics, 2 plus 2 equals 4, never 4 1/2 or 5. In logic, a is *either* A *or* not-A; it is never *both* A *and* not-A. It has been an all-or-nothing logic that admits neither contradictions nor shades of gray. In our daily lives, Aristotelian logic underpins definite statements. It is also the basis for the either/or thinking of classical physics—light is a wave *or* a particle; an object is here *or* there; cats are alive *or* dead. (See PROLOGUE.) In our computer culture, Aristotelian logic has been enshrined as the emblem of the digital computer with its high-speed, black-and-white binary strings of 0s and 1s.

With the beginning of the twentieth century, both general culture and the creative edge in science moved away from simplicity and certainty. Einstein's relative frames of reference stressed that there are many points of view, and HEISENBERG'S UNCERTAINTY PRINCIPLE emphasized uncertainty and partial truth. Looked at from the vantage point of classical physics and its Aristotelian logic, the new science made little sense. It was branded "absurd," although it clearly worked. Fuzzy logic is a move toward a conceptual scheme more suited to the "both/and," "a little of this, a little of that" logic employed by quantum physics. It also harkens back to more typically Eastern ways of thinking.

Like Buddhism and Eastern mysticism in general, fuzzy logic is at home with contradictions. It is a logic that stresses matters of degree and all those shades of gray in between black and white. It is about the possibilities that exist between 0 and 1, and how the new breed of "parallel" computers can be programmed to respond to them with something almost touching on creativity. (See NEURAL NETWORKS.)

A chief proponent of fuzzy logic, mathematician Bart Kosko, illustrates the principle of this kind of thinking with reference to an apple. An Aristotelian would look at a piece of fruit and say that it either is or is not an apple. But what happens when we take a bite out of the apple? Is it still an apple? Perhaps we take another bite, and still another, until there is nothing left. At some point the apple changes from a thing to a nonthing, from an apple to a nonapple. Where do we draw that line? What do we make of half an apple with all-or-none distinctions?

The half apple is a "fuzzy" apple, a key to understanding fuzzy technology. Suppose engineers want to make an intelligent traffic light that can time itself to change from red to green at different intervals, depending upon how light or heavy the traffic flow is. The binary switch of a digital computer is too crude to do this. Binary switches are either on or off. But fuzzy chips that allow traffic lights to readjust constantly have now been invented. They also delicately adjust subway control systems, the loading sensors of washing machines, the contrast buttons of TV sets, and a whole host of other "smart" machines. It took the Japanese to realize their vast potential first.

Fuzzy chips and machine circuits are modeled on the brain's own system of neural nets, in which each neuron is connected in a "messy" way to up to 10,000 other neurons. Like these brain circuits, fuzzy chips learn as they go, constantly readjusting themselves to varying data input.

# The Gaia Hypothesis

According to the Gaia Hypothesis, the entire earth is a living entity. Our bodies are composed of organs—heart, liver, kidneys, brain, digestive system, and so on—each performing its own specialized tasks. While individual organs are to some extent autonomous, they also act in concert to serve the whole body. In a similar fashion, an ant nest is composed of individual ants, yet each member acts in a highly cooperative way to serve the colony. Cooperative structures abound in nature (see COEVOLUTION), but can the concept be extended to an entire planet? Atmospheric scientist James Lovelock believes that each species, from bacteria to humanity, is part of the one cooperative system he calls Gaia. The earth is, in a sense, a single giant cell.

As part of NASA's project to detect evidence for life on other planets, Lovelock made a study of the earth and, to his surprise, discovered that its atmosphere and the composition of its oceans are quite different from those dictated by physics and chemistry alone. Take, for example, the oceans. Each year, millions of tons of salt are leached from the earth and washed down rivers into the sea. Despite this yearly addition, the salt content of the world's seas is surprisingly constant. Something must be monitoring and maintaining this level.

In a similar way, the amount of ammonia, carbon dioxide, and oxygen in our atmosphere should be quite different from their present percentages. Again, something is maintaining the earth's atmosphere in a stable, far-from-equilibrium state. The monitoring system, Lovelock argues, is life itself.

Our bodies and other living systems are characterized by homeostasis, the ability to maintain stable internal conditions in spite of external fluctuations. Even when the weather changes, our body temperature remains the same, as does the salinity of our blood. According to Lovelock, the earth itself is also in a state of homeostasis brought about by the cooperative action of life itself.

Just as the human body maintains a stable temperature in all weathers, so conditions on the earth's surface have not fluctuated too much even when the sun's temperature increased by as much as 30 percent. Plankton in the ocean emit gas that, through a chemical reaction, is transformed into aerosol particles in the atmosphere. It is thanks to these that water vapor condenses and makes clouds, which in turn reflect back much of the sun's heat and make life on earth tolerable. If the sun's output were to fall, plankton activity and in turn the density of the clouds would diminish. If it increased, a thicker cloud cover would be created. The result is a stable surface temperature.

Yet if Gaia exists in a state of homeostasis today, Lovelock believes it once evolved in such a way as to make abundant life possible. Long ago there was little oxygen in the earth's atmosphere, and life was particularly simple. During this stage, microorganisms called cyanobacteria pumped large amounts of oxygen into the atmosphere in a process that can only be described as large-scale planetary engineering. Increasing oxygen content made the atmosphere poisonous for the cyanobacteria themselves, but the sacrifice they unwittingly made allowed new and more complex life forms to emerge. Today, life forms all over the earth monitor its "vital signs" and take action to correct any deviation.

Certainly the planetary system (see SYSTEMS THEORY) is stabilized by many FEEDBACK loops, as is the physiology of a single organism. But the comparison cannot be pushed too far. Organisms have evolved by competition and natural selection; Gaia as a whole cannot have done so. Gaia may have many possible stable or metastable states, including the present one and the one before oxygen was abundant. (See COMPLEXITY.) We have no reason to regard our regime as uniquely privileged.

# Galaxies

Galaxies are large aggregates of gravitationally bound stars, plus an interstellar medium of gas and dust. There are some $10^{10}$ galaxies within the observable universe, ranging from dwarf galaxies containing about $10^7$ stars to supergiants of $10^{13}$ stars. Our galaxy, THE MILKY WAY, has about $10^{11}$ stars. Most galaxies are either spiral, because they rotate as a whole, or elliptical; a few are irregular in shape. The spiral ones, which make up about two thirds of the whole, resemble our Milky Way.

Galaxies are not evenly spread throughout the universe. They occur in clusters and superclusters, giving the universe a "lumpy" or foamlike appearance on the largest scales we have observed. Our local group has about twenty or thirty galaxies in it, the two largest of which are the Milky Way and the Andromeda galaxy. A really rich cluster of galaxies can have thousands of members.

Galaxies and stars began to condense out of the primordial gas about 2 billion years after THE BIG BANG. Their precursors, slight local variations in COSMIC BACKGROUND RADIATION from an even earlier era, have been detected recently. (See WRINKLES IN THE MICROWAVE BACKGROUND.) The contents of a galaxy are not homogeneous. There are many kinds of STARS, which often form open clusters, ranging from a few to thousands of members. The stars in a cluster formed from the same local gas at about the same time. There are also localized clouds of gas (mainly hydrogen) and dust (heavier elements). Stars enrich the interstellar medium by a constant stellar wind, or an explosion at a late stage of development. (See SUPERNOVAS.) The interstellar medium in turn condenses into a new generation of stars. It also contains strong magnetic fields, which accelerate elementary particles to relativistic velocities. These are cosmic rays.

A small proportion of galaxies are active ones, which produce vast amounts of energy at various wavelengths from a small galactic nucleus. There are several kinds of active galaxies, most notably Seyfert galaxies, which emit very bright light, and radio galaxies, which emit radio waves. QUASARS, discovered in the 1960s, appear pointlike. They may be the active nuclei of very distant galaxies, too faint for us to see. Nuclear physics cannot explain such violently energetic phenomena, which may represent supermassive BLACK HOLES onto which gas accretes. Our own galaxy may well have a much smaller active nucleus, hidden from us by dust.

## The Game of Life

An English mathematician, John Conway, devised a simple computer program that has scientists scratching their heads and asking if a computer could be alive.

The game of Life has, as its universe, a computer screen containing white and black squares. As you watch cells grow, multiply, and die, complex patterns form and parasites travel across the screen, consuming structures as they go. Each time, the patterns and behavior of the computer universe are quite different. The game of Life appears as rich and complex as life itself.

Surprisingly, in view of the nature of the display, the rules of the game are remarkably simple. The screen begins with a random pattern of black and white squares. White squares represent living cells; black ones, dead. In the next step, each square responds to what is happening in its neighborhood of eight adjacent squares. If there are too many white squares, it dies of overcrowding; too few, and life is not possible. Only if a square has two or three living neighbors (white squares) will it live in the next generation. Likewise, a dead square (black) will come to life if it has two or three white neighbors.

That is all there is to the game. As the computer goes through successive steps, colonies of white cells grow and oscillate or "breathe." Others break off and wander around the screen. Some shapes look like parasites that eat everything in their path. Using an extremely simple algorithm, Conway made a computer screen appear like the world seen when a drop of water is viewed under a microscope.

Conway's game of Life is a concrete example of Polish mathematician Stanislaw Ulam's abstract notion of cellular automata. Since the patterns on the screen reproduce themselves, move, and consume their rivals, it is tempting to ask if that world within the computer is in some sense "alive." (See ARTIFICIAL LIFE.)

# Games, Theory of

A mathematical analysis of games and their strategies was given by John von Neumann and Oskar Morgenstern in their famous book, *Theory of Games and Economic Behavior*. Von Neumann had earlier defined what he termed "two-person zero-sum games," games like chess or cards in which one player wins and the other loses. Game theory asks, What is the best way of playing such games? Von Neumann showed that each player has a best strategy, which reduces to a minimum the maximum possible losses that his opponent can impose. This is called the "minimax" strategy.

A simple example is the children's game in which two players simultaneously make gestures representing scissors, paper, or stone. The rule is that scissors cut paper, paper wraps stone, and stone blunts scissors. Payoffs are made accordingly. In this case, the minimax strategy is for a player in a series of games to make all three gestures in about equal proportions, randomly. Any more-patterned strategy can be exploited by the opponent.

The theory of games argues that rules developed for games have wide social applications to general competitive behavior in the social, business, and international arenas. By studying games and their strategies, it is assumed, it becomes possible to understand how rational decisions can be made in complicated situations. But this way of thinking has clear limitations. It assumes that each player is rationally trying to maximize his or her own return. But this applies only to certain competitive situations, i.e., two-person zero-sum games. Other situations of much greater complexity, known as non-zero-sum games, may involve two or more players. In these, it can be an advantage to co-operate and form alliances. And a purely rational strategy may no longer exist.

The most famous two-person non-zero-sum game is Prisoner's Dilemma. Two players are "arrested" on suspicion of a serious crime, and each is interviewed separately. If both independently deny all knowledge of the crime ("cooperate"), they will each receive three years in jail for a lesser crime. If each accuses the other ("defects"), they will each receive five years. If A remains silent ("cooperates") while B accuses him ("defects"), A will get seven years and B only one year.

Clearly, in this situation the best *total* result is obtained if both cooperate; the worst is obtained if both defect. But each may argue to himself, *Whatever my partner does is his business. I personally will be better off if I defect.* Both may thus defect, producing a poor result. This type of "game" occurs commonly in life, as when both members of a couple are tempted to unfaithfulness, or when not giving way in an argument becomes a matter of pride for both sides. The issue is the individual versus a community. How, in such cases, can the worst outcome be avoided? What is the best strategy to follow if the relationship is ongoing and trust can perhaps be built up?

Tournaments have been run to answer this type of question. In one tournament, anyone could enter a computer program embodying some definite strategy for playing Prisoner's Dilemma repeatedly against each of the other programs. "Always cooperate" (total altruism) and "always defect" (total selfishness) both did badly, as did some complicated programs. A winning program was called "tit for tat": It counseled "cooperate the first time you meet someone, and after that do what he did last." This strategy seems natural to many people, and may be a product of evolution. (See SOCIOBIOLOGY.)

One very important implication of the best strategies for "winning" non-zero-sum games (getting the best possible outcome for all) is that rational behavior is not always the most "rational." We know from studies of mind (see THINKING) that the fact that people are not always rational may be an important part of our survival makeup. Our simple emotional reactions, such as fear, anger, and parental behavior, are possible strategies programmed into our genes as instincts. (This is borne out by tests that show that stimulation of certain points in the brain results in predictable emotional responses.) The impulse to have children, to attack one's enemies, or to treat people equally may help or hinder group survival. The arbiter in these cases is not rational game theory but the long-term results. The "best" strategy depends on the context: Who else is playing, and what are the rules of that particular game? (See CO-EVOLUTION; DARWINIAN EVOLUTION.)

# Gauge Fields

Gauge fields, ways of connecting different points in space-time, explain the nature of physical force. Suppose you visit a planet that does not have a magnetic North Pole. Your compass needle spins and eventually comes to rest. It is entirely up to you to decide whether to call the painted end north or south. Move to a new location, and you are faced with the same ambiguity. There is no connection between the convention for north or south that you use in one locality and that in another; all directions are magnetically equivalent. In other words, space is isotropic. On earth, the convention matters very much. Since one end of the compass always points toward the magnetic North Pole, it is important to keep a consistent convention from location to location, always calling the painted end "north." Technically speaking, since like poles repel, the "north" pole of a compass is really the "north-seeking pole," a magnetic south pole.

A magnetic field can be thought of as nature's convention for orienting magnets throughout space. Physicists call this a gauge field, and it connects the orientation of a magnetic pole from one point in space to another. A similar convention applies to an electric charge. Once physicists decide to label the electron negatively charged, the gauge field ensures that a consistent convention between positive and negative charges is maintained throughout space.

The existence of a gauge field represents a special kind of SYMMETRY BREAKING. Empty space should be isotropic—the same in all directions—and homogeneous—the same from point to point. A gauge field selects certain directions in space, or an abstract isospace, as being privileged. Since the gauge field is a quantum field, its quantum vibrations are themselves BOSONS and, as it turns out, carriers of the force associated with each particular field. The quantum particles associated with the gauge field are massless, and the massless particles of the electromagnetic field are photons. In a case of weak isospin—that is, a gauge field of the weak force—additional theoretical considerations must be taken into account, and the carriers of the gauge field—that is, the particles associated with it called intermediate vector bosons—now have mass.

The first gauge theory was QUANTUM ELECTRODYNAMICS, a principle later shown to apply to the origin of all the other forces—gravitation, the weak force, and the color force. It is a fundamental concept of modern particle physics.

# General Relativity

In 1916, Albert Einstein proposed an interaction between mass-energy and the structure of space-time. Mass-energy produces "curvature" in space-time—i.e., it deviates from flat, Euclidean space. Curved space-time in turn changes the paths of mass-energy through it.

Consider the two-dimensional space of an even lawn. Where the lawn is flat, a ball bowled along it will proceed in a straight line at a constant velocity, if we ignore friction. This situation corresponds to empty space. But suppose the effect of a mass placed on the lawn creates a slight local depression, a "curvature" of the surrounding "space." A ball bowled near it will deviate from its previously straight path toward the mass. The effect is the same as that of gravitational attraction. Einstein worked out a theory, described in beautifully elegant equations, for the whole of four-dimensional space-time. Instead of gravitation, his theory gives us changes in the fabric of space-time geometry itself whenever it is near masses.

One test of the worth of any new scientific theory is whether it offers us additional predictions. How do the predictions of General Relativity differ from those of Newtonian gravity? One obvious one is that curved space will deflect not only matter but also light rays passing through it. Thus, light from a distant star should be deflected if it passes near the sun. Newtonian gravitation predicts a similar effect, but only half as much. During an eclipse of the sun in 1919, the measured deflection of light from a distant star won international acclaim for Einstein's theory. Given the technology of the time, one other experimental verification was possible, an explanation for an anomaly in the orbit of Mercury, and this, too, was achieved successfully.

In principle, the curvature of space predicted by General Relativity should be geometrically measurable, but in practice, real deviations from flat, Euclidean space are too small for us to measure under normal circumstances. The three angles of any small triangle add up to 180°, just as Euclid described. Thus, on the scale of the everyday, the predictions of General Relativity seem to have little application. But if we have a triangle on the earth's surface large enough to sit with its apex at the North Pole and its other two corners at the equator, it can have *three* right angles. We know that we can circle our "flat" earth by traveling straight forward, and the equations of General Relativity offer revolutionary predictions about the large-scale structure of the universe as a whole (see RELATIVISTIC COSMOLOGY) and about very high energy situations where the curvature of space-time becomes appreciably large (see BLACK HOLES). Still, without the "real-life" experience of its effects at everyday dimensions, Einstein's theory of space-time curvature didn't seem to bear on things that matter.

Since about 1960, improved technology has led to a much larger role for General Relativity in astronomy and physics, and to many more experimental tests of its validity. Radar ranging of the planets and of artificial satellites made high-precision measurements of their orbits possible. The gravitational redshift of light was measured accurately.

The discovery of QUASARS, pulsars (see NEUTRON STARS), black holes, gravitational lenses, and the COSMIC BACKGROUND RADIATION involved General Relativity. The most accurate test of the theory is now given by the binary pulsar. Here, observations match predictions to twelve or more decimal places.

Today, General Relativity competes with QUANTUM FIELD THEORY as one of the two most accurately confirmed physical theories ever. The two have not, however, been fully integrated. (See QUANTUM GRAVITY.) Also, gravitational waves, which are predicted by General Relativity and implied by the gradual slowing of the binary pulsar, are as yet beyond our ability to detect directly.

# Gestalt and Cognitive Psychology

Gestalt psychology and its later extension into cognitive psychology were founded on the belief that the inner life of the mind—our perceptions, attitudes, and beliefs—is important and can be studied scientifically. In contrast to the behaviorist (see BEHAVIORISM), who denies or ignores experience to focus solely on relations between stimulus and response, Gestalt and cognitive psychologists *concentrate* on experience. What human beings respond to, they argue, is not the stimulus itself, but our perception or belief *about* the stimulus. The way that we *see* the color red, and our inner associations with redness (our beliefs and attitudes), determine how we will behave in the presence of red things.

Gestalt psychology began in the early years of this century as a study of perception. Experiments on seeing and hearing demonstrated that human perceptions have holistic, self-organizing qualities. Certain data are always perceived *together*, or in association with one another—they have an inherent unity. Thus we hear collections of notes as melodies, we see arrays of dots as patterns, and we automatically fill in the missing links in incomplete circles or sentences.

A gestalt is, essentially, an organized whole whose parts have an intrinsic relationship. It is this relationship, the Gestalt theorist suggests, that structures what we see or hear. When we hear a melody, what we are perceiving is the relationship of the notes more than the individual notes themselves. Thus we recognize the melody as the same melody, in whatever key it is played. Our perceptions are assumed to have an innate synthesizing capacity, rather than being just passive reflexes. This is illustrated in Gestalt experiments on optical illusions, such as a flashing sign showing an arrow in two positions, which gives us the impression of movement. There is no actual movement in the arrow; we perceive it as such because of the relation between its two positions. Another experiment puts a gray patch first on a white background, then on a black background. In the first case we see the patch as dark, and in the second as light, but it is the *same* gray patch. What we really see is the relation between figure and background.

Later Gestalt experiments showed that we construct mental maps (the cognitive psychologists call them cognitive maps) of whole areas of our experience, and that these maps guide our behavior. In our language, for example, we form and understand sentences that have never before been used, implying that we have constructed an inner map of the kinds of words there are and how it is possible to combine them. Language is not just a set of push-button reflexes linking words, as the behaviorist would have it.

The role of cognitive maps in structuring experience is illustrated by a Gestalt twist on a classic behaviorist experiment. A rat is put into a maze and slowly learns to find its way through it. The behaviorist argues that the rat learns through a series of stimulus-response trial-and-error runs. But Gestalt psychologists took the same maze and flooded it with water. They found that the same rat could swim successfully through the maze on its first attempt, without hesitation or error, even though all its movements had to be different. They argued that, in fact, the rat had previously formed an inner cognitive map of the maze, which it used to guide itself through. Other experiments showed that human subjects can restructure their mental maps, through a sudden flash of insight for instance, without repetitive reconditioning.

As the Gestalt approach was extended to beliefs and attitudes rather than just to perceptions, and applied to whole personalities and to larger groups, it slowly grew into what today is known as cognitive psychology. In the 1950s, in his "personal construct" theory, the cognitive therapist George Kelly proposed that each person acts as a scientist putting together a set of generalizations from experience (his or her cognitive map), which can then be used to predict and control the world. These generalizations are holistic in that they hang together and are reflected in a set of coherent attitudes. A few bad experiences with one or two people may lead to the general attitude "people are untrustworthy."

Kelly proposed that, in some cases, these attitudes are maladaptive or "irrational," and he devised a form of cognitive therapy for changing them. Through talking about his or her attitudes, the patient would spontaneously come to see—perhaps in a flash of insight—that they were inappropriate, and in consequence would adopt a new cognitive map. There are some interesting relations between the rationale of this therapy and what actually happens when scientists undergo paradigm shifts in their work. Scientific theories, too, are holistic pictures of the physical world that include all the scientist's data. When data appear that do not fit a given paradigm, the end result may be that the whole paradigm is abandoned as "irrational" or "wrong," and a new one is adopted in its place.

Gestalt psychology and cognitive psychology accomplished two things. First, they demonstrated that the inner life must be taken seriously and that mental constructs, beliefs, and attitudes are causal factors in behavior. In doing this, they retrieved fragments of common sense and many insights of William James that the behaviorists had excluded, and brought these within the scope of scientific (methodical, experimental) investigation. Second, the experiments demonstrated clearly that the inner life has a holistic quality. Perceptions, beliefs, and attitudes "hang together." There is a unifying quality to consciousness. These achievements, however, raised scientific questions that so far remain unanswered, even by the new cognitive sciences. Newtonian physics contains no corresponding kind of unity. (See THE MIND-BODY PROBLEM.)

The tremendous advances made in the theory and construction of computers since the 1940s have created a new model of information processing and storage. Much of this has been taken up by today's cognitive psychologists to offer a new model in psychology. Contemporary cognitive psychology concerns itself with similar computerlike processes in human beings and animals, and is closely aligned with the field of ARTIFICIAL INTELLIGENCE. The drawback to this approach is that cognitive psychology may concentrate too much on mental processes (particular kinds of thinking and learning) that can be modeled on computers, and is thus in danger of neglecting other psychological processes. Raw experiences of redness, pain, emotion, humor, and creativity do not fall within the computer model, yet all of them are well-attested aspects of human psychology.

SERIAL PROCESSING and parallel processing (see NEURAL NETWORKS) in computer models of the mind can describe many mental processes, but insight, learning, the formation of new concepts, and the creative use of language—all clearly demonstrated in Gestalt experiments—are not accounted for in these models. The holism exhibited in Gestalt experiments has no physical correlate in existing Newtonian brain models, or in their complementary computer models of the mind. Where does the "holism" or unity of consciousness come from? Because of such questions, Gestalt work remains a challenge to the newest cognitive models.

# Gödel's Theorem

Gödel's Theorem, proved by the German mathematician Kurt Gödel in 1931, is generally considered the most important logical discovery of the twentieth century. It lies at the heart of the philosophical revolution this century witnessed, and it has sweeping implications for the nature of the mind and all the mind's claims to ultimate truth.

Gödel was a Platonist with faith in a realm of forms, a world of pure truth. But as a mathematician, he wondered whether we could ever express the whole of this truth in any language available to human beings. In mathematical terms, his question was: Can we ever formulate a rich or interesting mathematical system that could contain proofs of all its own truths? His answer was no.

The essence of Gödel's Theorem is expressed in the words "No finitely describable system, or finite language, can prove all truths. Truth cannot be fully caught in a finite net."

Gödel proved that any consistent logical or mathematical "formal system" rich enough to contain the natural numbers (1, 2, 3 . . . ) would also contain a statement that could be neither proved nor disproved from within the system itself. A formal system is a language or set of symbols defined in terms of a set of rules for manipulating elements of the language, or symbols of the set. All computer languages, for example, are formal systems. (See FORMAL COMPUTATION.) Gödel showed that the unprovable truths of any given formal system could be proved within an *expanded* system containing additional axioms, but then the expanded system itself would contain further true but unprovable statements. Anything that we can prove *within* such a system is always only a *partial expression* of some further truth.

According to Gödel's Theorem, any rich logical or mathematical formal system is always incomplete. This discovery flew in the face of the long-standing ambition of older mathematicians like David Hilbert, Gottlob Frege, and Bertrand Russell to derive all mathematical truths from a finite set of axioms. Gödel also proved a second theorem, that a *consistent* formal statement of this kind cannot be proved to be consistent by the methods formalized within the system itself.

These theorems seemed both surprising and problematic to mathematicians in the 1930s; they seem less so to contemporary thinkers schooled in the wider philosophical implications of today's science. It seems more natural to us now to wonder why we *should* think that an infinite sea of mathematical truth might be mappable onto a finite formal system. Since Copernicus and Darwin, we no longer think of ourselves as masters of the physical universe, but rather as finite creatures within its immensity. Physicists can calculate only with values measured in finite terms—definite amounts of mass, density, gravitational force, and so on. The equations of GENERAL RELATIVITY, which are used to describe the present universe, break down (produce meaningless, infinite values) when they try to describe the *origin* of the universe, THE BIG BANG. In QUANTUM FIELD THEORY, physicists must content themselves with describing the *manifestations* of underlying reality—they cannot do calculations about underlying reality (THE QUANTUM VACUUM) itself. Gödel's work shows us to be in a similar position with respect to the logical and mathematical universe.

Gödel's second theorem, on consistency, now seems more like common sense. We naturally feel that a formal system's proof of its own consistency is of dubious value. This, too, is part of the scientific ethos that demands independent verification of theory and replication of experimental data. Unless we had already tested a system and found it trustworthy, how could we trust an internal proof of its own consistency? It would be like trusting a witness who says, "I am telling the truth," without seeking to corroborate his or her story.

Some mathematicians and philosophers argue that Gödel's Theorem poses a serious challenge to the ARTIFICIAL INTELLIGENCE lobby. AI philosophy maintains that formal computation is the essence of all thinking. Since, ideally, any computing machine is capable of any formal computation, computing machines should be able to match any human capacity to think. (See THE CHURCH-TURING THESIS; TURING MACHINES.) But opponents of AI, like mathematician Roger Penrose (see PENROSE ON NONCOMPUTABILITY), argue that Gödel's Theorem proves otherwise. Since we human beings can *understand* the theorem, we have the capacity for some kind of thought beyond any formal system. We can, as it were, see beyond the program. Penrose and others argue that we do this through insight, or intuition, of which machines are incapable. (See THINKING.)

Penrose and others argue that Gödel's incompleteness theorem proves something about physics itself. We humans can understand the theorem, and our thinking originates in our brains. But brains are material substances and therefore subject to the laws of physics. Thus, the argument goes, the laws of physics themselves must be in some way richer than any given formal system. This line of thinking is topical at the moment, but very controversial.

Gödel's Theorem is in the spirit of the whole "postmodern" intellectual movement, embracing the philosophies of Wittgenstein and the existentialists, artistic movements like abstract expressionism, and some attempts to found a new pluralistic social philosophy. Nietzsche, a precursor, argued that we can achieve no God's-eye view of truth, that "we can never see round our own corner." Wittgenstein argued that we are always trapped within a "language game." Cubists and social pluralists argue that each point of view is just one perspective on truth. But whereas Gödel proved that any formal *statement* we make about truth will always be partial, his work leaves open the possibility of another kind of nonformal "knowing" through which we might gain access to the whole.

# Grand Unified Theories

The aim of Grand Unified Theories (GUTs), or Grand Unification, is to gather all elementary particles together under one theory. The quark theory explains hadrons as composite particles made out of QUARKS, which interact by means of the color force, a force carried by gluon particles. (See QUANTUM CHROMODYNAMICS.) Similarly, the electroweak theory combines the weak nuclear force and the electromagnetic force into a single scheme. (See THE ELECTROWEAK FORCE.) According to this scheme, forces are carried by the photon, the $Z^0$ boson, and two charged bosons, $W^+$ and $W^-$. These theories together make up the very successful STANDARD MODEL.

Is it possible to go further and unify the color and electroweak forces? Particle physicists believe that at high enough energies, this could indeed be the case. The strength of these interactions depends upon the amount of energy and momentum transferred in a collision. Since both the color force and the electroweak force vary with energy, physicists predict that at sufficiently high energies—enormously high energies, as it turns out—the two different interactions will come to have exactly the same strength. In such an energy range, nature should be more highly symmetric than in our own low-energy domain.

Grand Unifications can be understood in the following way: In the first instants, up to $10^{-34}$ second after its BIG BANG creation, the universe was highly symmetric with all forces equivalent in strength. During this period, the elementary particles mirrored each other through a variety of internal symmetries. But when, a fraction of a second later, the temperature of the universe fell to $10^{27}$ K, an act of SYMMETRY BREAKING occurred. The electroweak force began to separate from the strong color force, and the masses of hadrons from those of the leptons. Before this instant, there had been only one kind of force and one kind of matter; this was the GUT era. (See Table One in overview essay D, THE COSMIC CANOPY.)

The approach is based on the search for some more fundamental symmetry. The symmetries of the Standard Model of quarks and of the electroweak force must produce a new combined symmetry. It turns out that this new symmetry requires larger symmetry groups. The simplest is called SU(5). Physicists believe that SU(5), or a relative, must be the symmetry of Grand Unification. Accept this symmetry, and along with it goes the prediction of a new set of particles of force. In addition to the eight gluons of the color force, and the four carriers of the electroweak force, there are twelve new gauge bosons, called X particles, that help to carry the forces of nature, now unified under one scheme. An important prediction of this theory is that, under this grand unified force, the number of quarks is no longer conserved. This means that the proton, itself made up of three quarks, can decay into a positron and neutral pi-meson. However, this predicted effect has not yet been observed.

A Grand Unified Theory also allows the neutrino to have a mass, albeit a very small one. The implications of a massive neutrino are far-reaching. The total amount of observed matter in the universe can account for, at best, only 10 percent of the calculated mass. The question of the "missing mass" is crucial to cosmology and could possibly be resolved if the neutrino itself had a mass. Although the neutrino is far lighter than the electron, the enormous number of neutrinos in the universe may well account for this mass deficiency. (See DARK MATTER.) But no evidence has yet been discovered that points to the neutrino's having mass.

Grand Unification purports to bring cohesion into elementary particle physics. It makes interesting predictions, such as the mass of the neutrino and the finite life of the proton, and allows INFLATION THE-

ORY to be formulated. On the other hand, it requires the existence of yet more particles at energies that lie far beyond anything that can be produced by the present generation of elementary particle accelerators. And, like the Standard Model, the theory contains arbitrary features, such as a number of fundamental constants whose values are not determined by the theory itself.

A GUT theory still has three separate categories of entities: matter (FERMIONS), forces (BOSONS), and gravity (the curved space-time of GENERAL RELATIVITY). Such a theory cannot stretch back to the earliest instants in the history of the universe, THE PLANCK ERA, when things were so condensed that both quantum and gravity effects had to be treated together by a more unified theory. Physicists still dream of THEORIES OF EVERYTHING to describe the Planck era, but none has yet been fully formulated, and perhaps none will ever be.

# The Great Attractor

In 1977, American cosmologist George Smoot and his colleagues, looking for unevenness in COSMIC BACKGROUND RADIATION, made a surprising discovery. Our whole galaxy, indeed our whole local group of galaxies, is moving toward the constellation Leo at about 400 miles per second. This movement was shown as a slight Doppler effect on the background radiation, making the microwave "light" slightly "bluer" toward Leo and "redder" in the opposite direction.

The movement toward Leo could be explained only as the result of gravitational pull toward some "attractor." The Great Attractor, as it was named, cannot be nearby, or it would pull some galaxies harder than others. It has to be very large—about 150 million light-years across—and equally far away. This is a thousand times larger than our galaxy's diameter. The universe is far from uniform, even on this large scale. (See THE COSMOLOGICAL PRINCIPLE.)

In deducing the presence of The Great Attractor, scientists have taken the cosmic microwave background as the absolute standard of rest. It is not absolute space, nor the ether, but it plays some of the roles that the older physics attributed to those absolute constraints. In consequence, the victory of SPECIAL RELATIVITY has not been as complete as it once seemed.

# Hadrons

The matter particles called fermions are divided into QUARKS and LEPTONS. Hadrons are composed of quarks. They are either baryons, consisting of three quarks (e.g., the proton or neutron), or mesons, consisting of a quark-antiquark pair. Hundreds of kinds of short-lived hadrons have been produced by particle accelerators.

An older definition of hadrons was "particles that feel the strong nuclear force that holds protons and neutrons together in the atomic nucleus." But we now know that the strong nuclear force itself consists of pi-mesons, which, in turn, consist of quarks and gluons. So the older definition is no longer sufficiently fundamental. (See QUANTUM CHROMODYNAMICS.)

The terminology used to name these various particles is derived from Greek words: hadron ("bulky"), baryon ("heavy"), meson ("middling"), and lepton ("light"). This is generally accurate, although heavy, short-lived leptons ("muons") were discovered in 1937.

# Heisenberg's Uncertainty Principle

How much can we ever know about fundamental physical reality? Are ambiguity and uncertainty inherent features of the real world, or is our knowledge necessarily limited? Heisenberg's Uncertainty Principle addresses questions like these. It may also have wide implications for, and potential applications within, theories of knowledge and organization.

Quantum reality is a strange, uncertain, shadowy realm. The more we try to pin it down, the more it eludes us. The Uncertainty Principle asserts that it must always be so; we must always content ourselves with partial truth and ambiguity when dealing with fundamental physical reality.

A particle was always thought to have both position and momentum. A given particle should always be somewhere (have a location) and is always traveling at a certain speed. But we can never know both. If we measure, or focus on, the position, the momentum becomes unfixed; if we measure the momentum, we lose the position. It is the same with any of the other complementary pairs (see COMPLEMENTARITY) of which quantum reality consists, like waves and particles (see WAVE/PARTICLE DUALITY), energy and time, or continuity and discontinuity. Fixing one member of any pair in place always makes our knowledge of the other member become fuzzy.

A particle is fixed at some exact place in space and time, but this separates or "alienates" it from its neighbors. A wave is spread out over space and time and has an immediate, holistic relationship with its neighbors—and possibly with all waves in the universe—but it can never be located anywhere or anywhen. If we focus on the particlelike

properties of a quantum entity, we get a good sense of the isolated part at the expense of the whole; if we focus on the wavelike qualities, we have a sense of the whole but lose our ability to focus on the part or the particular.

Why does the Uncertainty Principle apply to nature? Why does nature seem to come in complementary pairs of determinate (exact) and indeterminate (fuzzy) aspects? The answer has to do with quantum theory's description of fundamental reality as a wavelike spreading out of infinite possibilities. A particle has the *possibility* of being anywhere or anywhen until it settles into one place or one time. The mathematical description of the particle in quantum theory is known as Schrödinger's equation. (See THE WAVE FUNCTION AND SCHRÖDINGER'S EQUATION.) It *is* a description of all the particle's possibilities. But when any *one* is *actualized*, when the particle settles into just one place or one time, all other possibilities disappear. This is known in physics as the COLLAPSE OF THE WAVE FUNCTION.

In our own lives, we experience something like the Uncertainty Principle. We can focus on the facts of a situation, or we can give ourselves over to the "feel" of it. Focusing on the facts costs us perspective, or a knowledge of the whole; gaining perspective distances us from the details of the situation. We can never be both detached observers and involved participants. In the same way, we find it difficult to focus on one clear and distinct idea (to be analytic) and to entertain a vague train of thought or pattern of loose associations (to be "poetic"). In our organizations we often find that we must choose between imposing rigid rules and tight structure, or allowing things to unfold creatively with a sense of self-organization. Tight structure gives us control but loses us the benefits of innovation. A marketing executive might gain a hard-and-fast knowledge of exact sales figures in the market at a given moment, but perhaps at the cost of understanding the overall drift of factors affecting market demand. In the same way, a pianist must often lose his or her sense of an overall piece of music temporarily in order to concentrate on improving technique for a difficult phrase.

For many years, physicists argued over whether uncertainty and ambiguity are actual features of the real world or merely constraints on our own knowledge and experience. In his "hidden variables theory," David Bohm argued that all variables have definite values, although we are unable to measure them all, but his thinking sits uneasily with the principles of Special Relativity. A series of experiments and complex arguments has led most physicists to believe that reality itself is inherently "fuzzy," or at least that it has both clear and fuzzy aspects at any given time. Trying to focus on it is like trying to grab hold of an evanescent dance.

The Uncertainty Principle has been taken to mean that there is uncertainty, indeterminacy, or unpredictability built into a situation. This use of the term catches the flavor of the actual meaning in physics, but in physics itself the Uncertainty Principle means that we have to *choose* between one or the other of a complementary pair of options.

# Holism

The old physics was atomistic; quantum physics is essentially holistic. What does this mean?

Holism as an idea or philosophical concept is diametrically opposed to ATOMISM. Where the atomist believes that any whole can be broken down or analyzed into its separate parts and the relationships between them, the holist maintains that the whole is primary and often greater than the sum of its parts. The atomist divides things up in order to know them better; the holist looks at things or systems in aggregate and argues that we can know more about them viewed as such, and better understand their nature and their purpose.

The early Greek atomism of Leucippus and Democritus (fifth century B.C.) was a forerunner of classical physics. According to their view, everything in the universe consists of indivisible, indestructible atoms of various kinds. Change is a rearrangement of these atoms. This kind of thinking was a reaction to the still earlier holism of Parmenides, who argued that at some primary level the world is a changeless unity. According to him, "All is One. Nor is it divisible, wherefore it is wholly continuous. . . . It is complete on every side like the mass of a rounded sphere."

In the seventeenth century, at the same time that classical physics gave renewed emphasis to atomism and reductionism, Spinoza developed a holistic philosophy reminiscent of Parmenides. According to Spinoza, all the differences and apparent divisions we see in the world are really only aspects of an underlying single substance, which he called God or nature. Based on pantheistic religious experience, this emphasis on an underlying unity is reflected in the mystical thinking of most major spiritual traditions. It also reflects developments in modern QUANTUM FIELD THEORY, which describes all existence as an excitation of the underlying QUANTUM VACUUM, as though all existing things were like ripples on a universal pond.

Hegel, too, had mystical visions of the unity of all things, on which he based his own holistic philosophy of nature and the state. Nature consists of one timeless, unified, rational and spiritual reality. Hegel's state is a quasi-mystical collective, an "invisible and higher reality," from which participating individuals derive their authentic identity, and to which they owe their loyalty and obedience. All modern collectivist political thinkers—including, of course, Karl Marx—stress some higher collective reality, the unity, the whole, the group, though nearly always at the cost of minimizing the importance of difference, the part, or the individual. Against individualism, all emphasize the social whole or social forces that somehow possess a character and have a will of their own, over and above the characters and wills of individual members.

The twentieth century has seen a tentative movement toward holism in such diverse areas as politics, social thinking, psychology, management theory, and medicine. These have included the practical application of Marx's thinking in Communist and Socialist states, experiments in collective living, the rise of Gestalt psychology, SYSTEMS THEORY, and concern with the whole person in alternative medicine. All these have been reactions against excessive individualism with its attendant alienation and fragmentation, and exhibit a commonsense appreciation of human beings' interdependency with one another and with the environment.

Where atomism was apparently legitimized by the sweeping successes of classical physics, holism found no such foundation in the hard sciences. It remained a change of emphasis rather than a new philosophical position. There were attempts to found it on the idea of organism in biology—the emergence of biological form and the co-operative relation between biological and ecological systems—but these, too, were ultimately reducible to simpler parts, their properties, and the relation between them. Even systems theory, although it emphasizes the complexity of aggregates, does so in terms of causal feedback loops between various constituent parts. It is only with quantum theory and the dependence of the very *being* or identity of quantum entities upon their contexts and relationships that a genuinely new, "deep" holism emerges.

## Relational Holism in Quantum Mechanics

Every quantum entity has both a wavelike and a particlelike aspect. The wavelike aspect is indeterminate, spread out all over space and time and the realm of possibility. The particlelike aspect is determinate, located at one place in space and time and limited to the domain of actuality. (See WAVE/PARTICLE DUALITY.) The particlelike aspect is fixed, but the wavelike aspect becomes fixed only in dialogue with its surroundings—in dialogue with an experimental context (see CONTEXTUALISM) or in relationship to another entity in measurement or observation. It is the indeterminate, wavelike aspect—the set of potentialities associated with the entity—that unites quantum things or systems in a truly emergent, relational holism that cannot be reduced to any previously existing parts or their properties.

If two or more quantum entities are "introduced"—that is, issue from the same source—their potentialities are entangled. Their indeterminate wave aspects are literally interwoven, to the extent that a change in potentiality in one brings about a correlated change in the same potentiality of the other. In the NONLOCALITY experiments done to test BELL'S THEOREM, measuring the previously indeterminate polarization of a photon on one side of a room effects an instantaneous fixing of the polarization of a paired photon shot off to the other side of the room. The polarizations are said to be correlated; they are always determined simultaneously and always found to be opposite. This paired-though-opposite polarization is described as an emergent property of the photons' "relational holism"—a property that comes into being only through the entanglement of their potentialities. It is not based on *individual* polarizations, which are not present until the photons are observed. They literally do not previously exist, although their oppositeness was a fixed characteristic of their combined system when it was formed.

In the coming together or simultaneous measurement of any two entangled quantum entities, their relationship brings about a "further fact." Quantum relationship evokes a new reality that could not have been predicted by breaking down the two relational entities into their individual properties.

The emergence of a quantum entity's previously indeterminate properties in the context of a given experimental situation is another example of relational holism. We cannot say that a photon is a wave or a particle until it is measured, and how we measure it determines what we will see. The quantum entity acquires a certain new property—position, momentum, polarization—only in relation to its measuring apparatus. The property did not exist prior to this relationship. It was indeterminate.

Quantum relational holism, resting on the nonlocal entanglement of potentialities, is a kind of holism not previously defined. Because each related entity has some characteristics—mass, charge, spin—*before* its emergent properties are evoked, each can be reduced to some extent to atomistic parts, as in classical physics. The holism is not the extreme holism of Parmenides or Spinoza, where everything is an aspect of the One. Yet because some of their properties emerge only through relationship, quantum entities are not wholly subject to reduction either. The truth is somewhere between Newton and Spinoza. A quantum system may also vary between being more atomistic at some times and more holistic at others; the degree of entanglement may vary.

# The Human Genome Project

The Human Genome Project is an extensive international collaboration in which scientists are attempting to read the entire human genetic code. The instructions that determine the growth and functioning of each cell in the body are written as chemical compounds (letters of the genetic message) strung along the chromosomes. While the basic alphabet and the "punctuation marks" of the code and some few of the genes have already been determined, it will take many years of concerted work to decipher the entire message.

The genetic material in a cell is arranged along one threadlike pair of chromosomes or more—one in a bacterium, twenty-three in a human being—which are visible under a good microscope. Each chromosome consists of thousands of genes, and each gene is a long string of DNA molecules functioning like a computer program. A virus contains only a handful of genes, while the human genome consists of about 100,000. Each gene is an instruction to make one particular kind of protein. All the genes are present in every cell, but only some are "turned on" in a given cell. Many human genes are never turned on in any cell, and scientists don't know why they are there.

The genetic code regulates the various metabolic processes that take place within each living cell. DNA carries information on cell growth and repair, instructing a cell to function as liver, heart, blood, muscle, and so on. Red hair, green eyes, height, athletic ability, hemophilia, schizophrenia, and a host of other characteristics are partly or wholly determined by DNA's molecular message.

Long before the discovery of DNA, it was known that eye and hair color, general height, and a variety of medical disorders (such as hemophilia) are inherited. Breast cancer, manic depression, and schizophrenia seemed to run in families, but it was not clear if these had a genetic basis or were influenced by environmental (family) factors. The eugenics movement in the early decades of the twentieth century went so far as to propose selective human breeding to eliminate inherited defects and enhance the intelligence and physical well-being of the human race.

It was only with the discovery that genetic instructions are coded in a four-letter "alphabet" along the DNA molecule that inheritance could be placed on a strictly scientific basis. Segments of DNA's molecular message were found to correspond to physical characteristics or disorders. It was discovered, for example, that sections of the code determine factors such as hair and eye color. In other cases, the code does not so much determine an outcome as produce an overall disposition. Thus we may try to blame our parents if we are overweight, but diet and lifestyle can do a great deal to ameliorate the condition.

Decoding DNA has been hailed by doctors and biologists as a way of curing a variety of disorders. Where a metabolic defect is the result of incorrect genetic instructions, it may one day be possible to introduce a corrected message into a patient's DNA. Parents with defective DNA could be counseled against having children. DNA testing could also indicate a person's likelihood to succumb to heart disease or cancer, for example. It is highly probable that some people have a genetic predisposition to lung cancer. A simple test would indicate those at high risk, who should then refrain from smoking. Specific genes may also give information on an individual's life expectancy.

In view of the enormity of the task, the Human Genome Project can be carried out only at the international level. In a project that will take years to complete, teams from across the world each take responsibility for a given section of the DNA molecule. Many scientists hail this as an example of what can be achieved by international collaboration and believe that the decoding of the human genome will be a fundamental scientific breakthrough. They point to enormous benefits to be gained—understanding the mechanisms of disease, discovering potential cures, eradicating a wide variety of ailments, and gaining a deeper understanding of the human organism. The project is sometimes compared, with possibly unconscious irony, to the Manhattan Project for the development of the atomic bomb.

Other scientists are more critical, pointing out that extensive funding and many teams of scientists will be diverted from what may ultimately prove to be scientifically more interesting problems. They argue that the business of decoding the human genome involves pedestrian science, and that it is not at all clear that anything deep or fundamentally new will be discovered from this work.

The project has also been criticized on ethical and sociological grounds. Being able to manipulate genetic material brings with it enormous power, and the past record of the human race is not that encouraging when it comes to the wisdom of its decisions and the way new technology is applied.

The Human Genome Project is also seen by some as an extreme form of scientific reductionism in which elements of behavior are linked to specific genetic messages. Some behavior does depend on complex nonlinear interactions of genes. It may indeed be possible to select genes for IQ, or to develop brains that have enhanced neuronal linkages. Still, all this is a far cry from being able to breed an exceptional mathematician or artist. Human abilities are complex and depend on more factors than intelligence alone. Intelligence without associated maturity, drive, willpower, imagination, creativity, and a host of other abilities is valueless. While genetic instructions may produce certain predispositions, this is a long way from understanding how an adult human being thinks, feels, and acts.

The Human Genome Project has inevitably precipitated scare stories about designing human armies for extreme aggression or producing psychopaths for certain tasks. Yet even its benevolent side can be questioned. What if genes for criminality are discovered? Should these be eradicated, or do social outcasts and misfits also perform a useful, albeit bizarre, function as critics of conventional society? And how would society deal with the ability to manipulate a hypothetical gene associated with sexual orientation?

How about the benefits, the miracle cures? Despite many medical advances in the twentieth century, lifestyle, general preventive medicine, and, where needed, conventional surgery continue to be the overwhelmingly important factors in general heath. Even the dream of producing more beautiful bodies involves troublesome questions about the way aesthetics is socially conditioned.

Most scientists regard these questions as side issues that detract from the overall importance of the project. Nevertheless the Human Genome Project raises issues that, in the end, reflect more about ourselves and our society than they do about a supposed objective of science itself.

# Humanistic Psychology

The humanistic philosophy that "man is the measure of all things" has its roots in ancient Greece. At various times through the ages, it has stood in opposition to theological or political absolutism, or to materialism and reductive science, all of which have posited some authority above that of the human or have reduced man to the less than human. In psychology, humanism became prominent in the 1950s, when it was regarded as a "third force" in contrast to the two prevailing psychological models, BEHAVIORISM and Freudian-inspired psychodynamic theory (see PSYCHODYNAMICS AND PSYCHOTHERAPY). In humanistic psychology, the emphasis is on the uniqueness and importance of each individual, who is thought to be able to structure his or her own life and capable, if given the opportunity, of "self-actualization." This is growth, or the self-fulfillment of an inner potential.

Carl Rogers was a leading figure in humanistic psychology. His thinking was applied to therapy, encounter groups, education, and management training. Rogers believed that individuals are damaged by the negative conduct and attitudes of others—parents, teachers, colleagues, work superiors, and so on. If the therapist could reverse this damage by providing a supportive atmosphere and "unconditional positive regard," the patient would gradually grow in self-esteem and effective capacity, without the more formal analysis and interpretations characteristic of Freudian therapy.

Abraham Maslow, a leading thinker in humanistic psychology, contributed the important idea that there is a hierarchy of human needs. People must have their lower needs satisfied in order to progress toward the more fulfilling gratification of higher needs. The lowest needs are physiological ones, for such things as food. Then come higher ones, for things like safety, love, and esteem. And finally there is a need for self-actualization. Maslow looked at the peak experiences of ordinary people and at biographical material from the lives of great men like Abraham Lincoln and Albert Schweitzer, whom he regarded as self-actualized. This more positive, growth-oriented end of the human psychological spectrum had been overlooked by earlier therapies and by behaviorism, all of which had focused on illness, maladaptation, or elementary abilities.

The humanistic vision of the self-actualized life was eagerly embraced by the hippies in the 1960s, in conjunction with radical politics and a general spirit of anti-authoritarianism. There were a great many therapies and practices introduced during that decade to expand consciousness and promote inner growth. Drugs, Eastern meditation practices, sexual experimentation, and communal living were a few of the things tried. With most, the long-term effect was less transformative for both individuals and society than had been hoped. This was probably at least partly because many individuals adopting these practices were less mature or psychologically developed than their self-actualization rhetoric implied. And apparently noble goals were too often used to rationalize a narcissistic attitude toward life.

Nevertheless, the 1960s humanistic therapy was a useful corrective to the more one-sided, impersonal, authoritarian therapies that had preceded it, and something of its ethos has survived. Management practices, educational philosophy, even psychotherapy itself have all become less formal and authoritarian, and more concerned with individual welfare and development.

Whatever the strengths and weaknesses of humanistic psychology, the question of whether it is possible to view it within a scientific paradigm remains. Its concerns and areas of focus are clearly about aspects of human psychology that are perennially important, but they are not as easy to quantify or to study experimentally as those that behaviorism or psychodynamic theory concentrates on. Concepts like responsibility, the self, and personal growth may need a clearer definition and might benefit from more connection with neurology or computational science, if that is possible.

# Identity in Quantum Mechanics

Quantum entities are not really "things" but rather patterns of active energy. Each has a wave aspect (a spreading out of shifting potentiality) and a particle aspect (something here and now that can be pinned down), but which aspect shows itself at any one time depends upon the surrounding circumstances. (See CONTEXTUALISM.) Given the "shifty" nature of quantum reality, can we distinguish one quantum from another, the temporary, contextual properties of a quantum entity from its "essence"? Do quantum entities indeed have an essence? The answers to these questions bear directly on the long-standing philosophical debate about the identities of persons and things in this world.

The philosophical tradition asks two kinds of questions about identity. The first, the question of substance, asks how or whether we distinguish the underlying reality of a thing from its properties at any given time. The second, which might be called the question of indiscernibility (Leibniz called it the identity of indiscernibles), asks whether, if two things have all the same properties, they are still two things or only one. Both questions must be rethought when we deal with entities in quantum physics.

## The Question of Substance

What persists through change? Is the bare winter tree we see from our window in December the same as the leafy green tree that we saw last summer? Is the man standing before us now the same boy whom we once knew? The early Greeks took extreme positions about such questions. Heraclitus maintained that everything changes ("We can never step into the same river twice"), while Parmenides argued that nothing changes (the All is One, always the same as itself). Atomic theory, from Democritus to classical science, has offered a compromise position: The indestructible atoms persist, but their arrangements, and hence the properties of larger wholes, change.

In classical physics we can always reduce a whole to its parts, and only those parts have a fundamental identity. Ultimately, they are the atoms of which the thing is composed. But quantum physics raises two objections to this view: It gives too *little* substance to complex, emergent wholes, and too *much* substance and permanence to the constituent "atoms" themselves, whether we understand them as the basic units of chemical elements or as subatomic particles like protons, neutrons, and electrons.

In quantum mechanics, complex entities acquire a *further* identity through the relationships of their parts. *A* and *B* may be constituents of *C*, but *C* is greater than *A* plus *B*. The whole has properties not possessed by either of its parts. (See EMERGENCE; HOLISM.) This may bear on the question of complex wholes in our everyday experience, such as the identities of persons and of states of consciousness.

In quantum mechanics, entities are wave/particle systems. Waves can overlap, coalesce, split in two, and transform themselves. Insofar as a quantum entity is a particle, it has only temporary identity. It can, for the moment and in a given environment, be pinned down and identified as a thing of a given sort, with a given mass and charge, but this does not last forever. An electron and a positron can annihilate each other if they meet, and become two photons. An atomic nucleus, which is a complex whole, can give off an alpha particle to become a nucleus of a different chemical substance. If we think of persons as complex relational wholes in the quantum mechanical sense, we think of personal identity as a thing that exists but is always changing. This is a middle way between a reductive position like that of Hume or Sartre, who argued that persons are nothing but their associations or experiences, and a more atomistic view, which holds that they are discrete, unchanging, and indestructible entities.

In the quantum sense, identity can exist on all scales, small and large, but it is not permanent. In quantum physics, only THE QUANTUM VACUUM, the substrate of all that is, has a permanent identity. So it is only at the level of this underlying vacuum that substance can be said to exist permanently as a thing distinguishable from properties. In this sense, quantum identity is similar to the Buddhist position that only the underlying Void (Sunyata) has substance, while all existing things are transient. But the quantum view regards existing things as *real* entities while they endure.

## The Question of Indiscernibility

In our commonsense Western view, physical objects have unique identities. Whether we can tell the difference or not, one blade of grass is distinct from all others, and a given girl is not her twin sister. At the same time, we do not accord this unique identity to copies of patterns. Things like computer programs, symphonies, or concepts (like "blue" or "true") have many examples, but no distinct identity unless we fall back on notions like Plato's archetypes or Karl Popper's World Three. In quantum mechanics, things have both a particle (object) aspect and a wave (pattern) aspect. What can we say about identity in this context?

Do elementary quantum entities like electrons and photons have discernible identities? Subatomic entities are too simple to have distinguishing marks (as, for example, one twin might have a mole where her sister does not), but we could still tell them apart if they were kept in different "boxes"—different places in space-time. But what if their wave aspects begin to overlap, to share the same place in space-time? What criteria are left to tell us whether they are more like individual particles or indiscernible waves?

There is a fundamental test to determine whether two entities have individual identities. If we exchange one for the other, is the new state in which we find them distinguishable in principle from the old one? If A and B were similar pennies or similar particles, the answer would be yes. It would be no if A and B were similar computer programs (i.e., similar patterns). If program A was copied onto computer B, and a similar program B onto computer A, the result would be indistinguishable in principle from the state things were in before the exchange. In the case of the pennies, *we* could not tell the difference between the exchanged pennies, but nature could. Each penny would, as it were, still know that it was itself. But in the case of the computer programs, even the programs would not know the difference. Patterns have no identity.

An analogy would be two triangular patterns made up of billiard balls. If we exchanged the balls making up triangle A with those of *B*, we would say that each of the two sets of balls was now in a different place and the total state had been altered. But it would be meaningless to say that we were going to exchange one triangular *pattern* for the other. The two patterns are indistinguishable—they *share* an identity, according to Plato, or they have no identity at all.

## Fermions and Bosons

Though the difference between pennies and programs, or billiard balls and the patterns into which they can be arranged, may be difficult to grasp, physicists have done actual experiments on exchanging quantum entities. Some behave more like pennies (particles), others more like patterns (waves), though neither has an exact classical equivalent. The two different kinds of quantum entities are known as FERMIONS and BOSONS, and their behavior in respect to identity illustrates the even-handedness of the quantum wave/particle duality. An exchange of two fermions, like that of pennies, results in things' being in a different state, whereas an exchange of two bosons does not.

Fermions include all the particles of which solid matter is made—electrons, protons, and neutrons. Desks and chairs and human bodies are made up of fermions. Bosons include all the quanta ("particles") of which the fundamental binding forces of the universe are made. Photons of light waves and electromagnetic forces, the strong and weak nuclear forces, and gravity, if it is quantum, are made up of bosons. Two similar fermions can never occupy the same state (Pauli's Exclusion Principle), which is why solid matter is solid, and why we have the diversity of chemical elements and forms of matter that we do. The fermions are "antisocial"; they push each other away and keep a space between them. Two or more bosons, on the other hand, can get into the same state. They are very "gregarious," and this is why large-scale forces can exist and bind things together. It also accounts for the special intensity of something like laser light: Millions of photons get into a single state—that is, share an identity—and the laser beam behaves as though it were made up of just one giant photon.

The medieval question "How many angels can dance on the point of a pin?" was about identity. Do angels have identities? In modern terms, the question would be posed as whether angels are made up of fermions or bosons. The perennial philosophical question about personal identity—"Would anything be different if I were you, and you were I?"—might also be asking whether persons are best described as fermions or as bosons. (See SPIN AND STATISTICS.)

# Implicate Order

According to the physicist David Bohm, the world we see around us is only a shadow of a deeper reality called the Implicate Order. Reality, in the Newtonian world, is clear and unambiguous. Material objects have specific boundaries and locations in space. Their properties are well defined and exist independent of any observer. Bohm refers to this as the Explicate Order and points out how deeply ingrained its associated world view has become in our thinking. An essential feature of Explicate Order physics is the Cartesian coordinate system, a grid placed over the world whose most fundamental elements are points and lines.

Our most fundamental scientific description of the world is quantum theory, which happens to be at odds with the language of the Explicate Order. Quantum entities, to the extent they can even be spoken of, are not well defined in space but delocalized and interpenetrating. Objects that are separated by large distances can be, in another sense, correlated to such an extent that normal (Newtonian) concepts of distance and separation no longer seem to apply.

Despite the radically new nature of quantum theory, much of its mathematical description is still within the older Cartesian and Explicate Order. It uses differential equations based on the mathematical notion of points, although the very notion of a point in space loses its meaning in quantum theory. Bohm proposes that quantum theory reveals an Implicate Order (or enfolded order) within nature. In this Implicate Order, object A can be contained within *B*, while at the same time *B* is contained within A. Explicate Order notions of boundaries, separation, space, and time are transcended within the Implicate Order.

Bohm suggests that elementary particles, which appear as objects in the Explicate Order, must be seen as processes within the Implicate Order. As these processes unfold, they give rise to what appears, for an instant of time, to be a localized particle. A moment later, this "particle" folds back into the overall Implicate Order. What is taken, in the Explicate Order, as two elementary particles in interaction becomes, in the Implicate Order, a single process that, for a moment, localizes within different spatial regions of the Explicate Order. What looks like an independent, localized object is, within the Implicate Order, an aspect of the whole. Bohm also proposes an underlying ground that he terms the Holomovement.

Bohm believes that the Implicate Order not only describes processes at the most fundamental levels of matter, but also describes the deeper nature of human consciousness. The reason the Explicate Order became so pervasive is partly that we live in a world of objects that appear, on our scale, to be stable and well localized. It is also the result of the hypnotic influence of the (European) languages we speak. These languages are strongly noun oriented and lend themselves to a perception of the world in terms of objects, categories, and fixed boundaries. "The cat chases the mouse" suggests independent objects in interaction, while the observer of such an event may be more aware of a scurrying movement. Experiments by the psychologist Jean Piaget point out that an overall perception of process and movement exists before a young child is able to comprehend independent objects. Piaget's developmental theories suggest that our primary perceptions are of an Implicate Order world.

Bohm suggests that a verb-oriented language (his hypothetical Rheomode) would accord well with our primary experiences of the world. It turns out that such languages do exist. In North America, for example, the Algonquian family of languages spoken by the Cree, Blackfoot, Ojibwa, Micmac, Cheyenne, and other peoples are strongly verb oriented and essentially reflect a PROCESS vision of the world. In keeping with his thoughts on the Implicate Order and Rheomode, Bohm has also attempted to develop algebras and other process-based mathematical forms that would be more appropriate to a description of consciousness and quantum theory.

Forms thrown up by the Implicate Order exist, for a time, within our Explicate Order world until they unfold back again. Thus, the Implicate Order (together with its underlying ground, the Holomovement) is the source and sustainer of the Explicate world. In later versions of his theory, Bohm introduces the notion of an even more subtle Superimplicate Order. In one sense, this forms a sort of feedback loop between the Explicate and Implicate orders. While the Implicate sustains the Explicate, it also responds to the Superimplicate. This Superimplicate Order is in turn responsive to what occurs in the Explicate Order. In this vein, Bohm speculates that below the Superimplicate Order there may be yet other orders of increasing subtlety, an infinite number of orders perhaps. Bohm's ideas are not easy to test by experiment; many physicists regard them as philosophy rather than physics.

# Indeterminacy

Can science rely on eternal, immutable laws of nature to predict the outcome of events? Classical physicists answer this question with a resounding yes, but quantum physics supports the likelihood that a radical indeterminacy, or contingency, limits any certainty surrounding the unfolding of physical events.

Since the most ancient times, there has always been a large element of contingency in human life. We have never been very good at predicting the weather, the onset of disease or disaster, or the behavior of others. Such uncertainty has been attributed to blind chance, the arbitrariness of the gods, or the simple vagaries of human experience. Those uncomfortable with so much contingency have often suggested there is a plan behind it all, that God at least, or Fate, knows or controls some scheme that determines things should happen as they do. (See DETERMINISM.)

In the physical realm, Democritus held that all events have a cause;

thus all atoms move in a fixed or determined way. Epicurus argued that some movements are without cause and some atoms swerve from their path without reason. He attributed human free will to such contingent movement of atoms in the mind.

Until the scientific revolution of the seventeenth century, how or whether physical events are determined was left to philosophical or theological speculation. But Newton's formulation of the three laws of motion, and his work on gravitation, seemed to offer a kind of scientific proof that causal laws really do determine events. Given knowledge of the initial state of any physical system, and a further knowledge of any force of change acting upon the system, the classical scientist could accurately predict exactly how that system would behave.

After Newton, uncertainty or apparent chance in physical events was attributed to ignorance—we might not *know* all the factors affecting a physical system, but underneath is a fully determined reason for its behaving as it does. It is only in this sense that, for instance, gambling casinos are said to be subject to "the laws of chance." We cannot *predict* into which slot a roulette ball will fall, but the ball's destination is fully determined. An omniscient Newtonian computer could forecast the correct number every time. The same is true of the weather. (See CHAOS AND SELF-ORGANIZATION.)

Until quantum mechanics turned everything on its head, uncertainty in physics meant simply unpredictability. Quantum theory is much more radical and, for many, unsettling. Events often happen without cause. (See NONLOCALITY.) An electron may leap into one energy orbit of an atom, or into another. A radioactive atom may emit a decay particle at any moment, or remain stable for thousands of years. Not only can we not predict these events, they are largely undetermined. There is no underlying set of factors setting out a program of events for quantum reality.

In classical physics, determinism is assured by forces of change acting upon the initial state of the system. But HEISENBERG'S UNCERTAINTY PRINCIPLE shows us that we can *never* know all the parameters of a system's initial state. If we know a quantum entity's momentum (its wave aspect), its position (particle aspect) is indeterminate. If we know its position, its momentum dissolves into a blur of uncertainty. Certain variables of a quantum system become unfixed or indeterminate when their complementary variable becomes fixed. (See COMPLEMENTARITY.) Unobserved or unmeasured, a quantum system is a spread-out of possibilities or potentialities that collapses when one of a complementary pair of possibilities is actualized. The collapse itself is random. (See ACTUALITY AND POTENTIALITY IN QUANTUM MECHANICS; COLLAPSE OF THE WAVE FUNCTION.)

Indeterminacy is one of the concepts that make quantum physics such a radical break with earlier ways of scientific thinking. The scientific enterprise itself, as defined by Newton and his successors, rested upon using observation and causal laws to predict the outcome of events. Now quantum theory tells us that our observations are necessarily limited by an underlying indeterminacy, and that the causal laws don't always apply. Randomness, or contingency, is a basic feature of the universe and its unfolding.

Quantum events are probabilities. Some are more likely than others to happen. Given a large enough number of them, we can predict certain patterns of outcome. Schrödinger's equation describes all *possible* observations of a quantum system, now and in the future, but it is really only a set of bookmaker's odds. (See THE WAVE FUNCTION AND SCHRÖDINGER'S EQUATION.) We can rarely say anything useful about the future behavior of a single quantum event. Newton's dream of a universal clockwork machine gives way in quantum physics to a universal roulette wheel or a game of dice.

Both the probabilistic and the random features of quantum events are revealed in an experiment commonly done on polarized light, that is, light that vibrates electrically in a selected plane. When such light encounters a sheet of Polaroid in the right orientation, all the light will get through. At a 90° angle of orientation, none of the light passes through, and at 45°, half gets through.

If we think of the polarized light as a stream of photons, at 45°, 50 percent of the photons will pass through. But whether any *one* photon will get through is problematic. Each has a 50 percent chance of being transmitted or absorbed, but nothing in either the photon or the Polaroid determines which will happen. It is a matter of chance.

For some physicists, indeterminacy has remained a primary objection to the whole quantum enterprise. Einstein could never come to terms with it and argued that something was deeply wrong or incomplete about quantum theory. David Bohm and others have proposed theories of "hidden variables" that lie beyond the quantum limits of our powers of observation but nonetheless determine the outcome of quantum events. There seem to be no scientific grounds for adopting these theories, however.

Within the categories and values of the old Newtonian world view, indeterminacy appears as a threat or negation of the scientific enterprise, but it lays the foundation for a great deal of what a new scientific paradigm suggests is positive. In quantum theory, indeterminacy underpins the potentiality, or the "what might be," of an evolving system. It is because many variables of a quantum system begin as unfixed or ambiguous that the system can evolve in creative dialogue with other systems, with its environment, or with human agents. Quantum HOLISM and the EMERGENCE of new relationships and new properties depend upon the unfixedness of initial states. So, too, does the possibility of free will require that something about the universe remain open to the influence of choice and action.

# Inertial Frames

Newton's law of inertia holds that any object continues to move with constant velocity (constant speed plus direction), unless acted on by a force. Hence the continuing motion of the planets, inertia plus gravitation, or of a rolling ball until frictional forces bring it to rest. Hence, too, by analogy, our own tendency to stay fixed in habitual (inertial) behavior unless acted upon by some inner effort or outside force.

In SPECIAL RELATIVITY, the laws of physics, including THE SPEED OF LIGHT, are observed to be the same within any given inertial frame: any space-time framework traveling at constant velocity. Thus we cannot tell whether a closed train or an elevator is at rest or moving unless it changes its velocity (accelerates), which we experience as a force. The perspectives given by different inertial frames use different book-keeping systems for measurements of mass, length, and time, but are all equally valid. (See RELATIVITY AND RELATIVISM.)

GENERAL RELATIVITY deals with noninertial frames, which are accelerating or rotating. In these frames, apparent forces (e.g., centrifugal force) are felt, which complicates the laws of physics. Einstein was able to unify the description of these noninertial frames and gravitational forces.

# Inflation Theory

Inflation theory is an elegant, although speculative, refinement of the BIG BANG model of the universe's creation. According to inflation models, the universe expanded enormously during the first $10^{-34}$ second of its existence, by a factor of $10^{50}$ or more. The first version of such models was put forward by Alan Guth in 1979, and several variants have been formulated since. All inflation models set out to resolve three problems that result from standard Big Bang theory.

The first is known as the flatness problem. Experimental observations have shown that the universe has a density parameter omega (average mass per cubic meter) somewhere between 0.2 and 2. If omega is exactly 1, the result will be a flat universe, delicately poised on the borderline between closed and open. (See RELATIVISTIC COSMOLOGY.) This is remarkable, because omega-equals-1 is an unstable point. If the universe had begun with the slightest deviation from this, omega would by now be very far from 1, and we would not exist. If omega were much less than 1, the universe would long ago have recollapsed. If omega were much greater than 1, gravitation would have been too weak to form stars. So why is the universe so flat? (See THE ANTHROPIC PRINCIPLE.)

The second problem with Big Bang theory is the uniformity problem. With the exception of tiny "wrinkles" (see WRINKLES IN THE MICROWAVE BACKGROUND), dating from the time when the universe was 300,000 years old, this background radiation is of the same intensity, in all directions, to one part in $10^4$. Yet regions in opposite directions have not interacted since just after the Big Bang. What physical mechanism has kept, or made, things so uniform?

The third problem is the antimatter problem. Every particle of matter has an ANTIMATTER equivalent, and when the two meet, both are annihilated. Yet our universe consists mainly of matter, not antimatter. How can this be? We can see various pairs of colliding galaxies, yet none displays the spectacular signs of the mutual annihilation of matter and antimatter. All matter has condensed out of the same hot, primordial sea of radiation. So what became of the antimatter? There must be some slight asymmetry in the laws of physics at that primordial stage so that more matter than antimatter was formed. What we see now, supposedly, is the surplus of matter left after the rest had been mutually annihilated.

According to inflation theories, a vast expansion of the universe took place when it was $10^{-36}$ to $10^{-34}$ second old. The effect was like blowing up a balloon with spots on it—the spots become farther apart. Though the distance between two spots may increase faster than the speed of light, this does not violate SPECIAL RELATIVITY. The spots have not *moved* on the balloon, but more space (balloon surface) has been created between them. The effect on a balloon the size of the universe is to make it "flatter" by a factor of perhaps $10^{50}$. Its observable flatness today (where omega is near 1) would have resulted from almost any initial conditions. Furthermore, all the regions whose microwave background radiation we can see today would have been in close contact before the inflationary era began. This solves the uniformity problem.

What physical mechanism could have caused such a massive inflation? An answer is found in the several GRAND UNIFIED THEORIES (GUTs), which unite the two components of THE STANDARD MODEL of particle physics (see overview essay D, THE COSMIC CANOPY), THE ELECTROWEAK FORCE and the color force. (See QUANTUM CHROMODYNAMICS.) In GUTs, these two disparate forces of nature were united at very high energies in very early times in the universe. As the universe expanded and cooled, when it was about $10^{-36}$ second old, the color force "crystallized out" from the mixture. At this point, a massive expansion would have taken place in the fabric of space-time. Shortly afterward, the baryons (protons and neutrons) would have been formed. In GUTs, unlike conventional quantum chromodynamics, there can be a slight asymmetry in the baryonic laws, resulting in a preponderance of matter over antimatter.

Inflation theories are elegant, but we have no hope of creating high enough energies in our particle accelerators to test the predictions of GUTs. In this case, we find ourselves near the frontier at which physics and astronomy meet philosophy, where a theory is accepted or rejected not so much on empirical grounds as because of its simplicity and aesthetic appeal. But even a Grand Unified Theory does not incorporate gravity, and so it cannot describe THE PLANCK ERA, the first $10^{-43}$ second of the universe's existence. For that, still more inclusive, and speculative, THEORIES OF EVERYTHING are needed.

# Information

We live in an "information age," travel the "information superhighway," and are surrounded by "information technology." But what *is* information? Merely something we can communicate from one person to another? Or is it embedded in the very structure of the universe? Why does it have the capacity to transform our lives? Why do we value it so much?

Within physics and biology, information relates to the structure of a thing. Any nonrandom structure represents some information. A maple leaf or a footprint in the sand contains information, but the random patterns of falling leaves or of sand deposited on a shore by the sea do not. The more complex its structure, the more information the thing contains. There is more information in a human fingerprint than in a simple drawing containing only a few lines, more information in the hologram of a jungle scene than in a black-and-white photo of the same. But structure is simply the *objective* dimension of information. It does not relate to how we can interpret it or put it to use.

The human concept of information has a *subjective* dimension. It raises the question of information *content* and what this content *means* to us. In human terms, information has meaning only to something that can interpret it—i.e., read what it is saying. This ability to interpret need not be resident in human consciousness. A computer program can interpret data fed into it and convert it to some practical purpose. A DNA decoding system in biological cells can convert the information in DNA structure into the manufacture of proteins. A boy scout can interpret the information latent in a map and find his way through the woods.

Our older, more familiar energy machines, like steam engines or gas-driven automobile engines, convert one sort of energy into another. An information machine, by contrast, converts structure into some sort of structured behavior. Thus a DNA decoder converts DNA structure into biological growth, a computer converts the structure resident in its program and data input into a screen display, and a telephone system converts the structure in sound waves into electrical patterns that can later be reconverted into sound waves. Although all these information conversions are physical processes, the important dimension in information "technology" is not how much energy the process uses but rather the *pattern* of that energy. This shift from energy to pattern reflects the growing shift in our culture away from the importance of muscle to the crucial importance of brains—the heart of the "information revolution."

Since information has both an objective and a subjective dimension, some psychologists (see FUNCTIONALISM) think that it is the key to THE MIND-BODY PROBLEM. Subjectively, we are aware of thoughts, perceptions, understandings, and actions, all of which have an information content. Objectively, the neural wiring in our brains, like the chips in a computer, can be said to process this information. But there is more to human experience and subjectivity than the mere processing of information and its conversion into action. The specific qualities of our sensations—the redness of red, the painfulness of pain—are experiences over and above the specific information content or formal structure of the things to which they relate. I may have the same amount of information about your toothache as I have about my toothache, but I am very aware of the difference. There is clearly more to the subjective aspect of information than can be understood in the current functionalist model.

Our understanding of information as nonrandom structure raises fascinating, though so far unanswerable, questions about the universe and its origin. High entropy is a state of zero information. It is completely random. The universe therefore could not have started from a high-entropy state, or it would not have its present incredible structure. It clearly began as a low-entropy, high-information state. But how did it get that way? Divine design? A previous vacuum (see THE QUANTUM VACUUM)? At the moment, these are questions outside known physics, which begins with THE BIG BANG and the creation of the first vacuum.

The mathematical theory of information, first developed by Claude Shannon, a Bell Telephone engineer, in the 1940s, demonstrates that errors in electronic messages due to noise or human error can be reduced by "redundancy"—i.e., through the duplication of messages. We can buy accuracy at the cost of increasing the message's length. The brain, too, employs redundancy in its information-processing procedures, duplicating the same task on different sets of neural wiring, which is why the loss of some neurons does not diminish actual performance. This is a feature of the brain's parallel processing ("graceful degradation"), though not so much of its serial, or one-on-one, neural connections. (See NEURAL NETWORKS.)

The basic unit for measuring information is the "bit," a binary unit. It represents a choice between two equally likely alternatives, as in tossing a coin, answering yes or no to a question, or placing an electronic switch inside a computer in the on or off position. In this measurement scheme, the alternatives have to be equally likely—not much information is provided by the happening of a fully determined or expected event. The "byte" is a larger unit of information—a string of eight bits. Each byte represents $2^8 = 256$ alternatives, enough to encode one character onto a computer program—a letter, a numeral, a punctuation mark, or some control instruction. Computer memories are measured in megabytes, i.e., millions of bytes.

# Intelligence in the Universe

Now that we know so much about the physical universe, and the fact that our earth is just one small part of it, it is natural to ask whether it contains other intelligent life. Are we alone, or are we one of many intelligent life forms? What conditions would be necessary to support intelligence elsewhere? Does the universe itself possess intelligence, or is it the product *of* intelligence? Possible answers are the standard fare of science fiction and of the wishful thinking that lies behind UFO sightings. But can we assemble any arguments on the basis of known science?

We do not yet know whether the universe is finite though large, or whether it is infinitely large. If it is infinite, almost anything that *could* happen *will* happen in some part of it. As in the many-worlds theory of quantum physics, there could be many earths resembling this one. (See THE MEASUREMENT PROBLEM.) Just as monkeys tapping on typewriters infinitely would eventually produce the complete works of Shakespeare, so nearly all physical events and histories would happen somewhere or other. But whether we find this scenario attractive or repulsive, it makes little practical difference to our lives. Most of the universe would be forever beyond our "event horizon," too far for messages from other galaxies to reach us even if they had traveled at the speed of light since the beginning of time. Those scientists who speculate about other intelligent life therefore concentrate on wondering whether it exists in our own galaxy.

Scientific speculations about intelligent life elsewhere in the galaxy usually assume that, if it exists, it would be based on a chemistry similar to our own and so would have to live on an earthlike planet, solid but providing water. If this assumption is wrong, we have little way of estimating the chances of life elsewhere. If it is correct, some probabilities can be calculated.

There are some $10^{11}$ stars in our galaxy. The probability of life like ours existing among them, and our knowing about it, depends upon a whole chain of probabilities:

1. Whether a star has planets. Though our telescopes are not good enough to see them directly, a few examples have been established indirectly.

2. Whether one of the planets in some other solar system is earthlike. Scientists think there is a more than 1-in-10 probability of this.

3. Whether the earthlike planet has developed life. Some dozens of simple organic molecules have been detected in space.

4. Whether at least one living species on this earthlike planet has developed intelligence.

5. Whether an intelligent species elsewhere has developed an advanced technology sufficient to communicate over interstellar distances. The easiest known way to do this is via radio waves. We have had sufficient technology to do this only in the twentieth century.

6. Whether this advanced civilization wishes to communicate and broadcasts messages we can recognize as messages and understand. If the civilization uses messages based on a technology as yet undiscovered by us, we might not notice the messages. Here on earth, dolphins behave intelligently and make about 30,000 different sounds that we think constitute language, but we have not managed to decode it. This is not encouraging.

7. Whether this advanced civilization and ours both survive long enough to send a message and receive a reply. Our advanced technological civilization might have destroyed itself with a nuclear war in the first century of its existence. At the other extreme, a civilization might survive as long as its parent star—$10^9$ years or more. What is the *average* lifetime of advanced civilizations?

Most estimates of probabilities 1 through 7 are pure guesswork at this stage, but some sample figures are illuminating. If we suppose that the combined probability of conditions 1 to 6 being met is 1 in 100, and that the average lifetime of an advanced civilization is 1,000 years, Frank Drake has calculated that there would currently be 100 such civilizations in our galaxy. The nearest one would be 10,000 light-years away, far too distant to make any two-way dialogue feasible.

If, however, the average lifetime of an advanced technological civilization is $10^7$ years, there would currently be about $10^6$ such civilizations, one for every $10^5$ stars in our galaxy. The nearest such civilization would then be about 300 light-years away, and some slow two-way communication would be possible. But the most likely estimate for that civilization's age since developing advanced technology would be $\frac{1}{2} \times 10^7$ years, compared with our own $10^2$ years. The intelligent beings there would probably have developed unimaginably far beyond us, both technically and politically. It is difficult to see what they would gain by communicating with us.

If all these estimates are correct, there may be between $10^2$ and $10^6$ other advanced civilizations, rather sparsely distributed, in our galaxy. These figures are sufficient to have motivated more than ten groups of scientists since the 1960s to listen in on radio frequencies from the stars, despite all the odds against their enterprise's bearing fruit. So far, there has been no successful detection of a message, but the enterprise continues.

# Intermittency

Is chaos a breakdown of order? Or could it be that order represents the breakdown of chaos? We tend to think of riots in the cities and stock market crashes as occasional lapses in an otherwise normal world, but what if chaos is as fundamental as order itself? (See CHAOS AND SELF-ORGANIZATION.)

Nonlinear systems, by virtue of their feedback and iterated loops, have a wide variety of behaviors, ranging from simple stability through repeated oscillations, extreme sensitivity, and even chaos. Some also exhibit what is called intermittency. This consists of periods of ordered calm interspersed with sudden bursts of chaos. Amplifiers can operate quite normally except for periodic eruptions of "noise." This noise is not the result of external interference, such as an unsuppressed electric motor, but is inherent in the nonlinear nature of the amplifier's electronics. Intermittency occurs within networks of coupled computers, signals in nerve membranes, convection currents in liquids, and even the length of the earth's day. It may also occur in biological evolution. (See PUNCTUATED EQUILIBRIUM.)

Maybe chaos and order are complementary aspects, so that enfolded within chaotic behavior is the potential for order and vice versa. (See COMPLEXITY.) It could be that periodic bursts of social unrest are just as normal as periods of quietness and calm. Rather than seeking to control, contain, or limit social disorder, politicians should view these outbursts as expressions of a society's underlying dynamics. The cure may be not to attack the symptoms but to rethink the entire social structure, including its values and even the design of its cities.

# Iteration

An iteration is the constant repetition of a simple sequence of instructions or processes, the result of one cycle's becoming the starting point of the next. Using iterations, highly complex shapes, structures, and behaviors can be built out of a sequence of elementary steps.

Computers operate with preprogrammed instructions called algorithms. Iterating the simplest programs over many cycles produces highly complex behavior. For this reason psychologists believe that many of the brain's cognitive processes are iterative in nature. The human brain is "hard-wired" to perform a variety of tasks based upon iterative processes, from recognizing faces to understanding language.

The fluctuation of insect populations, in which eggs laid in one generation hatch into insects that lay eggs for the next generation, is also an iterative system that can generate behavior of great variety, from static equilibrium to fluctuating population cycles and even chaos. (See CHAOS AND SELF-ORGANIZATION.) This sort of iteration describes the growth of rumors and the change of genes within a population.

Iterations occur in chemical reactions where the presence of a particular substance catalyzes its own production. These autocatalytic reactions generate stable structures within a chemical soup—order out of chaos. Their self-replicating forms suggest that the chemical basis of life could lie in such processes.

Iterative processes are responsible for the generation of FRACTALS. By repeating an algorithmic process on smaller and smaller scales, it is possible to generate shapes of inexhaustible richness. Fractals imitate the patterns of nature, such as clouds, coastlines, fractures in metals, interstellar dust, and noise in electrical circuits.

Chaotic systems are in the grip of strange ATTRACTORS, dynamic attractors with fractal dimensions. Again, there is a connection between the iterative processes that generate the strange attractor and the iterative feedback that produces chaotic behavior in the first place.

The way in which a wide range of shapes, behaviors, and structures can be simulated by iterations and algorithms suggests that these processes are ubiquitous in nature. It has even been proposed that human intelligence is based on, or can be reduced to, a series of algorithms. (See THE CHURCH-TURING THESIS.) By determining the nature of these algorithms, it is suggested, it may one day be possible to build artificial intelligence. Yet the human mind also has the ability to recognize shifting contexts and to transcend paradoxes. Our minds jump with ease through different logical types and, unlike a computer, do not become trapped in the infinite regress of Russell's paradox. (Russell's paradox is about the class of classes that are not members of themselves: "If the barber shaves all and only men who do not shave themselves, then who shaves the barber?")

Of course, a computer can be instructed to escape from one level of iteration to a higher type—but again, the instruction is itself algorithmic, an algorithm about an algorithm. The question remains: Can human intelligence be totally reduced to nests of algorithmic processes?

# Lamarckism

Do improvements in the plant and animal kingdoms occur by chance, or are they self-directed manifestations of an underlying life force or VITALISM? The chevalier de Lamarck believed that plants and animals respond in appropriate ways to changes in their environment and are able to pass on these characteristics to their offspring.

Darwinian evolution holds that mutations are the result of chance processes. Organisms do not seek to better themselves; it is fortuitous mutations whose competition within an environment selects those most fit to survive. Darwinian evolution, with its complex system of feedbacks, is a nonlinear system capable, in its normal phases, of extreme stability, so that species do not normally change. When, through some climatic alteration perhaps, the environment undergoes a sudden dramatic transformation, there ensues a more fluid period in which random mutations with advantageous features are selected by positive FEEDBACK and thus become the dominant form. Neo-Darwinism argues that even the perfection of such a complex structure as an eye can be accounted for by the selection of a series of random mutations.

Lamarckism, by contrast, holds that mutations are not random but the result of efforts by plants and animals to survive in a changing environment. One example is the evolution of the giraffe's neck. Where animals compete to eat the foliage from trees, it is of obvious advantage to have a slightly longer reach. Elephants achieve this with their trunks, giraffes by virtue of their necks. While Darwinian evolution argues that the giraffe's neck is the result of natural selection of genes for long necks among a population of random mutations, Lamarck believed that the giraffe constantly strains to eat the highest branches. As a result it develops a slightly longer neck, which is passed on to future generations. Another example is the horny pad on the knees of the camel. Constant kneeling produces a thick pad, which is inherited by the offspring. Modern genetics denies this possibility, arguing that acquired characteristics cannot be inherited. An athlete may train to run a fast mile, but the musculature acquired for speed will never be passed on to his or her children.

Lamarckism received a boost in the Soviet era when the biologist Trofim Denisovich Lysenko applied a somewhat corrupted version of the theory to his attempts to develop new hybrids of plants and animals. Ignoring conventional genetics, Lysenko adopted a holistic theory in which a plant or animal is considered to be one with its environment. In this way, the plant or animal is supposed to "feel out" environmental changes and adjust accordingly. Lysenko's experiments were based on the assumption that if plant hybrids could "learn" to survive in, for example, a cold climate, they would pass on those acquired characteristics to their progeny.

He reported startling successes in the production of new varieties by grafting plants or injecting blood from one animal into another. His ideas fit perfectly into the Soviet thinking of the time. Russia was badly in need of a general improvement in farming efficiency if its large population was to survive. In addition, Lysenko's holistic approach, together with the notion of a plant's improving itself by virtue of its own struggle, accorded well with the principles of dialectical materialism. However, Lysenko's experimental results proved to be fraudulent. The result was to discredit not only an individual biologist but the entire Lamarckian system on which his research program had been based.

# Language

Language is generally thought to be a uniquely human ability. It is one about which philosophers and scientists of the mind continue to disagree. Where does language come from? Why did it emerge so suddenly and in such complex form with human beings? Is language the same as thinking, or at least intimately bound up with thinking? Do the Chinese think differently than the English or the French? How do we learn existing languages, and why are we able to invent new language? These are just a few of the questions to which we lack certain, generally accepted answers.

Nonhuman life forms have obvious ways to communicate, though scientists do not consider these to be languages on anything like the human scale. Bees display a complex dance "language" when sharing information about sources of food. Birdsongs seem to communicate meaning about food, courtship, and danger. Chimpanzees can be taught sign language. Dolphins make about 30,000 different sounds, which may well be a language in the full sense of the word, but humans have not been able to decipher it.

Three sorts of theories have dominated scientific attempts to explain linguistic ability. Behaviorists like B. F. Skinner have argued that we learn languages by association, like conditioned reflexes. We hear "apple" when shown apples, and soon make the association. But the great linguist Noam Chomsky showed that this could not be an adequate explanation. Human beings can both generate and understand sentences they have never heard before, and much of our language refers to abstract concepts like truth or beauty. Chomsky argues that linguistic ability must depend upon some innate capacity that generates inner grammatical rules, very like computer programs. His criticism of BEHAVIORISM was one reason that it gave way to cognitive psychology. (See GESTALT AND COGNITIVE PSYCHOLOGY.)

Chomsky's own work has provided the most dominant influence in linguistic theory, but many of his theories remain controversial. He argues that we have a special language faculty, apart from general intelligence. This is partially supported by the fact that there are special language areas in the brain. Chomsky goes on to argue, very much in the spirit of Platonic rationalism, that this special language faculty is innate. We are born with it. This is partially supported by the fact that the grammatical rules of most languages are extremely complex, yet young children easily learn languages, and few people who successfully speak languages can state explicitly the rules by which they do so. But Chomsky's own attempts to formulate our innate linguistic rules have had only limited success.

A third element in Chomsky's work is the argument that we can apply the rules of any linguistic system correctly without knowing the meanings of the words involved. If true, this would lend great support to computationalist theories of mind, and to the argument that the brain functions like a computer, language being part of its software. But there are only limited cases where this argument seems to apply. We can sometimes construe the grammatical system of a sentence without knowing its meaning, as with Lewis Carroll's " 'Twas brillig, and the slithy toves / Did gyre and gimble in the wabe." But more often, the meaning and context of a sentence we hear are necessary to decipher it, and formal rules are of little help. In the sentence "He came in when John opened the door," we know that "he" does not refer to "John," but what rule is involved? In comparing the two sentences "Sailing boats can be enjoyable" and "Sailing boats can be green," we need to call upon our general knowledge to describe the grammatical structure. Examples like these have raised problems for attempts to develop computerized language-translation programs. To the extent that such programs work at all, they must rely upon an encyclopedia as well as a dictionary and knowledge of the grammatical rules of the languages concerned.

A third kind of linguistic theory has been proposed by Chomsky's MIT protégé Steven Pinker (*The Language Instinct*). According to Pinker, language is a human instinctual ability, acquired through the normal means of Darwinian evolution, just like the structure of the eye. Some of our ancestors found they could survive more effectively if they could speak, so those with the ability won the evolutionary race. It follows from this theory that all human beings have the same basic instinct, perhaps even a "language gene," and that thought and language are quite separate. Human minds work in the same way regardless of the language that we speak. Critics of Pinker, among them Chomsky himself, feel that evolution is an inadequate explanation of how language appeared so suddenly and so exclusively among human beings. Others argue that he has no means to explain associated abilities, like reading and writing.

Both Chomsky and Pinker rely heavily on the theory that minds work essentially like computing machines, and thus the many AI (see ARTIFICIAL INTELLIGENCE) attempts to apply their thinking across the board flounder wherever computationalist models show their limitations. So far it has not been possible to write a successful computer program for parsing natural English sentences. Language translation programs remain inadequate. And the goal of offering a purely computationalist account of linguistic meaning remains out of reach. Critics of AI say that its model is essentially flawed, basing their arguments on GÖDEL'S THEOREM and John Searle's CHINESE ROOM to cast doubt on our ability ever to derive meaning from purely formal systems. At this stage in the sciences of the mind, many features of our linguistic ability remain a mystery.

# Lasers

A laser (*l*ight *a*mplification by *s*timulated *e*mission of *r*adiation) produces an intense beam of light of a single color. A similar phenomenon operates in the *m*icrowave range of electromagnetic radiation using a *m*aser.

Light from a candle or light bulb is produced when excited atoms lose their energy by emitting photons. Since atoms emit their radiation at random, photons are given off at slightly different times, and the resulting wavelets of light are not in phase. Moreover, atoms emit photons of different frequencies, so that the resulting light is a mixture of colors. While such light can be focused by using mirrors and lenses, the beam is never very tight, so even the light from a searchlight will soon spread out. By contrast, laser light is coherent; it is all in phase and of a single frequency. Laser light consists of a single quantum state containing an astronomical number of photons. This state of affairs is permitted because photons obey Bose-Einstein statistics. (See BOSE-EINSTEIN CONDENSATION.)

Given time, and isolated excited atom will spontaneously get rid of its excess energy by emitting one or more photons. This process can be stimulated by using electromagnetic radiation. Expose an atom to a beam of light of exactly the right frequency (in tune with the atom), and it, too, will give off a photon. Use an ensemble of atoms, and each will add its own photon to the beam, in phase with the others. The result is to amplify the original triggering light beam. Stimulating an enormous number of atoms to emit their photons simultaneously can produce a coherent beam of light of considerable intensity and focus it into an extremely tight beam that does not spread out.

Lasers can be constructed in a wide variety of ways by using solids such as pure crystals, gases, and liquid dyes. All lasers use some means of exciting the atoms into higher energy states. This can be done with an initial beam of radiation (optical pumping), chemical reactions, electrical currents, heating, and so on. There are even the X-ray lasers of the U.S. "Star Wars" project, whose power source would be a small nuclear explosion! The radiation emitted is normally reflected back into the laser by means of mirrors, so that the beam traverses the excited solid, liquid, or gas, each time stimulating more atoms to emit photons. The result can be a beam of exceptional intensity but short duration. Other lasers produce continuous beams of coherent light, but of weaker intensity.

The scientific, medical, and commercial uses of lasers are manifold. Lasers are now widely used in everything from CD players to the store cash register that reads bar codes. In medicine, lasers attach retinas, unblock blood vessels, and remove birth marks. Industrially, they burn minute holes in the hardest material. They are also used in the production of holographs. In surveyors' equipment, lasers can determine long distances within a fraction of a wavelength of light.

A tight, coherent beam of in-phase light can be transmitted over great distances—to the moon and back—with very little divergence. Laser beams can also be conducted along fiber optic cables between cities. Since the amount of information carried along a communication channel increases with frequency, a single optical fiber can support many telephone conversations, television signals, or computer data links. In the laboratory, lasers are used in everything from making microscopic adjustments in living cells to freezing the movements of individual atoms to explore phenomena at the limits of quantum theory.

# Leptons

Matter particles (FERMIONS) are either leptons or QUARKS. Leptons are light and pointlike in experiments. Quarks have fractional electric charge and combine into the heavier HADRONS, such as protons and neutrons.

There are two kinds of leptons, electrons and NEUTRINOS, each with its own antiparticle. (See ANTIMATTER.) Electrons have mass, but neutrinos are of zero or *very* low mass. Each of these stable structures is duplicated by two more high-energy versions, giving six leptons in all. Ordinary electrons, with their electric charge of $-1$, are echoed by mu-electrons and tau-electrons. Ordinary neutrinos, with an electric charge of 0, are echoed by mu-neutrinos and tau-neutrinos. Nobody yet understands why this is so. The heavier tau-electrons rapidly decay into lighter mu-electrons and ordinary electrons in our relatively cool universe. Total lepton number is conserved in our present conditions, but electrons and neutrinos can interconvert via the weak nuclear force. (See THE ELECTROWEAK FORCE.)

The family of leptons, with its three different generations, resembles the three generations of quarks, though there are three times as many kinds of quarks, due to their red, blue, and green color charges. This parity is one of the attractions of GRAND UNIFIED THEORIES, according to which leptons and quarks at extremely high energies can be regarded as transformations of the same particle.

# The Mandelbrot Set

The Mandelbrot set has been called the most complex mathematical object in the universe. Displayed on the screen of a computer, it opens out into an endless world of FRACTALS. Following its discovery by Benoit Mandelbrot in 1980, mathematicians have investigated other sets that have similar properties of endless complexity.

The Mandelbrot set is the set of points on a plane that have a particular property. Each point corresponds to a pair of ordinary numbers—the point's distance from 0 along two axes (we could call them north and east). Mathematicians describe these two ordinary numbers as one "complex number." They have their own arithmetical rules. Each point $c$ corresponds to a process, going from any point $z$ to the point $(z^2 + c)$. Mandelbrot wondered what would happen if, starting at 0, we iterated this process over and over again. (See ITERATION.) He stipulated that if $z$ eventually zooms off to infinity, as it always will for a large $c$, then $c$ is not in the Mandelbrot set. But if $z$ remains near zero forever, then $c$ is in the set.

This simple rule gives rise to an extremely complicated set of points $c$. The computer is set on automatic, and the Mandelbrot set grows on the screen as a rather warty-looking blob. Now the fun begins. It is possible to "fly into" the Mandelbrot set and explore its richness. Zoom in on a particular "bud," and it is revealed to consist of endless buds upon buds, a self-similarity that recalls Jonathan Swift's comment on poets:

> So, naturalists observe, a flea
> Hath smaller fleas that on him prey;
> And these have smaller fleas to bite 'em,
> And so proceed *ad infinitum.*

Zoom into another region of the set, and gentle filaments of "pearls" appear. Magnify the details hundreds of thousands of times, and the filaments reveal new clusters of details. Because the act of "zooming in," or magnification, can continue indefinitely (computers can magnify these details thousands of millions of times), the Mandelbrot set is infinitely complex. In some regions it exhibits endless levels of self-similarity; in others, new shapes and structures are revealed. These are known as "surf," "knots," and "islands."

What is truly remarkable is that the abstract complexity of the Mandelbrot set and other fractals is deeply connected to the behavior of chaos, intermittency, turbulence, and other nonlinear systems. All of them depend on iterative processes. We can now see the shapes of trees, clouds, coastlines, and turbulent streams as possessing their own kind of complex structure and beauty, rather than as just failed straight lines.

# The Many-Worlds Theory

See THE MEASUREMENT PROBLEM.

# The Measurement Problem

Quantum reality is described as a bizarre world of both/and. Cats are both alive and dead; photons and electrons are both waves and particles, both here and there, now and then. The many possibilities carried by quantum SUPERPOSITIONS are spread out over space and time. Yet we live, at the level of ordinary experience, in a world of either/or. We see waves or particles, live cats or dead cats, and so on. For most of us, Newton's classical mechanics is an accurate description of how we find the physical world. So what is the relationship between the strange quantum world and the classical world of common sense? How do many possibilities finally become one actuality?

In quantum mechanics, these questions are known as the measurement problem, or sometimes the observation problem. Whatever the process that transmutes quantum both/and to classical either/or, it happens when we measure or observe a quantum system. The mere act of placing a detection screen or a photomultiplier tube in the path of a photon transforms the photon's wave/particle duality into a wave (screen) or a particle (photomultiplier tube). In Schrödinger's cat paradox, it is when we open the box and look inside that the alive/dead cat is suddenly alive or dead. But nobody understands why this should be so. How measurement or observation can so radically change the character of physical reality is the single most outstanding problem of modern physics. Until it is solved, many physicists believe, quantum theory will be incomplete.

The Schrödinger wave equation, which so accurately describes quantum reality as a superposition of possibilities, and attaches a range of probabilities to each possibility, does not include the act of measurement or its apparatus. There are no observers within the mathematics of quantum mechanics itself. Neither, therefore, does the wave equation describe the COLLAPSE OF THE WAVE FUNCTION, that moment when possibility gives way to actuality. In trying to come to terms with this glaring omission, physicists have resorted to six main theories, none of which has so far been considered the last, or even truly enlightening.

The first approach, the one still used by the majority of working quantum physicists, is "pragmatic": We shouldn't worry about measurement. According to this view, originally advocated by Niels Bohr as part of his Copenhagen Interpretation of quantum theory, the Schrödinger wave equation successfully makes accurate predictions about the outcome of experiments, and physicists shouldn't trouble themselves about how or why. Some people have likened this approach to treating quantum mechanics as an elaborate "sausage machine"—physicists put their data into one end, and out the other come their measured results.

Linked to this pragmatic "we won't talk about it" approach is Bohr's COMPLEMENTARITY, his view that the apparent contradictions of quantum reality—photons that are both waves and particles, cats that are both alive and dead—complement each other, as do the quantum and classical worlds. This, according to Bohr, is just one of life's deep mysteries, and physicists must accept it, not try to explain it.

A third approach to the measurement problem, suggested by the physicist Eugene Wigner, is known as the Wigner Interpretation. Since, according to quantum mechanics, anything described by the Schrödinger equation must exist in a state of superposed possibilities, this should be as true of measuring devices as it is of the photons they measure. Thus it appears that nothing physical *can* collapse the wave function. Wigner concluded that a nonphysical or extraphysical something must be responsible, and his candidate for this magical agent was the mind of the observer doing the measuring. Wigner's proposal has been widely adopted by journalists and popular books on quantum mechanics because it is dramatic, startling, and supportive of the sometimes popular notion that mind is the origin of matter. But very few physicists give this approach much credence.

Two other approaches try to solve the measurement problem by arguing that it is a false problem. David Bohm's hidden variables theory holds that every quantum particle is guided by a quantum wave, or field of potential, and that all the eerie effects quantum physicists describe—indeterminacy, wave/particle duality, superpositions—are really due to underlying causal factors too small for our crude measuring techniques to see. Some physicists like Bohm's approach, but it gives rise to even larger problems of its own when applied to other quantum effects, like NONLOCALITY.

The other famous approach that denies there is a measurement problem is the many-worlds theory. According to it, all the superpositions of the quantum world are real and remain real forever. Whenever we make a measurement, we create another possible direction that one of the possibilities might take, and the world branches. According to many-worlds advocates, reality consists of an infinite number of constantly branching worlds, in a great many of which there might be a given physicist getting any one of an infinite number of outcomes to his measurements. Every time the physicist makes a measurement or a decision, his world branches, and there is yet another copy of himself. Each of us exists in multiple copies on many of the world's infinite branchings. Nothing is ever lost. This theory, too, is appealing to popular writers because of its similarity to science fiction, but it is regarded as just that by the vast majority of physicists.

The final approach to the mystery of the measurement problem is a physical one. Increasingly, there are physicists who argue that the Schrödinger equation must be altered to describe a nonlinear process whereby quantum potentiality becomes classical reality as the result of a physical interaction between the wave function and the measuring device. In that case, some *physical* difference between noncollapsing quantum systems on the one hand, and measuring instruments and human brains on the other must be specified. Several differences have been suggested as the key one. The measuring device may collapse the wave function because it is larger, has more particles (the GRW theory, named after Italian physicists Giancarlo Ghirardi, Alberto Rimini, and Tullio Weber), is heavier (Roger Penrose's gravity theory), is capable of making permanent records of the results, and so on. Crucial experiments are hard to do in this area, but in time, a combination of theory and experiment may resolve this as yet unsolved problem.

# Meditation

A meditative state is one of alert relaxation. Induced by some form of deliberate practice, meditation has measurable effects on both conscious states and physiological processes, and has thus become an area of experimental research as well as something done for its own sake. Some form of meditation practice, such as yoga or Buddhist meditation, is common to the history of many cultures, particularly those of the East. Western scientists became interested in these practices' relevance to health and concentration in the 1970s.

A typical meditation procedure requires a person to sit comfortably for at least twenty minutes in a quiet place. Discomfort distracts concentration. On the other hand, lying down may produce so much comfort that the meditator falls asleep. Once comfortable, the meditator gently directs attention to some simple object, such as a repeated sound, a candle flame, a diagram, or the breath. Since this kind of attention does not involve the rational ego, that part of the mind becomes less active, often freeing attention for more subtle conscious experience. When meditators become aware of a distraction, such as getting lost in a train of thought, they gently return their attention to the object of their meditation.

If meditation is successful, it diminishes the "fight or flight" response of the sympathetic nervous system. Blood pressure, heart rate, breathing rate, metabolic rate, and electrical skin conductance all diminish. For this reason, meditation is an effective relaxation technique and is sometimes recommended in the treatment of heart conditions. At the same time, blood flow to the brain is increased during meditation, and EEG studies show that brain waves become more coherent. (See COHERENCE.) Alertness is maintained, and capacity for awareness actually increases. A "busy mind" becomes still, and later concentration often benefits.

States similar to meditative states are often induced by practices, such as religious chanting, jogging, sunbathing, counting sheep, or repetitive hypnotic suggestion. In all these states, stress is reduced, the rational mind is lulled, and practitioners become more suggestible as well as gaining access to normally unconscious material. The relaxed, meditative state can therefore be used as an adjunct to hypnosis, group suggestion, psychotherapy (see PSYCHODYNAMICS AND PSYCHOTHERAPY), or self-exploration undertaken for therapeutic or creative purposes.

# Memory

Memory is our capacity to store and retrieve information about the past. Sometimes this is in the form of words, images, sounds, smells, or feelings, but it may be more direct or more functional. Many bodily systems learn something from experience and store that information for later use, e.g., the enlarged muscles of an athlete or the immune system's ability rapidly to make antibodies to a virus it has encountered once before.

Our brains are the seat of most memory, and of our ability to learn at the highest level. In an average lifetime, a human being stores $10^{14}$ or more bits of information—the equivalent of one hundred thousand 4-megabyte hard disks or books. Some of this information accounts for skills like riding a bicycle or speaking a language. Some represents the ability to recall explicit incidents or facts, faces, and so on. And some is devoted to forming "the specious present," the binding together of approximately three seconds of mental life into a single experience, such as hearing a melody or a sentence rather than just individual notes or sounds. Having this specious present is essential to the unity of our conscious experience.

In the 1950s, an operation developed to relieve a kind of epilepsy shed light on memory and brain function. A part of the patients' primitive forebrain, the hippocampus, was bilaterally removed. Epileptic attacks became less frequent, but patients lost their ability to form permanent new episodic memory. After a few minutes, they could remember nothing of a conversation or a new face, though they retained their long-term memory of all that they had known before the operation. They were unable to learn new surroundings. However, both *very* short term memory (a few minutes) and an ability to learn new skills were still present. These experiences showed that the ability to form new episodic memories, but not the ability to retain old ones, involves the hippocampus. In people over forty, the hippocampus often suffers gradual deterioration, and older people lose the facility to form new episodic memory. The distant past becomes more real as the years go by, hence elderly people's tendency to dwell on it. (This is quite different from the global impairment of Alzheimer's disease, which affects the rest of the brain, intelligence, and many physiological functions.)

Unlike new episodic memory, the associative (knowing-how) memory needed to retain skills seems to be distributed all over the cortex, as is intelligence. These abilities are reduced only by widespread or massive cortical damage. Associative memory is described by connectionist models of thinking and learning (see CONNECTIONISM), based on NEURAL NETWORKS. Episodic memory, by contrast, is required by SERIAL PROCESSING models of thinking. The brain seems to have the structures necessary for both kinds of memory, and each has been studied at the level of single synapses.

It is generally thought that memory is stored in the brain via modification to single synapses. There are perhaps hundreds or thousands of synapses for each of the brain's $10^{11}$ neurons, enough to store significant memory. The first theory about this originated with the psychologist Donald Hebb in the 1950s, but it now has experimental support. There is also evidence from psychological tests that our memory of past events changes over time. The brain is not simply a storage and retrieval system; it is in a constant process of laying down new synaptic connections and diminishing old ones.

The infant's brain contains a surplus of neurons and connections, many of which die if they are not used. Even brain structure, therefore, depends upon early experience. For example, a North American investigation found that adults perceive horizontal and vertical lines more easily than diagonal ones. A later study showed that Native Americans who had been raised in wigwams could perceive diagonal lines equally well. Gerald Edelman refers to this principle of the survival of the brain's most-used neurons as NEURAL DARWINISM.

# Mesons

The meson, a boson (see BOSONS) occupying a position midway in mass between the baryons—constituents of the nucleus such as protons and neutrons—and the much lighter LEPTONS, was originally proposed as a way of explaining the forces that bind the nucleus together. In the early decades of the century, physicists knew only two of the forces of nature—gravity and electromagnetism. It was the electromagnetic attraction between orbiting electrons and a positively charged nucleus that bound the atom together. But why was the nucleus stable when it contained only positively charged protons and electrically neutral neutrons? Clearly some new attractive force had to be involved to overcome the mutual repulsion of positively charged protons and bind the nucleus together.

In QUANTUM ELECTRODYNAMICS, physicists explain the attractive force between an electron and a proton, or the repulsion between two electrons, in terms of an exchange of photons—the quantized units of an electromagnetic force. Just as two soccer players toss the ball back and forth as they run toward a goal, photons are constantly being exchanged between electrons and protons. Each interchange involves temporary borrowing and then paying back of momentum from the electromagnetic field. The result is that charged particles experience an attractive or a repulsive force. In 1935, the Japanese physicist Hideki Yukawa proposed that something analogous must occur in the nucleus between protons and neutrons in order to create a strong force of attraction. In electromagnetism, photons have zero mass, which means that the electrical and magnetic forces are long range. By contrast, the strong nuclear force is very short range and does not extend out of the nucleus itself. This led Yukawa to predict the mass of his force particle to be about 200 times larger than the electron's. It was not until 1947 that this particle, called a pi-meson or pion, was detected. Since then, many other mesons have been discovered.

Physicists now believe that, just as the nucleus is composed of neutrons and protons bound together by meson forces, neutrons and protons are themselves composed of triads of even more fundamental particles, the QUARKS, which are bound together by gluon particles. Mesons are in turn combinations of quark and antiquark pairs. (See QUANTUM CHROMODYNAMICS; THE STANDARD MODEL.)

# Methods of Studying the Brain

All sciences of the mind in the twentieth century focus on the relationship between mind and brain. The central questions are: Which areas of the brain are linked to and/or responsible for recognized areas of our conscious experience, and what is the nature of these links? In addressing these questions, modern technology has provided a variety of methods for studying the structure and function of the brain. Each technique yields part of the story. Structure itself, the anatomy of human and animal brains, can be studied with reasonable ease at all levels. Because functioning brains must be living brains, studying function is much more difficult, and necessarily indirect.

One indirect means of attempting to correlate brain function with mental ability is to compare postmortem studies of people who suffered brain damage—from strokes, violence, and so on—with the associated loss of function experienced during life. This is particularly effective if the brain damage has produced a clear deficit in a single ability. It is also possible to get an indirect picture of how brain function relates to mental capacity by making drugs that affect psychological function mildly radioactive. An X ray can then show which parts of the brain have absorbed the drugs. But both postmortem studies and radioactive drug tracing produce only fragmentary information to link mind and brain.

EEG (electroencephalograph) machines have given scientists access to neural activity linked to behavior or experience. First made usefully available by Hans Berger in 1929, EEGs are voltmeters that measure tiny electric potentials produced by functioning neurons. Scalp electrodes that can record the average activity of at least one million neurons can distinguish levels of sleep and arousal or record the average evoked potentials produced by sensory stimuli. But these electrodes are too crude to correlate neural activity with higher mental function. Much more detail can be seen by using microelectrodes placed on an individual neuron. As this involves opening the skull, it is possible on human beings only during necessary operations. The method is used on animals, but functions like language can't be studied by using monkeys or cats. A variation on the EEG, the magnetoencephalograph (MEG), which measures tiny magnetic fields generated by the brain's electrical activity, is more effective, but the equipment is extremely expensive.

There is an assumption that active parts of the brain show an increased blood flow, so technology has been designed to detect blood flow and correlate associated active brain regions with specific psychological capacity, such as various linguistic tasks. Both positron emission tomography (PET) and magnetic resonance imaging (MRI) use these techniques. Both are noninvasive and can be used on human beings, but they are very expensive. Both PET and MRI have produced valuable new data about brain function, but as each can read increased blood flow only over a time scale of a second or so, they cannot detect individual nerve impulses. Nonetheless, they are a useful complement to EEG and MEG, which reveal more temporal but less spatial brain-activity detail.

# The Milky Way

The Milky Way is our galaxy. (See GALAXIES.) Visually, it is the ribbon of faint light we can see in addition to individual stars, extending right across the night sky on a summer's night. In fact, it is a disk consisting of about $10^{11}$ stars. The word *galaxy* comes from the Greek word for milk. There are billions of galaxies like our own that can be seen through powerful telescopes; only three are visible to the naked eye.

The real size and shape of the Milky Way were first described by the American astronomer Harlow Shapley in 1917. Previous estimates had been distorted by large clouds of interstellar dust. Optical telescopes cannot see to the center of the Milky Way, though it can be observed by radio or X-ray astronomy. Shapley used a different approach. Our galaxy contains, as well as the main disk, about 200 globular clusters of stars. Using a new method for calculating the distance of these, based on a special class of stars called Cepheid variables, Shapley found that they lie roughly in a sphere. (See DISTANCE MEASUREMENTS IN ASTRONOMY.) The center of the sphere is the center of our galaxy.

The Milky Way has three main components. The central bulge consists of densely packed stars, both new and older. The disk, which rotates around the center, consists of newer stars, gas, and dust. The halo consists mainly of globular clusters, each containing about $10^6$ old stars. The whole galaxy is thought to have condensed by "self-gravitation" (it pulled itself together) from a vast, spherical cloud of gas about $10^{10}$ years ago. It first condensed around small irregularities, forming the globular clusters. The disk embodies the rotation of the original cloud. The stars in the central bulge are more crowded and move in orbits in random planes around the nucleus.

Most stars in the Milky Way are spaced a few light-years apart, except for binary stars that orbit each other. Thus, close encounters are very rare, and each star responds mainly to the combined average gravitational pull of the whole galaxy. Like the sun and planets, the

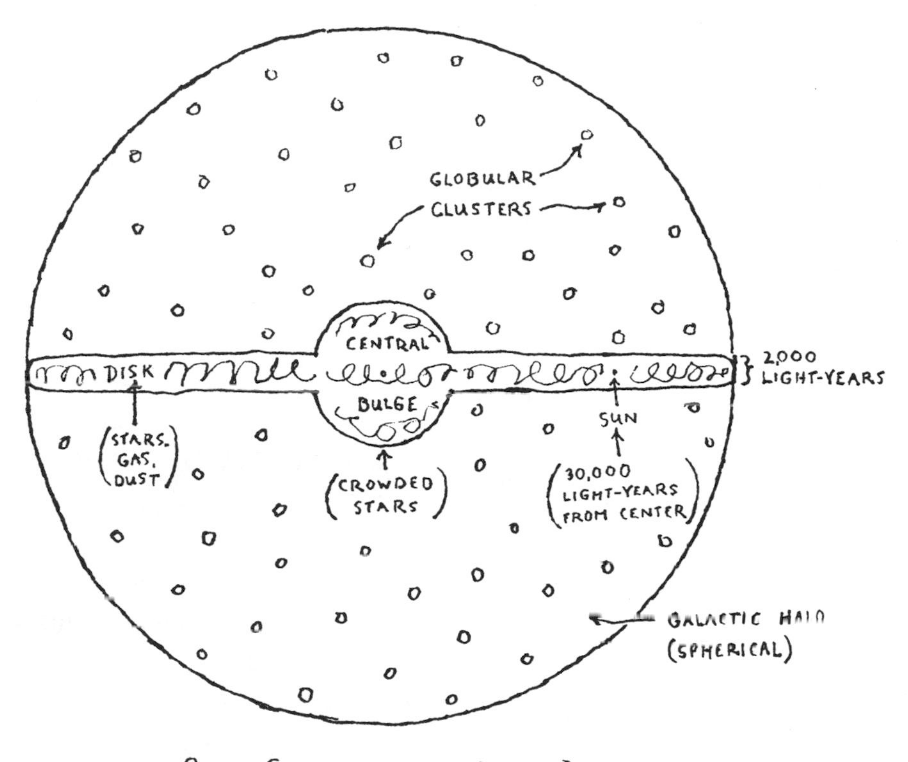


Our Galaxy Seen Edge On

inner regions of the Milky Way disk rotate faster than the outer ones, to balance the larger gravitational pull from the center. The sun completes a circuit of the galaxy about every 240 million years.

The speed at which any star moves in orbit around the galactic nucleus depends on the gravitational pull, which must balance the star's tendency to move off a straight line, i.e., centrifugal force. If we know the star's distance from the center of the galaxy, we can work out what mass there should be within the orbit to exert this amount of gravitational force. It turns out that the observed mass, in the form of visible stars, gas, and dust clouds, is only half the required gravitational mass. For clusters of galaxies, the discrepancy can be tenfold. There must, therefore, be some kind of DARK MATTER (invisible matter) in our own and other galaxies to make up the difference. There have been many speculations about what this could be.

Most stars in the Milky Way are fainter than the sun. Most of the galaxy's light is generated by relatively few heavy, bright stars. Our galaxy, like many others we can see in the sky, is thought to be spiral in shape. At its center there may be an active, high-energy nucleus such as we can see in other galaxies. This is relatively small and may perhaps be a black hole of $10^6$ times the mass of the sun. (See BLACK HOLES.) Dust prevents optical observation of the nucleus, but it is possible to observe at many other wavelengths—those of radio, infrared, X-rays, or gamma rays.

# The Mind-Body Problem

How are our minds related to our bodies or, more specifically, to our brains? Are the two entirely separate, made of different "stuff," following different laws, as dualists have long argued? Are minds wholly reducible to the matter and activity of brains, nothing but the end result of chemistry and firing neurons, as materialists argue? Or are both minds and brains different manifestations of a deeper, underlying common substance that has both mental and physical properties, as monists argue?

We know from both observation and medical fact that minds and brains are interconnected. Damage to the brain and drugs that affect the brain can make human beings unconscious, change moods, or even alter personalities. Yet minds and brains have very different properties. Brains are material substances weighing approximately a kilogram; they are gray and white, and are composed of neurons and other cells, ultimately of atoms. Minds have none of these material properties, but they have other, subjective qualities. "I" am conscious of myself and of an "inner" life. I am aware of colors, shapes, pleasure, and pain. I feel that I "know" certain things, and that I am responsible for my actions. Where does this "I" come from?

The mind-body problem is a collection of interrelated problems and philosophical issues about consciousness, the self, free will, meaning, knowledge, our experience of time, and so on.

Dualism stresses the differences between mind and body, or mind and brain. The Christian doctrine of the soul argues that mind and body have different causal properties and originate from different sources. This leaves dualists with the increasing problem of explaining how mind and brain can be so interdependent. The position fits poorly with the spirit of science, which always tries to unite different data and different substances within a few general theories. It is never comfortable with dualisms, feeling there is surely some deeper, unifying explanation. (Physicists are still hoping to discover THEORIES OF EVERYTHING.)

In contrast to dualism, monism regards mind and body as made of a single substance. This could be inert matter, as materialists argue, or pure mind, as held by idealists and some Buddhists, or it might be a third substance possessing both physical and proto-mental properties. This idea of a common, underlying substance inspired panpsychists (those who believe that every existing thing contains mental properties) and lay behind the thinking of Alfred North Whitehead. It is an increasingly popular theory today, now often associated with QUANTUM THEORIES OF MIND.

The relation of mind and brain is ultimately a scientific question that may have religious or spiritual overtones. Scientists now ask and investigate what parts or subsystems of the brain are associated with consciousness, memory, the experience of free will, and so on. They probe the possibility that each of these capacities is associated with a particular configuration of firing neurons, as cognitive scientists argue. Other scientists investigate whether something like consciousness originates from some quite other scale of brain organization, such as a molecular or quantum level. The origin of consciousness can be studied by seeing what systems of the brain are affected when anesthetics lead to a loss of consciousness.

The various theories that consciousness and its associated capabilities originate from the brain's neural activity are the most conceptually straightforward and most easily investigated by scientific means. These theories are supported by such contemporary figures as Douglas Hofstadter, Gerald Edelman (see NEURAL DARWINISM), and Francis Crick (see CRICK'S HYPOTHESIS). Neural explanations of conscious activity require no surprising new concepts, such as quantum or chaos activity in the brain, but they present their own problems. Why should neural activity have any subjective dimension? Where would the self-awareness of neurons originate? And how would a brain described solely in terms of its neural activity differ from something like the electrical activity in a computer or a television set? Are we to say that *these* are conscious?

The questions raised by neural theories of consciousness are deep philosophical problems, associated with a whole set of mind-body problems. If our minds originate from wholly material brains, which are themselves determined by the laws of physics, how can human beings have free will or be responsible for their actions? How can consciousness of the self be unified if its associated neural activity is distributed among billions of neurons, or among many neural systems all around the brain? How, exactly, could consciousness be related to matter anyway?

Even if various neural theories eventually provide scientific answers to these questions, we are left with a philosophical disquiet. There is too much clash between the objective, scientific paradigm in terms of which we understand the brain—mass, length, electrical activity, and so on—and the more subjective paradigm in terms of which we understand ourselves—self-awareness, phenomenal space and time, intentionality, free will, and such. As currently understood, the two paradigms seem like chalk and cheese. (See ARTIFICIAL INTELLIGENCE; HUMANISTIC PSYCHOLOGY.)

When paradigms clash in this way, it turns out that one or both need to be modified. The earth appears more or less flat, but knowledge gained from travel, observation, and scientific inquiry does not accord with this. Light appears to have wavelike properties, but we know that it can travel through a vacuum in which there is nothing to "wave." The philosophical conflicts associated with the mind-body problem diminish if we assume that the objective substrate of mind is a quantum system that can generate both mental and physical properties, or if there is some sort of emergent mental activity generated by the brain's COMPLEXITY. Such views are becoming more popular, but much further scientific research is needed.

# Nanobiology

Nanobiology is the branch of biology concerned with subcellular structure and function, the makeup and behavior of things smaller than single cells. The Greek prefix *nano* means "one billionth." Thus nanobiologists measure things on a space-time scale of nanometers and nanoseconds. These include relatively large cellular components, such as neuron synapses and microtubules, as well as very small things like DNA spirals and individual molecules.

Nanobiology is a new field of biology made possible by technological developments in the second half of the twentieth century. With the electron microscope we can view substances as small as an atom, opening up such things as the coordinated behavior of single-cell animals like amoebas and paramecia to study. Such creatures have no nervous system, but the electron microscope has revealed an internal structure of many interconnected, very fine tubes, known as microtubules or the cytoskeleton. These are now believed to transmit impulses like a mini–nervous system. Some scientist are even suggesting that microtubules may play some role in the emergence of human consciousness. (See QUANTUM THEORIES OF MIND.)

The development of very fine glass tubes known as micropipettes and a technique called patch clamping have enabled nanobiologists to study individual neural synapses, the junctions where neurons meet. Synapses are the site of action for chemical neurotransmitters and also for most drugs that affect the nervous system. Learning and memory are thought to depend upon the modification of synapses, so that some neural pathways are strengthened or inhibited. By patch clamping, a single synapse can be drawn up into the micropipette for investigation.

# Neural Darwinism

Neural Darwinism is a theory advanced by Nobel laureate Gerald Edelman (*Bright Air, Brilliant Fire; The Remembered Present*) to join neuroscience and psychology in a new physical theory of consciousness. The theory is speculative and controversial, but widely regarded as one of the most exciting and profound yet offered in this field.

In one of his books, Edelman sums up the traditional controversy in mind-brain studies by quoting an old play on words: "What is mind? No matter. What is matter? Never mind." Those in neuroscience who take their lead from Darwin see mind as a material object of biological evolution, a system of neural "wires" that necessarily results from a reductionist genetic code. Others of a more psychological bent, who take their lead from Freud, see mind as "psyche," a psychological entity in its own right, largely independent of brain structure. Edelman sees himself as influenced by *both* Darwin and Freud. He feels that the neurobiological workings of the brain *and* our psychological experience are formative, and he seeks a way to express both in a physical theory of mind.

Like many other critics of ARTIFICIAL INTELLIGENCE, Edelman wants to distance himself from notions that the brain is a computer. Though his own theory is "based remorselessly on physics and biology," he points out that minds are conscious; computers are not. Minds have experience; computers don't. And consciousness in the brain is associated with evolving structures that in no way resemble the point-to-point wiring systems in computers.

Edelman argues that the brain begins with far more neural capacity than it needs. This is subdivided into "neural groups," cooperating bundles of neurons that account for aspects of experience. As the brain engages in dialogue with its experience, both sensory and psychological, certain neural groups are chosen as most desirable, through a process of Darwinian evolution. These groups then constantly readjust themselves in a continuous learning and recategorization process. The rest die out. Thus, like William James, Edelman sees the human brain more as a dynamic process than as a thing—a marvelously flexible organ whose ever-changing structure originates in response to cellular movement within the developing embryo and continues throughout life. No two brains have the same experience; no two brains are alike. No brain is like itself from moment to moment, and the genetic code is more a set of constraints than a blueprint.

Edelman thinks that the key to the basic problems of mind can be found in the brain's complexity and in fundamentally new organizing principles that originate at the level of complex biological systems. The brain has many levels of structure and organization, each of which has evolved at a different time, in response to different needs. According to Edelman, each level and each neural "map" in the brain is linked by a system of interactive "reentry" loops, and these ultimately result in our capacity for consciousness.

Edelman works on the premise that if we can understand the brain's visual processing capacities, including THE BINDING PROBLEM, we will understand the emergence of consciousness. The bulk of his experimental research has been done on vision. A criticism of his theory, like all other such physical theories of consciousness, is that no amount of understanding how the brain processes visual data will amount to a full account of how we come to have conscious visual *experience*. There seems to be a missing factor still.

# Neural Modules

Neural modules are bundles or columns of nerve cells in the brain that cooperate when there is a need to perform some task. Each module contains about 120,000 neurons, many of which are active at the same time under the same conditions.

Much, possibly all, of the cerebral cortex, is organized into neural modules, so neurons with similar physiological functions are near one another. This has been demonstrated in experiments on vision, touch, and other sensory and motor areas. It might have been suspected from the anatomical fact that most neuron interconnections are local, within a range of about 1 to 2 millimeters.

A clear picture of the structure and function of neural modules has emerged from work with microelectrodes. These are small enough to record from one individual neuron; thus neuron performance can be correlated with an associated sensory stimulus or motor behavior. (See PERCEPTION.) The brain's cortex is in the shape of a folded-up sheet, 2 to 4 millimeters thick. A module or column of neural cells with a similar function is often about 0.5 millimeters square and runs through the whole thickness of the cortex. The function of the module is to extract information about some one feature from its input, for example a color recorded at a specific retinal position, and then to pass this on for some further level of processing.

# Neural Networks

Neural networks, also known as parallel processors, are a new variety of computer developed in the 1980s, although work on their as-yet-incomplete theory was begun twenty years before. Capable of pattern recognition—faces, voices, smells, handwriting, and so on—and simple learning, they have revolutionized the potential for computer technology.

Neural network theory was inspired by knowledge of the brain's own neural networks, hence the name. The brain has two sorts of neural connections—one-to-one "neural tracts" along which information is processed serially, as in a PC, and more complex "neural networks," in which many thousands of neurons in a bundle are all interconnected. Artificial neural networks are, of course, constructed from interconnected electronic processing elements (chips), not from biological neurons.

Most neural networks have three layers of neurons: input neurons, central neurons, and output neurons. Each input and output neuron is connected to central neurons, and the central neurons are connected to each other, usually in both directions. If one neuron "fires," it stimulates all the others to which it is connected. The connections themselves are of variable strength. The next unit will fire only if it has received a sufficient net stimulus from all its inputs. If a given *pattern* of input is received, the system will go through a number of stages and then deliver an output pattern. The whole effect is analogous to a pinball machine, with the variable connection strengths represented by the various obstacles and pockets met by the ball. For any given input, the system's state (i.e., the ball's position) will move around until it finally settles into one end result or another.

The performance of a neural network—what output results from each given input—depends entirely upon connection strengths. Thus the performance can change—i.e., the machine can "learn"—if there is a facility for changing these connection strengths. The first basic rules governing such learning were formulated by the psychologist Donald Hebb in 1949. They are a generalization of stimulus-response learning as described in BEHAVIORISM. (See also MEMORY.)

As a neural network begins to learn, the connection strength between any two units increases if they have fired simultaneously. Thus, if a group of input units frequently fire together, they become more strongly interconnected, reinforcing their tendency to fire together the next time. This procedure is very similar to the formation of habits or conditioned reflexes. Subsequently, if any *part* of the group is activated, it will tend to activate the whole group. Thus such machines can recognize a pattern when stimulated by only part of it, or by some variation of it—such as postal codes written in different handwriting.

Neural networks offer a model for pattern recognition, or learning by association, that has in turn become a widely accepted model for some types of human perception and learning. Human beings can recognize another person from a glimpse of the face, or from the voice, and then past associations with that person come to mind. (See CONNECTIONISM.)

Neural networks and the more standard serial computers, like our PCs, complement each other in their strengths and weaknesses. Neural networks are very poor at the rule-following operations that distinguish serial computation, such as arithmetic. On the other hand, their pattern recognition capacities are far superior to those of serial computers.

Both kinds of computation are probably done by the human brain, and each has useful machine applications. In the case of neural networks, they are only starting to be explored; the theory behind parallel computation is itself in the infant stage.

Existing applications of neural networks include the recognition of fingerprints and smells, rock classification, the detection of signature fraud, and company sales forecasting. Such networks have also recently become a powerful diagnostic tool in medicine, recognizing odors classically associated with conditions like diabetes. In each of these applications, many pieces of data are fed into the neural network, which in the early stages is "taught" how the data should be classified into groups. With practice, the machine often learns to do this for itself. Such applications offer an alternative to the EXPERT SYSTEMS based on serial computation.

There are still major drawbacks in neural network technology. The machines themselves are expensive to manufacture, although they are very much simpler than anything found in the brain. Each neural network chip can hold only a few hundred "neurons," while even the brain of an insect has about 100,000 neurons. A human brain is one million times as complex. Even though they are simple compared to real brains, neural networks are too complex to analyze theoretically. We have no means of knowing, for example, by what inner rules or processes such machines learn to distinguish the odor of one wine or perfume from another; thus we cannot be certain that they will operate successfully in new circumstances. This uncertainty rules out their acceptable use in applications such as flying a plane or judging legal cases.

Some drawbacks to parallel computation can be overcome by simulating neural networks on serial computers. Simulation engineers work out a program for what each "neuron" in a neural network would do, one at a time, but even here speed constraints limit the technology to networks of a few thousand neurons. These simulated machines are slow to learn and can store only a limited number of patterns. The complexities of the visual world are beyond them.

# Neurons

Neurons are the main type of cell of which the brain is composed and are responsible for much of its known activity. The brain contains other cells, called neuroglia, but they are so far less understood.

The human brain contains $10^{10}$ to $10^{11}$ neurons. Each makes hundreds or thousands of connections with others at neural synapses. Most such connections are local, within 1 to 2 millimeters, but many are long distance. Between sensory input to the brain and eventual motor output, there is a great deal of neural activity in the form of electrical signaling and chemical reaction at the synapses. There is also recent evidence that coherent oscillations among bundles of neurons, or among neurons that cooperate on a specific task, are another way that one part of the brain communicates with other parts. (See THE BINDING PROBLEM.)

A typical neuron has a treelike structure, and neurons, like trees, come in recognizably different types. From one end of the cell body there are extensions called dendrites, which are sensitive to sensory signal inputs. At the other end, a branching axon delivers output signals to other neurons. Dendrites meet axons at the synapses. When a neuron fires, an electrical pulse (the action potential) passes along its axon to the synapse. The pulse lasts about a thousandth of a second and travels at speeds up to 300 miles per hour. When the electric pulse reaches the synapse, a chemical neurotransmitter is released. This either stimulates or inhibits the relevant dendrite. (A few synapses communicate via electrical rather than chemical signals.) If at least a few dozen impulses reach a target neuron from adjacent axons, that neuron will in turn fire. Smaller numbers of stimuli cause transient electrical changes in the dendrite and raise local potentials. These are of some importance, as they affect neighboring neurons.

Since the 1940s, neuroscientists have learned a great deal about the subcellular events underlying this simple story. The first detailed study of action potentials was made in that decade by Alan Hodgkin and Andrew Huxley, working on giant neurons taken from squid. They later received the Nobel Prize for their discoveries. Because the squid neuron is so large, Hodgkin and Huxley could insert a fine silver wire to monitor electrical activity. Today's technology is so advanced that any neuron can be studied, and we now know of more than thirty neurotransmitters active in different parts of human beings' brains. Most drugs that affect human psychological states do so by increasing or decreasing the effect of some neurotransmitter. Tranquilizers, antidepressants, heroin, and LSD work this way. Further research into neurotransmitters and their psychological effects is extensive and rapidly expanding, funded by both universities and pharmaceutical companies.

Given the function and operation of neurons, comparisons between brains and computers are inevitable, with neurons represented by chips. Such models work up to a point and are the basis of ARTIFICIAL INTELLIGENCE efforts to build more intelligent machines, but there are many differences between real neurons or brains and their computer equivalents. Neurons work a million times more slowly than the chip components in a personal computer, although, to compensate, there are many more of them. The brain uses nearly one hundred different types of neurons, and its function is not greatly affected by the death of any one of them. The combination of slow performance and huge numbers means that the brain must do many of its calculations at the same time (in parallel), and these are distributed among many systems.

To discover how neurons function in PERCEPTION, MEMORY, problem solving, and other mental functions is the fundamental challenge of neuroscience, and comparison with computer elements could be misleading. It would be an oversimplification, for example, to regard a neuron as a simple on-off switch. Each neuron is more like a whole minicomputer in its own right. A better comparison might be to see one as on the scale of a single-cell animal, an amoeba or paramecium, capable of finding food, fleeing danger, and reproducing. Like these simple animals, neurons learn, and the brain constantly "rewires" itself in response to experience. (Some computers structured as NEURAL NETWORKS can do very simple versions of these things.)

Since there are more neurons in one human brain than there are people on earth, it is unlikely that we will ever know all about their structure and function. A full wiring diagram of the brain would be too extensive to be analyzed in a lifetime. At this stage, neurobiology is concentrating on finding a few general principles for the brain's structure and function, and hopes to build up a thorough understanding of a few simple areas and systems that can then become the basis for general models. One well-known philosopher of the mind, Colin McGinn, is fond of saying that the brain is far too complex ever to be understood by itself. Nonetheless, research continues, and our knowledge, however partial, grows.

# Neuroscience

Neuroscience, a study of the brain, is a very broad science. It asks such questions as: What are neurons, and how do they function? How are they grouped and interconnected? What neural activity accompanies our capacity for PERCEPTION, motivation, motor control, MEMORY, or learning? How are neural structures affected by brain damage through illness or accident, and what possibilities of repair exist?

Neuroscience is basically a study of neurons and a new science of this century. Even the ancient Greeks realized that the brain has a basic relation to our conscious life, but only the most indirect, general observations were possible in attempting to make specific links between human mental capacity and the kilogram of gray matter inside the skull. With staining techniques previously available, a microscope slide of brain tissue looked very like a slice of dense jungle containing small parts of innumerable plants. It was only toward the end of the nineteenth century that "Golgi staining" made it possible to stain a few neurons completely, leaving others untouched. This led to the discovery that NEURONS are the basic functional units of the brain. Camillo Golgi and S. Ramón y Cajal shared the Nobel Prize for this work in 1906.

With the constantly improving technology invented to study brain structure and function (see METHODS OF STUDYING THE BRAIN), neuroscience has led to an explosion of knowledge. The collection of facts now known about the brain is impressive, although it is only a very small part of what remains to be known. But facts alone don't make a mature field of science. The field must have a dominant paradigm that sets the questions it asks and frames the necessary experimental research. There is as yet no such unity in neuroscience. In terms of scale, necessary brain research ranges from tiny molecules or synapses up to looking at the whole brain. Neuroscientists usually specialize in one such level, and there is little integration with what is known about others. Scientists working on each level tend to seek explanations for everything, from motor control and vision to consciousness, in terms of activity on that level.

At Tucson, Arizona, in 1994, this lack of paradigmatic unity was illustrated at the first major cross-disciplinary conference devoted to seeking a scientific basis for consciousness. Each of the conference's five days was devoted to research and theories focused on one level of brain activity—the quantum, the molecular (see NANOBIOLOGY), the cellular (individual neurons), NEURAL MODULES, and the whole brain. No neuroscientist who spoke related his or her own level to others, and it was only the philosophers present who tried to integrate what was being said on the different days. (See THE MIND-BODY PROBLEM.)

The evolution of the brain itself is another reason why neuroscience is not yet a mature science like physics or chemistry. The human brain contains layer upon layer of nature's gradual adaptations, each grafted onto what came before. Just as the immune system uses several different weapons in its battle against disease, the brain draws on many levels of its evolved capacity when performing different functions. Usually these are complementary, but sometimes they are in conflict. Many critics of computer models of the mind point to the brain's complex, organic history as one reason why machines will never function like brains. When human beings think they do so with a cerebral cortex grafted onto the brain of a frog grafted onto the "nervous system" of a paramecium. It may be impossible ever to sum up human mental activity with simple generalizations.

# Neutrinos

Having no mass and no charge, the neutrino must be one of the most ephemeral particles in the universe. Ten of billions of neutrinos could pass through the earth before one of them happened to react with an atomic nucleus.

The neutrino was predicted, on purely theoretical grounds, by Wolfgang Pauli, one of the founding figures of atomic and quantum physics, in 1932, but not experimentally confirmed until 1956. Except for its lack of electric charge, the neutron is virtually identical to the proton. This led some physicists, early in the century, to believe that it must be a bound state of a positive proton and negative electron. Sure enough, the neutron is slightly heavier than the proton. Moreover, unlike the proton, which is either absolutely stable or has a life at least as long as the present universe's, the neutron has an immensely long life compared to other elementary particles, decaying after about fifteen minutes. Physicists identified the products of this decay—called beta decay—as an electron and a proton.

The problem with decay arose when the accounting was done on the reaction. Conservation laws dictate that both total energy and momentum must be the same before and after the disintegration. By contrast, every experiment indicated that energy was being lost in the process. Did this mean, as some physicists speculated, that quantum matter did not obey the fundamental conservation laws of nature?

Pauli came to the rescue, arguing that energy (and momentum) must still be conserved. The only way a failure to balance the energy books could be explained was that a third, invisible particle was being emitted in beta decay. Pauli predicted that this particle, the neutrino, must have zero or nearly zero mass, no charge, and a spin of ½. In 1956 this particle was experimentally observed and its predicted effects on other particles were confirmed. (See THE ELECTROWEAK FORCE.)

Neutrinos may have a very small mass. If so, there are so many that they would be a significant fraction of the total mass of the universe. (See DARK MATTER.) This could also solve the problem that the sun appears to radiate only about one third as many neutrinos as our theories predict. If they had any mass, they could oscillate into forms our neutrino detectors would fail to register.

# Neutron Stars

A neutron star is one possible end product of a normal star that has run out of fuel and collapsed. (See STARS.) It is so dense that all normal atomic structure has broken down.

The collapse of a normal star has three possible outcomes. As well as producing a neutron star, collapse can produce a white-dwarf star or a black hole. (See BLACK HOLES.) A white dwarf has a mass comparable to that of the sun, but its size is more like that of the earth, so it is denser than a normal star. In white dwarfs, normal chemical elements and reactions are preserved. In a neutron star, the elements undergo dramatic transformation.

When a parent star is sufficiently massive, its collapse squashes the component chemical elements out of existence, perhaps in a supernova explosion. (See SUPERNOVAS.) Electrons and protons collapse into neutrons, and the whole star becomes like a gigantic atomic nucleus. At this point, the star is about the size of a large city, and, according to theory, its interior is a superfluid and a superconductor. (See BOSE-EINSTEIN CONDENSATION.) Neutron stars have very strong magnetic fields. As they shrink, their speed of rotation increases to compensate, until the star is rotating several times per second. Over millions of years, this fast rotation is braked by the star's magnetic interaction with its surroundings.

Neutron stars have been observed in two contexts. Radio pulsars, neutron stars that emit very regular pulses of radio waves a few times a second, were discovered in 1967. The waves are generated by the stars' magnetic poles, as material falls into them from the surrounding gas. If the stars rotate obliquely, their radio emission can sweep the earth once during each rotation, like the rhythmic beam of a lighthouse. At least one binary pulsar is also known. GENERAL RELATIVITY predicts that this should lose energy very slowly by radiating gravitational waves as it orbits its companion. The tiny predicted reduction in the binary pulsar's pulsing rate has been observed, adding one more confirmation to Einstein's theory.

Other neutron stars, though optically invisible, are known to radiate X rays via a different mechanism. They, too, are members of binary star systems. Matter from the companion star rains down on the neutron star, producing great heat and X rays as it does so.

# Nonlinearity

The equations that describe much of nature, as opposed to simplified models used by scientists, are called nonlinear. The range of solutions associated with them is rich, varying from extremely stable behavior to exceptional sensitivity, sudden jumps, bifurcations, chaos (see CHAOS AND SELF-ORGANIZATION), and repetitive cycles. The equations offer such serious difficulties to mathematicians that in most cases exact solutions are unobtainable. The best that can be done is to map out general observations about the different classes of solutions, or use high-speed computers to work out approximate answers.

High-speed computers and modern mathematical methods didn't appear until the second half of the twentieth century, and for the preceding 200 years scientists generally ignored this rich but intractable nonlinear behavior, looking only at those regions of nature in which things change gently and slowly—smooth rivers, systems close to EQUILIBRIUM, tiny flows of energy, and so on. While it was known that bizarre new forms of behavior could take place outside these regions, it was generally considered unimportant for scientific understanding, and the whole phenomenon of nonlinearity was sidelined.

Linear systems are characterized by things that change slowly in expected ways. Knowing one solution makes it easy to work out another. Extrapolations from one range of behavior to another are straightforward. A small influence produces a small effect, and doubling the size of the influence doubles the effect. Linear systems are well behaved, analyzable, predictable, and controllable. Linear science complemented the dreams of a nineteenth-century society that was rule-bound, reliable, predictable, and unlikely to shock, a society that believed in eternal progress through the manipulation and exploitation of natural and human resources.

It was only during the latter part of the twentieth century that modern techniques enabled scientists to venture outside regions of linearity. They learned of systems in one region of which a small influence might produce a small change in behavior, while in another a small change might precipitate radical new behavior, repeated oscillations, or even chaos. THE BUTTERFLY EFFECT spoke of the extreme sensitivity of such systems. Bifurcation points were discovered, crossroads at which the system had to choose between two radically different forms of behavior.

Nonlinear systems could be thought of as a rich landscape of hills, valleys, mountain peaks and passes, fast-running streams, and water meadows. Over certain regions of this landscape, the system is predictable and controllable, but in other regions it becomes resistant to change and external influence, and in yet others it engages in bizarre chaotic fluctuations.

Nonlinearity is found in a wide variety of natural, social, and economic systems. Not only are nonlinear studies transforming physics, chemistry, and biology, they are making inroads into the social sciences and politics. What are policy makers to do when the systems they study are infinitely richer and more complex than the models being used? Or when systems are so sensitive that any intervention causes them to swing into radically different forms of behavior? How are stock market decisions to be made when a sudden fall in prices turns out to be an intrinsic feature of a nonlinear stock market, not the result of external influences?

Specific models have been created in CATASTROPHE THEORY, PHASE TRANSITIONS, chaos, and COMPLEXITY. Some theorists hope that universal, predictive laws will emerge; others are more pessimistic.

# Nonlocality

Nonlocality means action in the absence of local forces, action without causation. It is a prediction of quantum mechanics that distant events should be related nonlocally, but this is a violation of some of the most fundamental principles of classical or Newtonian physics.

Classical physics rests squarely on the principle of locality—correlated events are related by a chain of causation. If we push against a door, and it opens, we say that the door opened *because* we pushed against it with a force. If we hear a voice on a telephone wire, it is because some electronic signal carried that voice from its source. Similarly with a radio program, the radio broadcasts a distant voice because a set of electromagnetic waves carries that voice from a studio.

In all such locally caused events, some form of energy is transferred from A to *B*, and *B* is later in time than A. The energy moves continuously through space from A to *B*, at a speed no greater than the speed of light. By contrast, in quantum nonlocal events, two distant events are correlated without any transfer of energy from A to *B*, and *B* can happen at exactly the same time as A.

The simplest instance of nonlocality in quantum physics relates to the behavior of just one particle, although that particle in its wave aspect has multiple possibilities, and the nonlocal correlations are really between these. Imagine a very large two-slit experiment, one that is actually done in radio astronomy. A single photon leaves a distant star and, in effect, passes through two slits, A and *B*, located miles apart on earth. A beam of such photons creates an interference pattern because, although we are dealing with photons one by one, the wave function of each is spread out to cover all the possible paths each photon might take through the slits. Thus the detected interference pattern is an interference of possibilities—the possibility of going through slit A, and the possibility of going through slit *B*. (See the discussion of the two-slit experiment in CONTEXTUALISM.)

We know from that discussion of the two-slit experiment that if we add particle detectors in front of the two slits, the photon is detected as a particle and seen to have gone through only one slit. But at the very instant that we detect the particle at slit A or *B*, the *possibility* that the particle will also have gone through the other slit disappears. So does the interference pattern at the screen. Once we have pinned down the particle, its wave function disappears from everywhere else. We can never detect the same particle twice.

This disappearance of the other branches of the wave function, or of the full range of possibilities associated with the photon, is a nonlocal effect. Detection of the particle at A changes to zero the probability that the particle will be found at *B*. Thus events or probabilities at A and *B* are correlated; as the probability at A becomes one, the probability at *B* becomes zero (or vice versa) at exactly the same moment. This is not causal. No force or signal passes from A to *B*.

The effect, though instantaneous, does not violate relativity because it has not employed a faster-than-light signal. The two probabilities are correlated simply because they were aspects of the same larger whole (the wave function), and any change in one is automatically a change in the other. (See HOLISM.) Quantum physicists say that the probabilities were interlinked.

From a Newtonian perspective, one might be tempted to say the race was fixed—the photon was already on its way to slit A or slit *B* before we decided to put up our particle detector. But this is just a particle interpretation of the photon, which takes no account of its wave aspect. It can't account for the interference pattern that is seen when no particle detector is present and that shows the photon goes through *both* A and *B*. To understand the whole experiment, we must keep in mind wave/particle dualism, quantum indeterminacy, *and* nonlocality. Quantum physics is all of a piece.

There are also nonlocal correlation effects between two or more particles separated in either space or time. (See BELL'S THEOREM.)

# Observational Astronomy

Observational astronomy is the factual basis upon which all our subsequent cosmological models and theories are based. Nearly all its knowledge is gathered via observation of electromagnetic radiation of various wavelengths. (See Table Three.) Originally, this was done with the naked eye, but since the seventeenth century it has been supplemented by increasingly powerful telescopes.

Since 1945, astronomers have had much better instruments that function over a wide range of wavelengths. The first developed were radio telescopes, which have contributed greatly to cosmology. Apart from visible light, radio waves (1 centimeter to 10 meters in wavelength) are the only form of electromagnetic radiation that reaches the ground in large quantities. Other wavelengths are absorbed by the atmosphere or reflected by the ionosphere. Radio telescopes are huge, revolving dishes, hundreds of feet across, or fixed arrays of ground-

TABLE THREE

THE ELECTROMAGNETIC SPECTRUM

| TYPE OF RADIATION | WAVELENGTH (MILLIMETERS) | |
|---|---|---|
| radio | greater than 100 | radio "window" |
| microwave | 0.1 to 100 | in the atmosphere |
| infrared | 0.001 to 0.1 | |
| visible | $2 \times 10^{-4}$ to $10^{-3}$ | visible "window" |
| ultraviolet | $10^{-6}$ to $2 \times 10^{-4}$ | |
| X-ray | $10^{-8}$ to $10^{-6}$ | |
| gamma ray | less than $10^{-8}$ | |

The shorter the wavelength, the higher the energy per photon.

based aerials extending over miles, even thousands of miles. Due to their large size, they can "see" very far into space. GALAXIES, pulsars, QUASARS, SUPERNOVAS, the sun, and the planet Jupiter act as radio sources. The COSMIC BACKGROUND RADIATION was detected by radio telescope.

Observations at other wavelengths have been made in the last few years by raising instruments above most or all of the atmosphere, on mountains, balloons, planes, or satellites. The Hubble Space Telescope has vastly increased our knowledge at both ultraviolet and optical wavelengths. Infrared telescopes, which now incorporate better detector materials, see through the clouds of obscuring dust in THE MILKY WAY and elsewhere. There are also X-ray and gamma-ray telescopes that can be mounted on satellites. Neutrino detectors measure the flux of neutrinos from the sun and registered the neutrino outburst of Supernova 1987A. And instrument-carrying space probes have visited the surface of the moon and Mars and have flown close enough to other planets to give us a great deal of information about the solar system.

All telescopes are set to measure the brightness and direction of various sources in the heavens. If the "light" detected has well-defined absorption or emission lines at certain "colors," spectroscopy reveals the chemical composition or character of the source. The Doppler effect, the shift in the natural position of these lines, reveals whether the source is approaching or receding, and how fast.

The distance of each observed source is not directly revealed by telescopes. It is difficult to know whether a given star is bright and far away, or faint and near. Such information is essential for interpreting many observations, and a great deal of effort has been expended in measuring distances. (See DISTANCE MEASUREMENTS IN ASTRONOMY.) We still remain unsure of whether our distance scales are accurate by a factor of two, although progress has been made. Hence the uncertainty about the exact age of the universe since THE BIG BANG, which is in the range of $10 \times 10^9$ to $20 \times 10^9$ years.

A huge limitation to the success of observational astronomy lies in the fact that most—perhaps 90%—of the mass of the universe is not visible to any sort of telescope. This is the so-called DARK MATTER whose existence can be deduced from the movements of stars within galaxies, and of galaxies themselves within clusters of galaxies. They move too quickly to be gravitationally bound simply by the matter that we can see. The origin and nature of this dark matter is one of the great unsolved puzzles of astronomy. There are probably many more phenomena in the cosmos awaiting our discovery as our technology improves.

## Olbers's Paradox

Olbers's paradox, named after the nineteenth-century German astronomer Wilhelm Olbers, relates to the fact that the night sky is very dark when, given the number of stars it contains, it should appear bright.

Given an infinitely large universe, or even a very large finite universe, we would expect there to be a star in almost every direction. Many would be faint because very far away, but their numbers would make up for their faintness. On the average, the night sky should be as bright as the surface of an average star, like the sun. But clearly it is not, and this is the paradox. Though it was named after Olbers, earlier astronomers, including Kepler, had commented on it.

The paradox is not removed by supposing that parts of the universe are filled with dust or other dark matter. This would eventually have heated up to the same temperature as the stars, and should be equally bright. But two effects predicted by the BIG BANG model of the universe resolve the paradox.

Since the universe is of finite age (about $10^{10}$ years), light from stars more than $10^{10}$ light-years distant has not yet had time to reach us. The stars appear sparsely distributed, whatever telescope we use. Since the universe is expanding, light from distant stars within this event is redshifted, which has reduced the temperature of the COSMIC BACKGROUND RADIATION from about 4000 K, when it was formed, to 3 K now. Thus it is not hot enough to raise dust or dark matter to the temperature of stars.

Any good cosmological model of the universe incorporates one or both of these two features—the universe's finite age and its expansion. Olbers's paradox is one of the fundamental observations that constrains any such model building. The Big Bang model incorporates both.

## Olfactory Perception

The olfactory system (the sense of smell) is simple and anatomically well understood. EEG measurements can be made on it fairly easily, so it has been studied extensively. Some of the most significant information was produced in the 1970s by Walter Freeman's work on the olfactory system of rabbits. This has led to a connectionist model of the system—one based on parallel processing—and to a significant understanding of the role played by chaos in the brain. (See CONNECTIONISM.)

The brain has an outgrowth known as the olfactory bulb where smells are processed. This is connected to chemical sensors in the nose. It is to the nose a little like the retina is to the eye, and like the retina, the bulb carries out the first stage of processing on its sensory input. By attaching sixty-four surface electrodes to different areas of the bulb and separately recording their EEG responses to various situations—no smell, familiar smell, and unknown smell—Freeman and his coworkers were able to build up a picture of what takes place during this first stage of processing and how the information is then passed along the olfactory nerve to the brain's cortex.

When Freeman's rabbits were exposed to a familiar odor, the EEG patterns recorded from all over the olfactory bulb were coherent—they oscillated in synchrony. (See COHERENCE.) What distinguished the response to one smell from another was not due to which neurons fired or what part of the bulb was affected, but rather to the *relative amplitude* of the response in different parts of the bulb. This led to the theory that chemical receptors in the nose are spatially clustered by types, and each type of smell is associated with a corresponding amplitude profile in the olfactory bulb. The whole bulb acts as a neural network (see NEURAL NETWORKS) in which all the neurons are interconnected, and it is the collective amplitude pattern of this network that is transmitted along the olfactory nerve to the cortex. Each of the transmitted patterns remains constant if the rabbit is exposed to familiar odors; new patterns emerge when new odors are learned through conditioning.

The role played by chaos in olfactory perception emerged in connection with the absence of odor. In this situation, the corresponding EEG patterns in the bulb wandered irregularly through all possible frequencies and local amplitudes. This was a chaotic state, poised to respond quickly to any change in input. When the rabbits were exposed to a familiar odor, their EEG patterns moved out of chaos and immediately into a coherent state. Thus chaos is associated with no perception, and coherence (order) with a definite perception. Similar phenomena have been noticed in studies of visual perception, and some scientists speculate that chaos may play a similar role in states of consciousness. (See CHAOS THEORIES OF MIND.)

# Open Systems

For more than 200 years physicists treated nature as if every system were enclosed in a box. The laws of thermodynamics dictated that the entropy in a box increases as the system runs down toward featureless equilibrium. Today, however, scientists realize that most natural and social systems are open systems, capable of ordering, structuring, and regulating themselves.

The world around us is filled with complex structures, growing plants, and evolving societies, all of which appear at first sight to defy the inexorable march of ENTROPY. A seedling grows into an oak tree that lives for centuries. Human beings, cities, patterns in heated water, and self-catalyzing chemical reactions spontaneously structure themselves and continue to sustain high degrees of order. They are all open systems, in which matter and/or energy is free to enter and to leave.

Water placed in the freezer compartment of a refrigerator decreases in entropy, making a phase transition into solid ice. (See PHASE TRANSITIONS.) The reason this can happen is because a refrigerator is an open system, one into which electrical energy is pumped and entropy (or heat) is extracted. Thanks to a flow of energy through an open system, it is possible for entropy to decrease internally and for water to move from a relatively chaotic into a more orderly state. Of course, the total amount of entropy in the system plus its environment increases. Order is able to prevail and internal entropy to decrease by virtue of the entropy that is being pumped out into the environment.

Put a pan of chicken soup on a stove. Heat at the bottom warms the liquid, causing it to move upward in a series of convection currents. (Hot water is less dense than cold and attempts to rise.) At the top of the pan, the water cools, releasing its heat into the air above. The result is a constant flow of heat from the bottom to the top of the pan.

Initially there are random disturbances in the soup. Hot soup attempts to rise; at the same time cool soup at the top is trying to fall to the bottom. The result is a complex, chaotic motion, filled with many different sorts of fluctuations. But as the heating continues, some of these fluctuations amplify. Hot soup in one region begins to flow upward, taking some of the neighboring soup with it. At the top, the hot soup spreads out and helps to force cooler soup to fall to the bottom. Soon there are well-defined columns of rising and falling soup. If you were to observe it carefully, you would see a pattern of cylindrical rolls or hexagonal cells with hot soup flowing upward and cooler soup flowing downward. The result is a stable system, with the appearance of order out of chaos that results from the fact that energy (heat) is flowing through an open system. A similar pattern of cells can be seen from an aircraft flying over desert sand; hexagonal cells of hot air rise and disturb the sand beneath to form regular patterns.

Open systems can also involve the flow of matter. Examples are the stable pattern of a fountain in an Italian piazza and the vortex formed as water flows down the drain of a bathtub.

A tree is an open system that is able to structure itself by virtue of a flow of both matter and energy. Water is sucked in through its roots and carbon dioxide breathed in through its leaves. Energy enters the tree in the form of sunlight. Thanks to the special processes of photosynthesis, the tree is able to synthesize sugar (carbohydrates) for its growth out of the basic components of water and carbon dioxide. If any part of this system were to be closed off from the environment by shielding the tree from light, blocking its leaves from transpiring, or cutting its roots, the internal ordering process would cease. The tree would die, and order would move back toward chaos.

Open systems can also be social and economic. Self-ordering occurs in traffic flows on highways. A city is an example of an open system whose internal order is maintained at the expense of the entropy that is pumped into the external environment in the form of garbage, sewage, and the city's heat. Indeed, all life on our planet, like convection patterns, depends on a flow of energy through it, which is ultimately supplied by the sun.

# Parallel Processing

See NEURAL NETWORKS.

# The Participatory Universe

What role do observers play in the unfolding of physical reality? Is the physical world any different than it would be if there were no conscious beings? Do human purpose and intention in any way alter the direction taken by the physical universe?

A classical physicist would give a negative response to all these questions. He might not even understand them. In the Newtonian paradigm, there is a sharp dichotomy between the observer and the observed. Physical reality has its own nature and obeys its own laws, quite apart from what observers do, and even apart from whether there *are* observers. A stone is a stone is a stone, and it remains the same kind of stone with the same physical characteristics and propensities in whatever environment it finds itself and regardless of whether anyone knows it is there. Making this clear, freeing the material world from all supposed psychological influences, was an important step in distinguishing the Newtonian scientific world view from the earlier medieval notion that spiritual and psychological forces influence matter. The observer/observed distinction is crucial to what the Newtonian scientist means by scientific objectivity. We stand back from our observations, make them dispassionately, and keep ourselves and our goals out of the procedure.

The shift to a quantum physical understanding of the material world changes much of this. The quantum observer stands *inside* his or her observations, which themselves play an active role in bringing about the very reality they then look at. In a sense not yet fully understood, the quantum observer helps to *make* the world of his or her observations. (See PERSPECTIVE AND INTERACTION.)

The unobserved world of quantum reality is a plethora of possibilities. The Schrödinger wave function describes an infinite array of possible states into which a quantum entity might settle. All this possibility becomes an actuality only when the wave function collapses—and whatever else may make the wave function collapse, we know that measurement or observation does. The mere act of measurement converts possibility to actuality, and the *kind* of measurement that is made determines *which* kind of actuality will be plucked from the infinite sea of possibility. To some extent, the observer sees what he or she looks for.

A photon, for example, has both position possibilities (a particle-like aspect) and momentum possibilities (a wavelike aspect). A physicist can set up an experiment to measure, and hence fix, either of these, although in fixing one the other is lost (HEISENBERG'S UNCERTAINTY PRINCIPLE). The physicist's interference—the measurement or observation—seems to influence which side of its dual nature the photon will exhibit. An experiment illustrates this graphically.

If we have a photon emitted by a source, and that photon has the option of traveling through either one or two slits in a screen (being quantum mechanical, it has the option to do both), the physicist's planned experiment will have the following result: If he or she places two particle detectors in front of the slits on the opposite side from the photon source, the photon behaves like a single particle—it follows a definite path through one slit and strikes one particle detector.

If, on the other hand, the physicist replaces the particle detectors with a single screen that "sees" both slits, the photon behaves like a wave—it travels through both slits and contributes to an interference pattern on the detector screen. Physicist and photon are involved in a creative dialogue that transmutes one of many possibilities into a fixed, everyday actuality. Wheeler claims that this dialogue shows that we live in a "participatory universe." Another leading physicist, Nobel laureate Ilya Prigogine, says, "Whatever we call reality, it is revealed to us only through an active construction in which we participate." (See CONTEXTUALISM.)

One very important note of caution is required, however, as we consider the meaning of observer-participancy in quantum mechanics. Physicist Eugene Wigner offered a famous interpretation of the collapse of the wave function that argued that the collapse is brought about by human consciousness. Wigner's logic was that nothing physical can collapse the wave function; since anything physical must exist in a state of superposition, according to quantum mechanics, it must be collapsed by something nonphysical—i.e., consciousness.

This way of thinking requires a dualist interpretation of consciousness, placing it outside the physical world and at the same time placing a great deal of importance on conscious observation. In fact, though much has been made of Wigner's hypothesis in popular books on quantum theory, very few physicists lend it much credibility. The majority believe that if the wave function collapses, it does so for physical reasons, and if consciousness is one thing that can collapse it, consciousness, too, must have some physical basis. It is usually supposed that many things other than consciousness bring about collapse—photographic plates, Geiger counters, and so on.

Though observer-participancy is an indisputable fact of quantum mechanics, and one that makes us reassess the relationship between humans and the rest of the physical world, it requires subtle understanding. It will be fully understood one day in the context of the whole larger question of THE MEASUREMENT PROBLEM and COLLAPSE OF THE WAVE FUNCTION.

# Penrose on Noncomputability

Sir Roger Penrose, who has done distinguished work on the physics and mathematics of General Relativity, has in recent years devoted his attention to the problem of consciousness. He is one of the leading proponents of QUANTUM THEORIES OF MIND but is best known for his argument that human thinking has a noncomputable element—that is, a quality that cannot, in principle, be replicated or simulated by a computing machine.

Penrose bases his argument on the theorem made famous by Kurt Gödel in 1931. GÖDEL'S THEOREM states that any formal mathematical system of rules will always be incomplete, because there will always be truths not provable by those rules. It takes a vision of the system to create the system of rules, and that vision is never encompassed within the rules themselves. Penrose argues that this is a fair description of the relationship between a computer, its program (system of rules), and its programmer (us). The computer itself can never get outside its own program, but we can.

According to Penrose, no machine can understand what it is doing and act on that understanding. Human beings can. We detect the meaning of things and act on that meaning, rather than blindly following rules (algorithms). Human beings are conscious; machines are not. We have insight, sensitivity, and "soul." A crucial aspect of our creative thinking draws on these capacities. No machine can do likewise. No matter how clever the simulation, an intelligent person will always know whether he or she is dealing with a mind or a machine.

Penrose is one of the strongest opponents of both strong and weak AI (see ARTIFICIAL INTELLIGENCE). He is opposed to the notion that *all* thinking is nothing but computational, rule-bound thinking (strong AI) and also to the argument that a computing machine can *simulate* any kind of human thinking (weak AI). (See THE CHURCH-TURING THESIS.)

Penrose is also a self-declared Platonist. He believes that, in mathematics at least, there is a God-given realm of pure ideas, which have some sort of timeless, independent existence. This realm of ideas is *there,* waiting to be discovered, and when mathematicians communicate their insights, this is made possible by each one's having a direct access to it. According to Penrose, consciousness plays a mediating role between the realm of ideas and the realm of material reality. His project is to describe the physics that would make this possible. He has suggested that consciousness is crucially dependent upon the effects of QUANTUM GRAVITY acting on the brain.

# Perception

According to Nobel laureate Francis Crick, "The overwhelming question in neurobiology today is the relation between mind and brain." Crick won the Nobel Prize for his contribution to the discovery of the DNA molecule. Today he devotes a large part of his scientific life to studying perception, because he, like many others, believes that visual perception can give us clues to the physical basis of the relation between mind and brain. This larger mind-body question gives the study of perception a more exciting cutting edge. Ultimately, it is hoped, understanding perception will help us to understand the physical basis of consciousness itself.

Our perceptual abilities—sight, hearing, smell, touch, and the others—are windows into the mind, the vehicles through which our subjective sense of self meets data from the outside physical world. Semir Zeki, professor of neurobiology at London University, says, "Our inquiry into the visual brain takes us to the very heart of humanity's inquiry into its own nature."

Perception is one of the most promising areas of psychology or brain function to study. Unlike language, it is a function of both humans and animals and can easily be studied in both. Researchers can initiate a visual or other stimulus and ask the subject what he or she has perceived, or they can measure the physical response in a human subject's or an animal's brain. Unlike attempts to investigate motor behavior or thinking, studies of perception permit the same stimulus to be given repeatedly, under the same or altered conditions, until a general picture of response is built up; therefore a great deal of work has been done on various aspects of perception.

Human beings have about twelve perceptual senses. In addition to sight, hearing, touch, taste, and smell, we have a sense of balance, a sense of the positions of our joints and the tension and relaxation in our muscles, a sense of the fullness or otherwise of our stomachs, and so on. Other animals have still other senses. Bats use a form of sonar, and some desert insects can see a fourth primary color in the ultraviolet region. Perception researchers operate on the hope that all sensory systems, however different, have enough in common to make detailed studies of one system yield general principles about all. (See OLFACTORY PERCEPTION; VISUAL PERCEPTION.)

Much of the information gathered by our sense organs is unconscious. Blood pressure receptors in the neck arteries supply information to the lower brain, which in turn corrects blood pressure if it is too high or too low. We are unaware of this. Far more information flows along the optic and auditory nerves to the brain than we are conscious of. BLINDSIGHT is a graphic illustration of this. The brain carries out a great deal of computation to yield such useful information as distance perception from raw perceptual data *before* we become conscious of the result. Ideally, scientists would like to trace the neural mechanisms involved at each stage of perception, up to the final conscious result, but we are still very far from this goal.

# The Perfect Cosmological Principle

The Perfect Cosmological Principle states that the universe looks more or less the same, not only at all places (see THE COSMOLOGICAL PRINCIPLE), but also at all times—past, present, and future. Such a view, except for a possible initial creation, was common to most ancient cultures. It was also held by Newton and, originally, by Einstein. But Edwin Hubble's discovery in 1929 that the universe is expanding (see THE EXPANDING UNIVERSE) seemed to refute it. From Hubble's discovery, cosmologists went on to the BIG BANG hypothesis that all the galaxies we can see diverged from an initial point some $10 \times 10^9$ to $20 \times 10^9$ years ago.

Efforts to save the Perfect Cosmological Principle were made in 1948 by Fred Hoyle and his colleagues, with their proposal of the steady-state hypothesis. According to this, the expansion of the visible universe is compensated for by the continuous creation of new matter in the consequent spaces, so that the overall density of matter in the universe remains the same. With the new observational evidence of the COSMIC BACKGROUND RADIATION in 1965, and the knowledge that there was a higher proportion of radio galaxies in the early universe, most astronomers abandoned the steady-state hypothesis.

It now seems clear that our own universe is evolving, but diehard steady-state theorists still hold that this is only part of a static larger pattern—for example, a succession of universes that itself exhibits some constant features. (See THE PLANCK ERA; THEORIES OF EVERYTHING.)

# Perspective and Interaction

The new physics, particularly relativity theory and quantum physics, stresses the crucial importance of the perspective, or context, from which or within which we observe something, and the interaction between ourselves and what we observe. The paradigm for truth that results from this stands in sharp contrast to the earlier views of both direct realism and relativism.

Direct realists argue that the world is pretty much as it appears to be. There are objects, such as trees and buildings and people, and also, possibly, an abstract world of logical and moral truths. All who look at these objects (or ideas) see the same thing, if they look clearly and without error. Relativists, on the other hand, argue that what we see or what we believe is conditioned by our culture and individual experience. They say there is no objective standard by which we can adjudicate between conflicting points of view.

Both relativity theory and quantum mechanics offer a more subtle view of truth. Both hold that there *is* an objective reality, but its features are very different from those of the world of human experience. Any particular observer in our world perceives an aspect of underlying reality evoked by the interaction between the observer and the environment. (See THE PARTICIPATORY UNIVERSE.) In SPECIAL RELATIVITY there is a unified, four-dimensional space-time, which is divided into separate space and time by any given observer, according to that individual's motion. In QUANTUM PHYSICS, objective reality is described by the abstract Schrödinger's equation, which contains an infinite array of differing potentialities. (See THE WAVE FUNCTION AND SCHRÖDINGER'S EQUATION.)

What the quantum or relativity observer may actually *see* depends upon how he or she interacts with ("measures") the underlying reality, or the overall context of the experimental apparatus. (See CONTEXTUALISM.) Thus the "real" world contains a multiplicity of possible (and partial) viewpoints, one of which we experience, depending upon our situation. These possible or partial viewpoints are not arbitrary.

They are related by the abstract laws of the whole. There is a diversity-within-unity that is a constant theme in the new science, and this, in turn, can be used to reflect on much of twentieth-century thinking. (See RELATIVITY AND RELATIVISM.)

# Phase

The phase of a wave is a measure of its time of arrival at a given point. Waves are characterized by their size or intensity—technically called their amplitude—and by their frequency—the number of oscillations in a given time. A wave's length is an additional characteristic, but this is directly related to frequency. For a given speed, the longer the wave, the lower its frequency.

Two waves can have identical amplitudes and frequencies, yet be in or out of phase with each other. When the peaks of two waves arrive in the same location at exactly the same instant, they are said to be "in phase." But if the peak of one arrives with the trough of another, they are said to be exactly "out of phase."

When waves in water and air, or electromagnetic waves like light, meet, their effects add up or subtract. This is called interference. If the peaks of two identical waves arrive in phase, their size is doubled. If they are exactly out of phase, the effects will cancel out through destructive interference.

When a boat is docked in a harbor, incoming waves are deflected by its hull and meet in the region behind. The result is a complex interference pattern, with waves of various phase differences meeting. Some waves add while others subtract their effects.

The notion of phase applies to oscillatory behaviors in general; thus one speaks of the moon as having phases. In quantum systems, because of HEISENBERG'S UNCERTAINTY PRINCIPLE, the phase and the number of quanta in a wave are complementary (See COMPLEMENTARITY): Both cannot have definite values at the same moment. So quantum COHERENCE is unlike its classical analogies in subtle ways.

# Phase Transitions

A phase transition occurs when a substance undergoes a dramatic change of internal order, as when ice melts or water boils. Within a solid, atoms are arranged regularly in the form of a lattice. Local forces between atoms operate like little springs that pull an atom back to its normal position as it vibrates. Heat the solid so that its temperature rises, and the atoms vibrate even more. But when the temperature reaches a certain critical point, a phase change occurs. As additional heat is added, the solid's temperature no longer rises because the incoming heat energy is being used to break the bonds that hold atoms in their lattice positions.

During this phase, atoms separate and begin to move more freely, although they remain somewhat influenced by their neighbors. The result is a new phase, liquid. Once the phase transition is complete, the temperature rises again as more heat is added. A second phase change occurs when the liquid begins to boil and becomes a gas. Again, incoming heat excites the atoms, breaking the weak bonds that attract them to their neighbors.

In addition to using heat, phase changes can also be brought about by pressure. Apply pressure to a gas and it liquefies. A sufficiently high pressure will cause liquids to solidify. Phase transitions between solid, liquid, and gas are all transformations of order, movements across the boundary that separates one order from another.

There are several different kinds of phase transition. The element carbon, for example, is found as a hard, shiny diamond in which atoms are arranged in a crystal structure. It can also occur as soft, black graphite with atoms aligned in flat planes. The extremely high pressures that occur inside the earth induce a phase transition from graphite to diamond. Other substances like sulfur and tin are found in several different phases. Each phase is characterized by the particular external physical form. The magnetization of iron and the transformation of normal metals into SUPERCONDUCTORS and of fluids into SUPERFLUIDS are also phase transitions.

# The Planck Era

The Planck era was the first $10^{-43}$ second in the life of our universe. It was effectively "time before time," because it was an era that preceded the creation of time itself. During this period, the whole universe was no larger than an elementary particle.

INFLATION THEORY, which describes the expansion of the universe from a point $10^{-36}$ second after THE BIG BANG, pretty well explains the evolution of the universe without regard to whatever may have come earlier. As a result, there seems to be little chance that we will gain observational evidence of any earlier stages. Any sound theoretical speculation would have to rest on a full theory of QUANTUM GRAVITY, which we lack. Nonetheless, those first moments in the history of the universe remain of great philosophical interest.

Everything that we can know began $10^{-43}$ second after the Big Bang—that is, *after* the Planck era. It was only at that moment that space-time, matter, and energy (forces) as we know them were created. Thus, within our time coordinates, there is no point to asking what came "before" $10^{-43}$ second. Yet because we think in terms of causality, we find it impossible not to ask what happened just after the Big Bang or, indeed, what preceded it.

One speculative idea about this inaccessible "before" is presented in cyclic Big Bang theory. According to it, the universe is like a giant heartbeat. It begins point-like, expands from a Big Bang, eventually collapses into a Big Crunch, then reexpands from this point, again and again. A related idea is the theory of a branching universe, which suggests that new universes are forever breaking off from some expanding universe, each originating in its own Big Bang. Both these possibilities bear some relation to the old steady-state hypothesis, which holds that the overall density of matter in the universe remains always the same. A further variant is proposed by Lee Smolin, an astrophysicist who proposes that physics evolves through a kind of Darwinian evolution among universes. A new universe is created through the collapse into a black hole (see BLACK HOLES) of a star within an existing universe.

It would have slightly varied values of physical constants, thus allowing the universe to evolve. (See THE ANTHROPIC PRINCIPLE.)

Stephen Hawking has yet another tack for getting around the problem of the Planck era. He suggests that the universe did not begin as an infinitely hot, dense singularity, but rather as a kind of "blur," like the North Pole on earth. If we leave out the problem of an infinite singularity, we can apply inflation theory from the beginning, except that a Heisenbergian blur would require a theory of quantum gravity to deal with the uncertain blotches of space-time, and we don't have one. Thus, though Hawking's approach would allow the whole history of the universe to be discussed in terms of the finite laws of physics, it is in practice untestable, and therefore more philosophy than science.

A further refinement of the Hawking idea is the theory that the universe was created out of "nothing." The hypothesis here is that the origin of the universe resulted from a quantum fluctuation of zero total energy. The energy released by expansion during the inflationary era was precisely equal to the mass-energy of the contents of the universe today. This idea has its attractions, but the initial state could not really have been "nothing." It may have had zero total energy, but there was still a quantum field in which the fluctuation could take place, and there would still be laws of physics governing this quantum field.

All difficulties considered, perhaps the sort of scientific explanation for the origin of the universe that requires some *further* fact or theory can *never* be complete. All such explanations break down somewhere. Might we, at some point, have to fall back on philosophy or theology for our explanations? Or simply admit that some things are beyond our capacity to understand? (See THEORIES OF EVERYTHING.)

# Planck's Constant

Planck's constant is one of the two most important constants in the whole of modern physics, the other being the speed of light.

Max Planck was one of the early founding fathers of quantum physics. His main contributions were the theory that electromagnetic radiation happens in discrete quanta, and the discovery that the size of each quantum is associated with a universal constant, a physical ratio or proportion that stays the same in all circumstances and all frames of reference. Today we associate quanta with any periodic or wavelike process, e.g., oscillations or the orbits of electrons in atoms.

Planck's constant is the ratio of a particle's energy to its frequency. Mathematically, this is written $\hbar = E/f$, where $\hbar$ is the symbol for the constant. Thus, if a particle's frequency increases, its energy must also increase. If its frequency decreases, its energy will decrease. But Planck's constant always stays the same and is always equal to one quantum.

The constancy of $\hbar$ can be appreciated in the effect known as the Doppler shift. This is observed when, say, a train whistle appears to drop its note as it passes us because fewer vibrations per second pass. If we view a photon traveling in one frame of reference, say the earth, from a vantage point in another, a plane, the apparent color of the photon will depend on the relative speeds of the reference frames. The perceived color will be more red if they are moving apart, more blue if they are approaching each other. The perceived frequency will change in the same way. But Planck's constant, the ratio of color (energy) to the frequency, will remain absolute, providing a stable reference point in an otherwise relative world. (See PERSPECTIVE AND INTERACTION; SPECIAL RELATIVITY.)

# Plasma

Plasmas are found in the atmospheres of stars as well as in interstellar space, electrical arcs, nuclear fusion devices, and hydrogen bombs. They have been called "the fourth state of matter" to contrast them with solids, liquids, and gases. Although on earth they are found only in the artificial conditions of the laboratory, the vast majority of matter in the universe exists in the plasma state.

The temperature inside a plasma is so high that outer electrons are stripped from atoms, leaving interpenetrating clouds of positively charged nuclei and negatively charged electrons. Because of these electric charges, plasmas behave quite differently from gases. The molecules in a gas are essentially free, experiencing their neighbors only when they collide. The electrical forces in a plasma are long-range and have a profound effect on its behavior.

Scientists attempting to control nuclear fusion as the power source of the future use hot plasmas. Nuclear fusion occurs when nuclei, such as deuterium (heavy hydrogen), collide with sufficient violence to overcome their mutual repulsion. They fuse together to form a new atom and release energy in the process. These fusion reactions take place only at the extremely high temperatures within a plasma. The difficulty in harnessing nuclear fusion as a power source is keeping the high-temperature plasma confined long enough for the fusion reaction to take place. Because the plasma is electrically charged, one solution is to confine it within a "magnetic bottle"—a specially shaped magnetic field that deflects the plasma particles as they try to escape, focusing them back into the region where the fusion reaction is taking place. (See also COLD FUSION.)

In the 1950s, the physicist David Bohm suggested that the "gas" of free electrons within the positively charged lattice of a metal acts as a plasma. Bohm developed a way of describing plasmas that emphasizes their dual collective and individual natures. The electrical force between two electrons acts over an extremely long range, but in a plasma electrically charged particles rearrange themselves around an electron to screen out the effects of its charge. At one level, therefore, electrons experience only the short-range, screened influence of their neighbors and behave like free electrons, or molecules in a gas. Yet their freedom arises from the fact that they are also able to group themselves so as to screen out the effects of electrical charges. Not only are electrons free individuals, they also contribute to an overall collective behavior—the plasma vibration. (See COHERENCE.)

The unique properties of plasmas arise because free individual motion is made possible by collective response, and collective vibrations are made possible in turn by the motions of many free individuals. In their complicated order, plasmas are like a human society, one in which the individual is free yet responds to the overall values of the group, values in turn created by free individuals.

# Predator-Prey

The constant duel for existence between predator and prey is a striking example of the way interacting systems can settle down into the stable, oscillating behavior called a limit cycle. (See SYSTEMS THEORY.)

Imagine a lake stocked with trout. After many years the trout population settles down into a stable equilibrium determined by the size of the lake and the amount of food available. Introduce a few pike into the lake, and they gorge themselves on a plentiful food supply of trout. The pike and many of their offspring survive. Within a few years the number of trout has dropped significantly.

With a diminishing food supply, the pike begin to die out. With fewer predators, the trout begin to grow back. As the trout population peaks, the number of pike begins to grow. Given suitable rates of birth and death, pike and trout follow each other in an endless circle of growth and decline.

Such a nonlinear limit cycle (see NONLINEARITY) is extremely stable under certain conditions. If a few more trout are added, they will be eaten by the pike; the overall perturbation is soon smoothed out, and the same oscillations in population occur year after year. It is as if the system is being attracted back to its habitual cycle. For this reason its behavior is said to be determined by a limit-cycle attractor. (See ATTRACTORS.)

Limit cycles of the predator-prey type occur whenever two systems compete together. The records of the Hudson's Bay Company show oscillations in the number of snowshoe hare and lynx pelts brought into factors' posts. Again, the two species were locked together in a limit cycle. The behavior of electrical activity in the brain and fluctuations in economic systems are also governed by stable limit cycles.

Additional factors can be added to a limit cycle to produce even more complicated cycles within cycles. As anglers begin to visit the trout lake, they compete with the pike. When the lake's stock falls, persistent anglers still turn up, but eventually the lake gets a poor reputation. It is only some time later, when the trout population has been reestablished, that the anglers return. The result is a cycle within a cycle. Another case would occur if the pike were efficient enough hunters to exterminate the trout. Then they, too, would die out. This is not a limit cycle. Overfishing sometimes has this kind of result.

# Process

Recent movements in the natural sciences lean toward a process philosophy of nature, a movement from Being to BECOMING. Newtonian science has a mechanistic, reductionist view of the universe in which transformations are explained in terms of rearrangements of material building blocks.

Quantum theory, with its emphasis on the role of the observer and the way that certainty is replaced by ambiguity, has done much to counteract this mechanistic viewpoint. In addition, the new sciences of complexity, nonlinearity, and self-organization embrace a vision of nature in which fixed forms are replaced by flux and change. Some scientists have even begun to question the inviolate nature of physical law. Traditionally, the laws of nature are absolute, existing before space, time, and matter and imposing their form on an emerging universe. But could these laws also be part of a wider evolutionary process? Is it possible that not only systems but the laws that describe them are self-structuring?

Earlier thinking pictured the world as built of fundamental entities, with complexity arising through the conjunction of simpler parts. Order was fundamental; chaos was the breakdown of order. By contrast, NONLINEARITY and chaos theory (see CHAOS AND SELF-ORGANIZATION) offer images more in accord with a process philosophy; order emerges out of chaos, and the creation of novelty is characteristic of OPEN SYSTEMS.

Process philosophy is as old as recorded thought. Near Eastern creation myths picture the deity creating order out of primordial chaos. While some Greek philosophers sought the origin of the world in a fundamental element, such as earth or fire, Heraclitus viewed the world as emerging out of flux. Early in the twentieth century, the philosopher Alfred North Whitehead argued for a process philosophy in his book *Process and Reality.* Our bodies are processes that retain their structure, although all the individual molecules are replaced every few years. More recently, process philosophy has been the basis of the approaches of such scientists as David Bohm and Ilya Prigogine.

Bohm stresses the process nature of quantum theory and the need for related mathematical approaches. His IMPLICATE ORDER is based on pure activity and transformation—the Holomovement. Bohm also draws attention to the way our predominantly noun-oriented languages—all European languages—incline us to see the world in terms of objects, boundaries, and categories, rather than processes and transformations.

Ilya Prigogine emphasizes the dynamic nature of time, which he sees as an animating force within nature. His approach to physics, biology, chemistry, sociology, and economics draws upon notions of self-organization and the emergence of novelty.

QUANTUM FIELD THEORY and SPECIAL RELATIVITY redescribe particles as evanescent processes, like ripples on a river.

# Psychiatry

Psychiatry is a branch of medicine practiced by qualified doctors who are concerned primarily with the physical causes of mental disturbance and with physical methods for treating them. Psychiatry differs from both psychology, which is an attempt at pure science, and psychotherapy, which tries to help people by way of analysis or conversation. (See PSYCHODYNAMICS AND PSYCHOTHERAPY; PSYCHOLOGY IN THE TWENTIETH CENTURY.)

The ancient Greeks were aware that some people suffer episodes of insanity or depression. They usually attributed these to possession by gods or evil spirits, and they felt that the appropriate treatment was prayer, dedication to some ritual, or physical restraint. The Greek doctor Hippocrates held the much more modern notion that mental disorders are diseases of the brain, to be treated with rest and quiet or with herbal remedies. There was little theoretical advance on either of these positions until the past century.

The great advance in medical knowledge in the twentieth century has shown that certain mental disturbances have clear physical causes. Brain injuries, Alzheimer's disease, epilepsy, mental deficiency, and the mental inertness due to an underactive thyroid gland all have physical origins. Other conditions—schizophrenia, paranoia, severe episodes of mania or depression, a tendency to become addicted to drugs or alcohol—certainly have a physical component, as is shown by studies of identical twins reared in different environments. In all these disturbances, an inherited tendency does not ensure that disease will follow. Environment, lifestyle, and even something as vague as attitude or willpower are known to play a part.

If we ask whether psychiatry is a science, we have to take the combination of physical and more general factors into account. In most psychiatric illness, physical treatment is only part of the answer. This is clearly borne out by antidepressant drugs, which treat some forms of depression successfully but often need supplementing by other measures.

Roughly 1 percent of the population suffers from severe "clinical depression"—that is, periodic depression with no clear emotional cause. This is usually dealt with effectively by antidepressant drugs, but the same drugs have little effect on depression resulting from bereavement, broken relationships, or business failure. In these cases, counseling (or psychotherapy), the attention of a sympathetic friend, or a good holiday seems more to the point.

While some successful psychiatric treatment depends upon drugs that affect relevant brain systems, too little is known about the brain to make the direct search for new treatments very feasible. Most existing treatments were discovered by chance. ECT (electroconvulsive therapy), which helps some severely depressed people, was originally tried as a treatment for schizophrenia. The first antidepressant drug was conceived as an antibiotic against tuberculosis and accidentally found to help clinical depression. The first useful antischizophrenic drug was developed as a remedy for hay fever.

## Psychodynamics and Psychotherapy

Psychodynamics is a model of personality that assumes there are several layers or "divisions" of the psyche that interact with each other through the push and pull of forces. Opposing forces lead to internal conflict that can give rise to neurosis or illness. Psychotherapy is a method for talking people through these conflicts in the hope of resolving them.

The first psychodynamic model was articulated by Sigmund Freud, who called it a "hydraulic model," in the spirit of Newtonian physics. In Freud's early work the psyche was divided between the conscious and the unconscious, which he later expanded to a division between the id and the ego (with its attendant superego), either of which might be partly conscious or unconscious. After Freud, there were scores of variants on his original model, stemming from the work of Carl Jung, Alfred Adler, Wilhelm Reich, and others. But all psychodynamic models have certain characteristic in common.

All of them hold that there is a primitive mental system, which

Freud called the id or "primary process." This is assumed to be the only mental system developed in infancy, and its dynamics are thought to function according to instinct and association. The baby has an instinctive need for food and associates the mother (or the breast) with the fulfillment or denial of this need. Freud himself recognized only two instincts, sex and aggression, but other psychodynamic theorists broadened the range to include many other sources of motivation. Animal ethologists recognized traits like maternal instinct and territoriality, while humanistic and transpersonal psychologists added social and even spiritual instincts or motivations. (See HUMANISTIC PSYCHOLOGY; TRANSPERSONAL PSYCHOLOGY.) All assume that the instinctive level of the psyche has its own "logic," or set of motives and associative principles. These are often unconscious but can become more, or less, conscious.

The id, or primary process, can now be described successfully in terms of the latest computer models of mind as functioning according to the dynamics of NEURAL NETWORKS, electronic processing circuits modeled on the brain's own neural networks.

All psychodynamic models also recognize a more conscious part of the mind, which Freud called the ego or "secondary process." This layer of the self controls speech and voluntary behavior and is assumed to function according to rational rules. In recent computer models of mind, the ego's rational functions are described in terms of FORMAL COMPUTATION, or serial processing, associated with the brain's one-to-one neural tracts and used to program personal computers. (See THINKING.)

Freud, and others following his example, assumed that, while some unconscious but potentially conscious id material is simply not noticed, other such material is actively *kept* unconscious, because the ego finds it too painful to face. In ignoring the painful material, the ego uses various defenses—e.g., repression, rationalization, or fantasy. There may be a temporary gain from such defenses, but ultimately a person can pay a high price in terms of restricted abilities, neurotic symptoms, or prolonged immaturity. The repressed material may appear indirectly in dreams, impulsive behavior, neurotic symptoms, or slips of the tongue.

While there was some idea of unconscious motivations before

Freud, the concept of active defenses against painful unconscious material was his most original contribution to psychodynamic theory. He arrived at the insight by studying patients suffering from hysteria and those who had undergone hypnosis. In both cases, an experience that might normally be conscious can be made unconscious through psychological means. Freud generalized these observations to apply them to the whole of mental life.

By now, many laboratory experiments have confirmed Freud's theory of active defenses. They have shown that subliminal (unconscious) stimulation can affect behavior, and events not consciously registered can raise blood pressure or alter skin resistance, both of which are signs of emotion. Other experiments demonstrate that unpleasant memories are often more quickly forgotten than neutral or pleasant memories. But Freud's more specific ideas about the primary importance of sexual or aggressive instincts, or of particular childhood experiences leading to the Oedipus complex, are not borne out by experimental research. Human motivation and learning seem to be as varied and complex as the proliferation of later psychodynamic models has suggested.

Psychotherapy can use any psychodynamic model or a combination of many as its theoretical foundation. The range of psychotherapies—there were more than 250 at last count—is greater than the range of psychodynamic models, but they all have in common the belief that a person can be talked through his or her conflicts. Classical Freudian or Kleinian analysis requires that the patient attend sessions daily for years, but the revolution in psychiatric drugs such as tranquilizers and antidepressants now offers powerful competition.

The classical psychotherapeutic method of dealing with defenses against unconscious material encourages the patient to "free-associate"—to let one thought follow another. From this, the therapist perceives gaps or distortions in the patient's thoughts or feelings—about other people, about dream material, or about the therapist ("transference"). The therapist comments on these gaps (makes an "interpretation") in a way intended to lead the patient toward insight.

Research suggests that psychotherapy is a helpful treatment for some people, but by no means a cure-all. The result seems to depend as much on the patient's determination to make progress and on the personality, sensitivity, and caring attitude of the therapist as on any particular theory. Some therapists achieve good results; others make their patients worse.

The authoritative Freudian analyst, who always knows better than his patient, is an increasingly unbelievable figure. A more two-way therapeutic process, based on dialogue between patient and therapist, is used increasingly in humanistic psychology. Greater numbers of counselors and eclectic therapists use a mixture of methods drawn from various kinds of therapy and from humanistic and transpersonal psychology. All seem to give pretty much similar results.

The psychodynamic discovery that parts of the mind can be split off from ego consciousness while continuing to function and to influence behavior raises problems for some philosophical ideas about personal identity and consciousness. (See CONSCIOUSNESS, TOWARD A SCIENCE OF.) How can we account for such events if the self is a unified, particlelike soul or psyche, as in some philosophical (and religious) models? Identifying psychodynamics with certain kinds of brain function and with the different kinds of neural processing now known to exist in the brain perhaps makes more sense. (See also SPLIT-BRAIN PHENOMENA.)

# Psychology in the Twentieth Century

Psychology, in one important sense, is as old as the human race. It is used for the implicit or intuitive understanding of other people, and its insights appear throughout world myths and literature. But in this ancient sense, it was not a systematic science. (See overview essay C, THE NEW SCIENCES OF THE MIND.)

Psychology as a natural sciences began in 1879 when Wilhelm Wundt opened the first psychological laboratory in Germany. It studied perception and association experimentally, though with a large reliance on introspection. The basic question was to ask: What are the simplest elements (John Locke's "ideas") of which consciousness is composed? Soon afterward, Ivan Petrovich Pavlov's physiological studies of conditioned reflexes in dogs offered a more precise and objective way of investigating one way in which stimulus and response are associated. Experimental psychology had begun.

The two towering figures of twentieth-century psychology are William James and Sigmund Freud. With his own training and background in chemistry, physiology, and medicine as well as philosophy, James was keen to see psychology develop as a natural science. His rigorous collection of data about both behavior and inner experience was a real contribution toward giving this new science a solid, factual foundation. But perhaps more significant still, James was not a reductionist. He adopted the scientific method, based on observation and experiment, but at the same time he rejected the atomism and determinism of Newtonian physics.

James felt that mechanistic physics was applicable neither to human experience nor to the physical world. The basic stuff of the world, in his view, was neither mental nor physical but rather something prior to such distinctions. Since, he argued, we have an experience of wholes as well as an experience of flowing time, reality itself must have some holistic, processlike aspect. Since we experience freedom of the will, the world must have at least some indeterminate aspect. And since we experience a sense of purpose, basic reality must have an evolutionary aspect. James believed, in short, that the categories of human experience reflected features of the outside world. Many of the features he ascribed to physical reality (holism, indeterminism, evolutionary process), though incompatible with Newtonian physics, were later to emerge in the new physics of quantum mechanics. In important ways, therefore, James's philosophy and psychology could be seen as harbingers of the new paradigm in physics. (See QUANTUM PHYSICS.)

Freud had a much darker, decidedly older paradigm vision of human psychology, though his work in exploring the unconscious and the structure and dynamics of the psyche was undoubtedly groundbreaking. Freud very much admired Newton's work in physics and consciously set out to emulate it in psychology. He wanted to articulate a new "scientific psychology" whose laws would mirror those of physics and chemistry. The basis of psychodynamics, he believed, was the blind and deterministic forces of the instinctive id. (See PSYCHODYNAMICS AND PSYCHOTHERAPY.) Human beings were driven by forces of sex and aggression, and rigidly determined in their later development and behavior by early childhood experience. The psyche was a battlefield in which the dark forces of the id were locked in battle with the repressive forces of the ego. This mechanistic, reductionist vision left Freud no scope to accommodate reason, ethics, or higher spiritual values.

Freud was to be the dominant influence in the development of the whole tradition of psychodynamics and psychotherapy and the enormous pool of "popular psychology" they have spawned. His creative vision altered, perhaps forever, the way we think of ourselves, although many of those who have followed him disagreed vehemently with his specific ideas and methods. William James, on the other hand, directly inspired much that was to come in the development of experimental psychology, as well as being a strong influence on TRANSPERSONAL PSYCHOLOGY and HUMANISTIC PSYCHOLOGY.

The dominant tradition in academic psychology today is experimental and remains as close as possible to the thinking and traditions of hard science. Some branches study the effect of human beings acting on the environment, by way of motivation and behavior, while other branches concentrate on the effect of the environment on human beings, by way of PERCEPTION, MEMORY, and the structure of THINKING made possible by the physical brain. (See also BEHAVIORISM; COMPUTATIONAL PSYCHOLOGY; GESTALT AND COGNITIVE PSYCHOLOGY; PSYCHIATRY.)

The question of whether psychology as a whole can ever be a fundamental science like physics remains. The material with which it deals will always be partly subjective. Most psychological findings about perception or meaning, for example, could be explained by a wide range of differing brain mechanisms. To describe which mechanisms are relevant, we need a knowledge of NEUROSCIENCE. In fact, to become proper subjects of hard science, all psychological findings must be unifiable with the discoveries of neuroscience, biochemistry, and even, perhaps, fundamental physics.

For some aspects of psychology, this union with the brain sciences could be difficult. Things like consciousness and intention, for example, look very familiar, even simple, but from a purely scientific viewpoint it is difficult to imagine how any known brain mechanism could

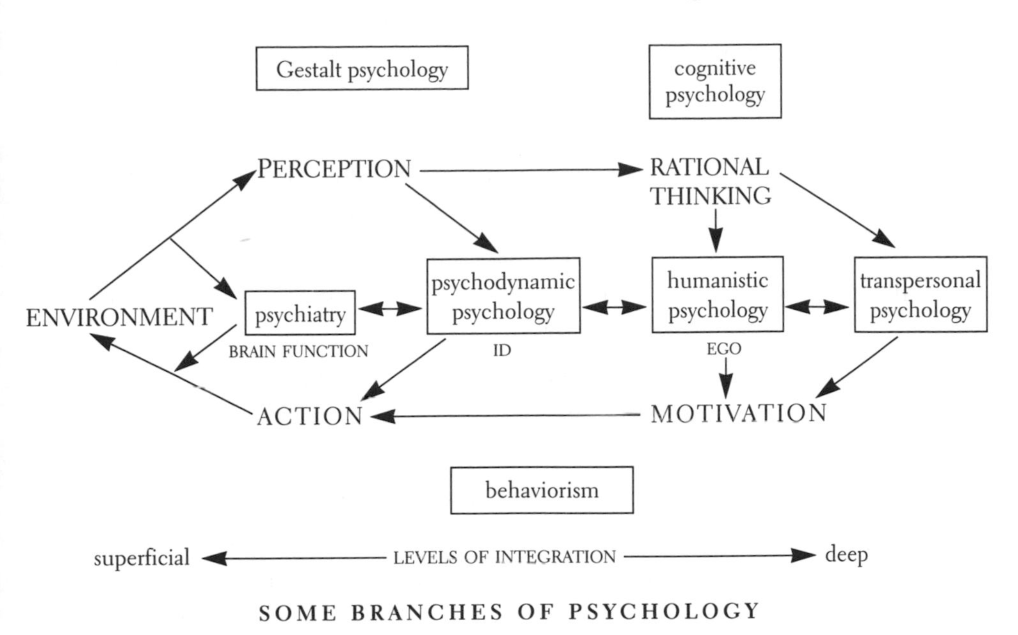


SOME BRANCHES OF PSYCHOLOGY

produce them. Without this ability to rigorously correlate experience and behavior with brain function, psychology remains analogous to natural history. Its practitioners usefully collect and classify "specimens" of experience and behavior but lack adequate tools for theoretical explanation. Their data "from the field" frames the questions that the harder sciences of mind must answer.

# Punctuated Equilibrium

Does evolution proceed in a gradual fashion, or through a series of sudden leaps called punctuated equilibrium? Neo-Darwinians believe that individual species change through the accumulation of a series of small adaptations. After England's industrial revolution, for example, a dark variety of peppered moth began to supplant the previously more common light variety. According to the neo-Darwinian view, while the genes for the dark variety had already been present, protective coloration had previously given the light variety an evolutionary advantage by allowing it to merge with tree bark. But in regions where trees had become blackened by industrial soot, the dark variety now had the advantage and came to predominate.

According to conventional evolutionary theory, the gradual accumulation of many such adaptations allows plants and animals to change slowly, to the extent that an entirely new species may eventually develop. Not everyone accepts this explanation. Looking at the fossil record, the biologist Stephen Jay Gould pointed out that individual species look more or less the same, with only minor variations and adaptations, from their first appearance until their eventual extinction. By contrast, new species appear relatively rapidly and more or less in their final forms. There appear to be no "missing links" between species, no long evolutionary periods in which plants or animals intermediate between one species and another had time to flourish. This led Gould and his coworkers to propose punctuated equilibrium, the idea that the evolution of a new species occurs in a sudden burst. Critics of the theory argue that no convincing explanation has been given for the biological mechanism whereby these evolutionary jumps are supposed to take place. There is, however, a possible model in terms of self-organization in biology. (See CHAOS AND SELF-ORGANIZATION; DARWINIAN EVOLUTION; FEEDBACK; INTERMITTENCY.)

# Quantum

The word *quantum,* etymologically derived from *quantity,* refers to a small bundle or packet of action or process, the smallest unit of either that can be associated with a single event in the microworld.

When an electron moves from one energy orbit to another within the atom, it does so in units of action that can be measured as so many quanta. Quanta are not divisible. No movement of a particle from one state to another ever uses up one and a half quanta, or three quarters of a quantum. Thus the term *quantum leap*—an abrupt movement from one discrete energy level to another, with no smooth transition in between. It is like a person jumping down a mountainside on a path into which discrete steps have been carved. He or she is forbidden to place a foot anywhere in between. Indeed, to make the analogy accurate, there can't *be* any in between.

Nearly every constituent of the microworld—and hence of the physical world as a whole—is spinning like a tiny top. Anything that spins has angular momentum, that is, the amount of energy that would be required to stop it from spinning over a period of time. Literally, a quantum is a unit of spin or of angular momentum.

We can get a feel for quanta and quantum leaps by imagining the action of a spinning top. It does not simply spin slower and slower in a smooth, continuous way. It goes *click, click, click,* as though jumping from one cog on a wheel of descending angular momentum to another. This is true whether the "spinning top" is an ice skater, a child's toy, or a tiny electron. All spinning motion is quantized, or "lumpy." The popular use of the term *quantum leap* has come to mean any sort of abrupt or gigantic change, a change out of all proportion to what came before. This is fairly true to its origins in the microworld, but quantum leaps need not be gigantic, just discontinuous. They are also indeterminate, unpredictable from what came before. (See INDETERMINACY.)

The number of quanta used to spin is proportional to the mass of the spinning object. A child's toy or an ice skater uses many trillions of quanta while spinning, whereas a tiny electron spinning around the lowest-energy orbit of an atomic nucleus uses just one quantum. Thus the *click, click* descending motion of the top or the ice skater is imperceptible because it is so small compared to the overall angular momentum. But the quantum leap of a single electron would be very dramatic if we could see on that scale.

Quanta are associated with the wave nature of matter, and we can visualize them by thinking of an undulating wave. A bound wave, e.g., an electron in an atomic orbit, with just one undulation—one peak and one trough—represents one quantum of action. A wave with ten undulations represents ten quanta. An everyday example of a bound wave would be a plucked guitar string, which is held still at its two ends and thus can vibrate only at certain "quantized" frequencies. Every photon, or freely moving particle of light, has an action of just one quantum per undulation, and this indeed is part of the definition of a quantum—the capacity for action of a single photon. One quantum is also equal to PLANCK'S CONSTANT, the ratio between the energy of a spinning particle and the frequency of its rotation.

The quantization, or "lumpiness and jumpiness," of action as depicted in quantum physics marked one of the new physics' sharp breaks with the Newtonian paradigm. Classical physics represented motion as smooth, continuous change over time, and energy as increasing or decreasing in a continuous spectrum. The existence of quanta now makes it possible to understand color (see COLOR—WHAT IS IT?), which could not be understood in classical terms.

Conceptually, quanta are in the same tradition as ancient Greek ATOMISM, which argued that atoms are the smallest discrete building blocks of nature. Quanta represent the discrete or atomistic aspect of matter, a complement to its periodic or wavelike aspect. (See WAVE/PARTICLE DUALITY.) Indeed, in the QUANTUM FIELD THEORY, an extension of QUANTUM PHYSICS into high-energy phenomena, "atoms" or particles *are* regarded as the quanta of an underlying vibrating field.

# Quantum Chromodynamics

Quantum chromodynamics (QCD) has arisen since discovery of a "color" force that binds quarks together to form heavier elementary particles.

Fifty years ago, the atomic nucleus was thought to be composed of two kinds of fundamental particles, protons and neutrons. It was believed the nucleus does not fly apart, despite the mutual electrical repulsion of protons, because protons and neutrons attract each other through the strong nuclear force, whose quanta are pi-mesons. Since this force is short range and fades away at distances larger than the diameter of a large nucleus, there is a natural constraint on the maximum size of a stable nucleus.

But since this earlier picture, more than 200 additional "elementary" particles have been discovered in particle accelerator experiments. Details of the "strong nuclear force" have proved to be very complicated and a more fundamental, simpler kind of analysis has been formulated to bring some order into all this diversity. HADRONS, heavy "elementary" particles like protons and neutrons, are now known to consist of three QUARKS, held together by the color force, which acts between them. There are six "flavors" of quarks in all, arranged in three "generations," the two lightest flavors being "up" (electric charge $+2/3$) and "down" (electric charge $-1/3$). (See overview essay D, THE COSMIC CANOPY.)

Quarks also have another kind of charge, like an electric charge, but more complicated. Each quark has one of three color charges, arbitrarily called red, blue, and green. The rule is that like charges repel; different charges attract. Thus the most stable combination is one red, one blue, and one green quark of the two lightest flavors, up and down, making up a baryon like a proton (two up and one down) or a neutron (one up and two down). Antiquarks have anticolors.

QCD was named by analogy to QUANTUM ELECTRODYNAMICS (QED). In QED there are only two electric charges: positive and negative. In QCD there are the three color charges: red, blue, and green. This terminology is used to suggest that the only stable combination is "white"—i.e., made of the three primary colors.

Actually, even this picture is too specific. Quarks do not have fixed colors or flavors; the color force and the weak nuclear force continually transform and exchange their attributes. A quark has no individual properties at all; it is totally caught up in a kind of quantum "dance" with its partners, existing only in relationship. Quarks are not found in isolation ("quark confinement"). But in collisions a proton or neutron appears to contain lumps called partons, which are thought to be the quarks. The mathematics of QCD quickly becomes horrendous.

When QCD is put together with electroweak theory (see THE ELECTROWEAK FORCE), the result is THE STANDARD MODEL of particle physics.

# Quantum Electrodynamics

The quantum theory created by Heisenberg and Schrödinger explained the behavior of matter, like electrons and protons, in atoms and molecules. While the behavior of an electron was quantized, it was still assumed to interact via classical electrical and magnetic forces. The next step was to extend quantum theory to fields of force themselves, beginning with the electromagnetic field.

In Maxwell's picture of the classical electromagnetic field, magnetic oscillations produce electrical oscillations, and these electrical oscillations in turn create magnetic fluctuations. The field is self-sustaining, and its vibrations travel with the speed of light. Depending on their frequency, they take the form of radio waves, infrared, visible light, ultraviolet waves, and so on. The quantum version of this theory involves quantizing the vibrations to produce photons—quanta of energy.

In classical physics, charged bodies interact via the electrical field. In the quantum version, two electrons are constantly interchanging photons, like soccer players running downfield and passing the ball to each other. Each interchange produces a small shift in the electron's momentum, and the net result of this interchange is that two negatively charged electrons repel each other. An electron is likewise attracted to a positively charged proton. The photons exchanged in this fashion are called VIRTUAL PARTICLES since, unlike light, they lack the energy to propagate far across space. They are, in a sense, borrowed from the quantum field and then paid back. The study of this total system of electrons plus photon field is called quantum electrodynamics, or QED. Its new predictions are well supported by experiment.

In a classical world, the lowest energy state of an oscillating string or, for that matter, a vibrating field is one of absolute rest. In this case, the energy of oscillation falls to zero. Things are quite different for a quantum system, since the lowest energy level of the quantum field still exhibits what is called zero-point vibration. Even at absolute zero, with all available energy removed, the quantum field remains in a state of quantum vibration. Because of HEISENBERG'S UNCERTAINTY PRINCIPLE, the field's position and momentum cannot both be exactly zero. Nevertheless this is still called its ground or vacuum state, since the system has nowhere else to go. (See THE QUANTUM VACUUM.) It can never get rid of this final energy and come to rest.

When the energy of the zero-point vibrations is added up over all possible frequencies of vibration, it turns out to be infinite. This infinite energy of the vacuum state represents one of the outstanding problems facing quantum field theory. There is more energy contained in even 1 cubic centimeter of the vacuum than exists in all the matter of the observable universe. The material universe is no more than an insubstantial cloud compared to the underlying ground state. But is this really the case? Are the infinities an inevitable consequence of the quantum world, or simply a defect within a theory of only limited validity?

Infinities within QED also arise in yet another way. To calculate the energy of an electron, physicists have to take into account the electron's interaction with its own electromagnetic field. This also turns out to be infinite. For several decades, physicists have been working on a variety of theories to eliminate the infinities and divergences from QED. (See SELF-ENERGY.)

# Quantum Field Theory

Quantum field theory (QFT) is an extension of quantum theory into the realm of high-energy physics. Instead of treating particles as solid substances, as in standard quantum theory, quantum field theory regards all particles as excitations of some underlying field. All entities resemble patterns or ripples on a "pond" of energy, rather than being seen as solid, individual lumps. The universe itself is described as consisting of so many patterns of dynamic energy.

In any system in elementary quantum physics, the number and type of particles present are fixed. All that changes is their movements.

But high-energy phenomena do not work in this way. According to SPECIAL RELATIVITY, mass and energy are interchangeable. An electron and a positron can meet and annihilate themselves into two gamma-ray photons. Removed from the nucleus, a neutron decays in about ten minutes into three other particles. To accommodate such transformations, quantum theory must be extended to provide a conceptual framework in which particles are not basic.

Quantum field theory is now the established theoretical framework in all branches of particle physics. Since all entities are viewed as excitations of fields, a photon is seen as a quantum of the electromagnetic field, an electron as a quantum of the electron field, and so on. A particular state of the electromagnetic field might, on measurement, be found to contain zero, one, two, or any number of photons, each possibility with its own contribution to the wave function (a mathematical description of a spread of possible quantum states). (See THE MEASUREMENT PROBLEM.)

A field is analogous to a guitar string or a jelly; its possible excitations are its possible modes of vibration. These modes can be superimposed, giving several at once. If we add some form of interaction with other fields, corresponding to plucking the string, we have a (classical) field theory. This becomes a *quantum* field theory if we add quantization—the WAVE/PARTICLE DUALITY. This is a framework in which given particles can be transformed into other forms of energy.

Schrödinger's equation, though a good description of ordinary quantum phenomena, cannot correctly describe the high-energy domain. (See THE WAVE FUNCTION AND SCHRÖDINGER'S EQUATION.) In consequence, there are several quantum field theories for different sorts of high-energy particles—QUANTUM ELECTRODYNAMICS (QED) for photons, Paul Dirac's equation for fermions (particles of matter), the Klein-Gordon equation for massive spinless bosons (particles of force), and so on. All these are compatible with Special Relativity. The first quantum field theory to be fully worked out was QED, for which the American physicists Richard Feynman and Julian Schwinger and the Japanese Tomonaga Shin'ichiro received the Nobel Prize in 1965. All subsequent quantum field theories have modeled themselves after QED.

From Dirac's equation in quantum field theory it follows that all

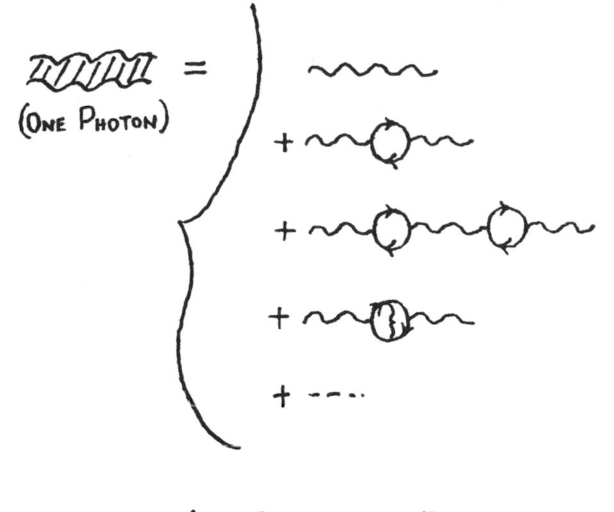


A FEYNMAN DIAGRAM

fermions should have antiparticles with the same mass but opposite electric charge—i.e., electron and positron, neutrino and antineutrino, proton and antiproton. This prediction has been amply confirmed. It can also be proved that all particles with integral spin obey Bose-Einstein statistics, whereas all particles with half-integral spin obey Fermi-Dirac statistics. (See SPIN AND STATISTICS.)

Most quantum field theory equations cannot be solved precisely. They allow an infinite number of possible virtual subprocesses within any given process. (See THE QUANTUM VACUUM.) However, a very good answer can often be obtained by considering only the most important subprocesses. This is easier to represent by using Feynman diagrams rather than equations. For example, the diagram above represents some of the processes in which a photon (the wavy line) can "propagate." This in turn can become an electron-positron pair and then change back again. Meanwhile, one of the pair may emit and reabsorb a photon, and so on.

The infinities that can be generated by the equations are a worrying feature of quantum field theory. They are dealt with by ignoring or subtracting them and concentrating simply on the *changes* in an electron's energy—a mathematical procedure known as renormalization. But the infinities remain troublesome and suggest that we may one day have a still more accurate theory. The theory of SUPERSTRINGS is an attempt at this.

Philosophically, quantum field theory has much in common with Alfred North Whitehead's PROCESS philosophy, with the early descriptions of matter by Heraclitus, and with Buddhism. Where entities resemble ripples on a pond, we begin to ask questions like: Is the ripple now lapping the shore the same ripple made by the passing boat? Questions of identity for which there are no firm answers arise. (See BECOMING; IDENTITY IN QUANTUM MECHANICS.) In this sense, quantum field theory moves physics into a new paradigm, away from the more fixed philosophies of Plato and Aristotle that have dominated Western thinking.

# Quantum Gravity

Physicists hope to account for all the particles and forces of nature within a single scheme. Gravity, a ubiquitous force of nature, must naturally be included in the final theory. At this point, GRAND UNIFIED THEORIES must move into the much deeper waters of THE PLANCK ERA and cope with the problem of reconciling the pillars of modern physics, QUANTUM PHYSICS and GENERAL RELATIVITY.

In the early days of quantum theory, physicists hoped that it would be possible to bring gravity into the quantum fold, to write down the equations for the classical gravitational field and set about quantizing it the way physicists had quantized the electromagnetic field. Because of a number of serious technical and mathematical difficulties, this simply would not work.

Considerable technical difficulties are being encountered today as physicists attempt to push ahead with a new generation of unified theories. Have we now reached a level where nature can be read only in terms of complicated mathematics, or are we simply no longer asking the correct questions?

The problem with "quantizing gravity" is that it puts the cart before the horse. It tries to distill a quantum extract out of General Relativity without ever confronting the deep incompatibility between the two theories. Despite half a century of work by some of the best minds in physics, science seems no closer to bringing quantum theory and relativity together. Placed side by side, the two theories contain such mutual inconsistencies that they can never be welded together. The solution must lie in a new, much deeper theory, out of which both quantum theory and General Relativity emerge.

A key concept in relatively is a signal—a light ray connecting one event to another. Space, both theoretically and in practical astronomy, is mapped out or defined by the crisscrossing of light rays. Without this concept it would be difficult even to begin discussing relativity. Yet the notion of a signal evaporates in quantum theory. For a signal to make sense, there has to be a distinct difference between a well-defined sender and a receiver, together with a message that passes between them. In every quantum event, observer and observed are so irreducibly linked that the notion of a signal between them becomes dubious. Defining space in relativity depends upon having accurate clocks and measuring rods. But the most accurate clocks are atomic devices, and the most accurate systems of measurement use radar and lasers. So relativity seems to demand the preexistence of quantum theory. But to define a quantum state, it is necessary to have a rigid, well-defined laboratory apparatus. Quantum theory presupposes the existence of such classical objects, yet when one moves to relativity, the notion of a rigid body cannot be maintained. Each theory seems to depend on the other, but in ways that are deeply incompatible.

Relativity is formulated by continuous differential equations—it assumes the continuity of space and time down to the level of the dimensionless point. Quantum theory indicates that space-time must break down into a foamlike structure before this limit is reached. (See THE PLANCK ERA.) Another difficulty is that nonlocality is an essential feature of quantum theory, yet its spirit is incompatible with relativity.

Considerations like these suggest that it will never be possible to unify quantum theory and relativity in their present forms. A new and deeper theory is needed. But it cannot simply be thought up by an enterprising physicist. It demands new physical insights beyond anything in the current theories.

The basic order of physics remains Cartesian, an order that predates even Newton. The laws of physics are expressed as differential equations, fields are defined on a continuous space-time, and Cartesian coordinates are the lingua franca for all equations.

There is general agreement that physics can no longer assume continuous space-time as the backdrop for all its theories. Rather, space-time and gravity must emerge out of an underlying theory. As John Wheeler puts it:

> Day One: Quantum Principle
> Day Two: Space-time

Day Two still seems a long way off for a true Theory of Everything. (See THEORIES OF EVERYTHING.) It will not be merely a matter of making adjustments to the current Standard Model. For any comprehensive theory to accommodate gravity, the face of physics itself may well have to be changed.

A semiclassical theory of gravity has been patched together. The background space-time is treated in the usual classical way, while matter and gravitational waves are treated as quantum fields. The resulting approximation is good enough, so long as gravity and its consequent distortion of Euclidean space-time are not too strong. Useful results have been achieved from it—for example, Hawking's radiation from BLACK HOLES. (See VIRTUAL PARTICLES.) But this hybrid theory turns out not to be subject to renormalization. (See SELF-ENERGY.) It cannot become a full quantum theory of gravity. For that, an entirely different approach is needed. Current candidates are SUPERSYMMETRY and SUPERSTRINGS.

# A Quantum Hussy

Think of a sheltered young lady who is about to be introduced to society for the first time at her coming-out ball. The girl is excited to find herself surrounded by dozens of would-be suitors, each of whom presents his card and requests that she accompany him on a date. A whole new world of possibilities has suddenly opened for her, and she naturally wants eventually to realize her greatest potential for a happy marriage to the man of her dreams.

In the real world, the world of everyday reality, the girl would have to establish a social list to explore each of her possibilities, one by one. She would date each of her suitors in turn, perhaps dating some several times before feeling certain she could settle on the right one. But for

a quantum debutante, things would follow a very different course. The dizzy girl would date *all* her suitors at once. She might even decide to set up house with them all simultaneously, and if her scandalized parents wished to protest, they would have to contact her at all her many addresses, all at once, because she would be living at all of them. If the girl wished, she could stand on the balconies of all of her many love nests and wave at herself!

In the end, having explored all her possibilities, the girl would eventually settle down with just one of her suitors for a lifetime of monogamous bliss, but not without having left traces of herself. Many people living in the city would remember having met her in the street, or would at least have a sense of déjà vu about her, although they could not remember exactly where she had lived or what she had been wearing. If nature followed its course, any of her simultaneous liaisons might have led to offspring, a composite image of which would also be vaguely remembered by many, although a more definite version was now at the toddler stage.

The case of the quantum hussy may seem extreme or farfetched, but it in fact illustrates the *both/and* nature of quantum reality. When quantum systems evolve, they throw out possibilities in every direction at once. Each possibility is a future direction the system might take, or a future state into which it might change. The possibilities or "trial runs" into a future direction or state are called SUPERPOSITIONS, and there may be an infinite number of them, some contradicting or canceling others out, but all happening at once.

When an electron in one of an atom's energy orbits is about to move into another orbit, it first "tests the waters" by making temporary transitions into *all possible* alternative orbits, all at once. For a small but nonetheless discrete amount of time, the electron is smeared all over space and time and is to be found everywhere and everywhen. Similarly, Schrödinger's equation, which describes any quantum system, contains an infinite array of simultaneous but often mutually contradictory possibilities, each of which describes a future state into which that system might collapse. (See COLLAPSE OF THE WAVE FUNCTION; THE WAVE FUNCTION AND SCHRÖDINGER'S EQUATION.) In the PROLOGUE, we saw this illustrated by the case of Schrödinger's cat—the cat's wave function carried both the possibility of aliveness and the possibility of deadness *at the same time*. And for a time, both possibilities were *real*—the cat in its quantum state was alive *and* dead.

The coexistence of many, often mutually contradictory, possible movements or possible states is a characteristic feature of quantum reality. But so, too, is the "real" nature of the possible. Just as the fantasies or temptations we entertain in our imaginations often have a real effect on our own or others' behavior, the quantum hussy's "possible" liaisons might lead to real children, as an electron making a temporary transition may collide with another particle, and that particle will remain off course ever after. (See VIRTUAL TRANSITIONS.)

In the two-slit experiment, each photon goes through both slits at once but ends at one place on the screen. (See CONTEXTUALISM.) The quantum hussy's toddler, likewise, resulted from multiple liaisons superposed.

# Quantum Physics

Quantum physics is as much a new way of looking at the world as it is a new science. It makes very accurate and very unexpected predictions about the behavior of the physical world, predictions that make sense only in terms of a larger set of new assumptions and expectations about things we find in the world and how they behave and relate to one another.

Elementary quantum mechanics, which was created in stages from 1900 to 1930, was largely the work of six men: Albert Einstein, Niels Bohr, Paul Dirac, Erwin Schrödinger, Max Planck, and Werner Heisenberg. Its first achievements were piecemeal theories formulated to make sense of odd experimental results that could not be fitted into the old classical paradigm. All its early thinking was focused on the microworld, and quantum theory is often misunderstood as a science that applies only to the behavior of very small things. This is untrue.

Quantum theory applies to physical reality on every scale—the very small, the everyday, and the very large. Without it, we cannot make sense of how stars produce nuclear power, why chemical compounds produce the range of colors that they do, why solids have strength and often the capacity to bend (solid-state physics), why electron currents can move along wires, or phenomena like superconductivity and laser light. The whole technology of the microchip is a quantum technology, and quantum effects are increasingly seen as important in biology.

*Quantum* refers to a little bundle or packet of energy, the smallest discrete amount that can be associated with a single event in the microworld. When an electron moves from one energy orbit to another, it always takes on or gives out an amount of energy that can be measured as so many quanta. Quanta are not, however, divisible. No movement of a particle from one state to another ever uses up one and a half quanta or three quarters of a quantum. Thus the term *quantum leap*, an abrupt movement from one discrete energy level to another. One physicist has described quantum physics as a physics of "lumps and jumps."

Quantum physics' "lumpiness and jumpiness" mark one of its sharp breaks from the Newtonian paradigm. Classical physics represents motion as smooth, continuous change, and energy as increasing or decreasing in a continuous spectrum. The existence of quanta explains why.

During the late 1920s, the piecemeal theories and predictions of quantum mechanics were systematized into a coherent mathematical picture. Quantum theory was born, elegant and complete and able to predict a wide range of physical phenomena accurately to a great many decimal points. But the kind of things, events, and relationships it describes seem to violate all common sense.

Where the old physics describes the world as made of two separate kinds of things, particles and waves, quantum theory postulates a WAVE/PARTICLE DUALITY. The basic building blocks of the universe, whatever form they may take, are "wavicles," indeterminate things with the potentiality to behave like waves in some circumstances and particles in others. Like children who behave well with some adults and badly with others, they manifest one property or the other, depending upon their context or environmental surroundings. (See CONTEXTUALISM.)

A quantum entity is *both* its capacity to manifest itself as a wave, in which case it has momentum, *and* its capacity to manifest itself as a particle, which has position. We can never know the position and the momentum of the entity simultaneously. Indeed, it doesn't even possess them simultaneously. If one becomes definite, the other becomes hazy. This is the nub of HEISENBERG'S UNCERTAINTY PRINCIPLE. Trying to view quantum reality is like looking at indistinct figures through blankets of fog.

Classical physics is rigidly determinist, and therefore predictable. The laws of Newton's universe mean that *B* will always follow A in the same predictable manner if all other conditions affecting them remain the same. But quantum physics has shown that this is only an approximation of the truth. In quantum theory, *B may* follow A, and one can assess the probability that it will do so, but there is no certainty. Quantum events often happen "just as they happen," and there is no way to know what will happen next, or why, or how. (See INDETERMINACY.)

Classical physics reduces all complex things to a few simple components and stressed their absolute, unchanging nature, their actuality or "what is." Quantum physics, by contrast, sees that new properties emerge when simple things combine or relate. The whole is greater than the sum of its parts. There is always the possibility of becoming other or more than what is. Every quantum bit has the potentiality to be here *and* there, now *and* then, a multiple capacity to act on the world. (See ACTUALITY AND POTENTIALITY IN QUANTUM MECHANICS; A QUANTUM HUSSY.) Underlying quantum reality itself is the ground state of being, a "sea of potentiality" described mathematically as a wavelike spread of possibilities. (See THE WAVE FUNCTION AND SCHRÖDINGER'S EQUATION.)

In the old physics and in common sense, things move and events happen as part of a chain of cause and effect. Something is acted upon by a force, or communicated with by a signal, and it responds accordingly. Without such localized action or causation, things remain stationary. But quantum events are often "nonlocal"—that is, they happen without apparent cause, in the absence of any known force or signal. The constituents of quantum reality are somehow correlated; they respond to one another and move harmoniously, as though they are all undulating parts of some larger but invisible whole. (See NONLOCALITY.)

The classical atomistic picture of a world consisting of tiny separate parts, each isolated in its own corner of space and time and linked only through force, is outdated by quantum mechanics. In the quantum universe—and this is the *whole* universe—every "part" is subtly linked to every other, and the very identity—the being, qualities, and characteristics—of constituents depends upon their relation to others. It is impossible, except as an approximation, to apply the part of the scientific method that calls for isolating an entity from its environment when one is investigating quantum entities or systems. The part comes to *be* fully only in the context of the larger whole. (See HOLISM.)

It is also impossible to isolate the observer (or measuring device) from what he or she (or it) observes. Observers have no place in the equations of classical physics. They play no "active" role in the deterministic chain of causal events. But in quantum theory, the observer is *part* of what gets observed. The observer's body and position, his or her choice of experimental design or measuring apparatus, perhaps even his or her conscious mind, are in a mutually creative dialogue with the way quantum reality manifests itself. The phrase "It all depends on how you look at it" takes on a powerful new meaning. The observer actively *changes* physical reality, actively evokes one or another of its underlying potentials. (See PROLOGUE on Schrödinger's cat; THE PARTICIPATORY UNIVERSE.) Exactly how or why this is so, and how it is that quantum reality changes radically to the more familiar reality of everyday experience when it is observed or measured, is the outstanding problem of quantum physics. It is known as THE MEASUREMENT PROBLEM or the observation problem.

Because many features of quantum reality seem to violate common sense, quantum physics has a reputation for being bizarre, an *Alice in Wonderland* physics. Einstein said that it struck him as "the system of delusions of an exceedingly intelligent paranoic." More recently, Nobel physicist Richard Feynman declared that it is impossible to *understand* quantum physics and useless to try. But all this is beginning to change.

In what is almost a third stage of quantum theory's development, philosophers of physics are beginning to understand the wider implications of the theory. Scientists are beginning to see how this physics relates to developments within chaos theory and complexity physics, and contributes to a new overall scientific paradigm. Nonscientists are increasingly aware of how the categories of existence and patterns of relationship described by quantum theory serve as meaningful models for our attempts to understand human psychology and relationships. Philosophers of the mind find parallels between quantum reality and the nature of consciousness. Changes in the cultural paradigm, new emphases on holism, and a greater need for a creative dialogue between human beings and the natural world all contribute to bringing quantum physics within the scope of a renewed common sense and everyday concern.

At the high energies of nuclear reactions, particles can be created or destroyed. (See SPECIAL RELATIVITY.) Here, elementary quantum physics must be extended into QUANTUM FIELD THEORY. At still higher energies, physical theories are still provisional. (See THEORIES OF EVERYTHING.) But nobody doubts that the principles of quantum physics will be a part of any future syntheses.

# Quantum Theories of Mind

Quantum theories of mind stem primarily from philosophical motivations, but they have a scientific aspect and are increasingly the subject of experimental research. Philosophically, they are exciting because they offer a new paradigm for cognitive science, one that seems better suited to our actual mental experience than the dominant mechanistic theories.

Mechanistic theories of mind are necessarily reductionist. Mental activity is reduced to brain activity, and brain activity is modeled on computers. It is difficult to see how such "mind machines" could be conscious, could exercise intention or free will, or could display the unity of experience that we take for granted. Quantum theories offer an alternative physical theory of mind that many proponents believe gets around these objections.

The first suggestions that human mental life bears many similarities to the properties of quantum systems were expressed by the biologist J.B.S. Haldane in the 1930s and drawn out in more detail by

David Bohm in the 1950s. Since then, the quantum view has been made more popular by the work of Roger Penrose, Danah Zohar, Ian Marshall, and others.

Newtonian machines are fixed; they remain the same in all circumstances. Both quantum reality and many features of human language and human nature are "situational," or context-dependent. An electron behaves like a wave in some experimental contexts, like a particle in others. Human beings have many different sides to their characters, which are drawn out by different associations or circumstances. The tone and the context in which we make a statement affects its meaning.

Newtonian physics is exact and stresses an either/or vision of truth. Both quantum physics and the human imagination allow for nuance and SUPERPOSITIONS. They contain several possible realities, often mutually contradictory and all juxtaposed one on top of the other. Both throw out "feelers toward the future" in exploring the viability of these possible realities. Quantum systems do this to test the most stable future energy state; the human imagination does it to test the best possible future-life scenario.

In many other ways, Newtonian physics, and thus computationalist mental models, seems too restrictive to describe our experience. Quantum models offer more scope. Newtonian determinism allows no room for free choice, but quantum indeterminism offers at least the possibility. HEISENBERG'S UNCERTAINTY PRINCIPLE limits what a quantum physicist can know about a physical system, and a kind of uncertainty principle operates when human beings must choose between vague thought and concentration, between the kind of answers given by one sort of question and those provided by another, or between being efficient and being creative. Higher mental faculties, like artistic ability, morality, and spirituality, seem unlikely products of mere neural activity. Quantum models of mind suggest they could be emergent phenomena, based upon neural activity but not sufficiently explained by it. (See EMERGENCE.)

Computationalist models of mind are based on Newtonian separate parts—the activity of individual neurons, neuron bundles, and their connections. They cannot account for how the whole brain manages to unify its conscious and perceptual experience. If there was a quantum system active across the brain, quantum HOLISM and NON-

LOCALITY might account for this unity. Quantum particle/wave COMPLEMENTARITY offers more viable social and psychological models for how both individuals and groups can be equally important. Roger Penrose (see PENROSE ON NONCOMPUTABILITY) believes there are crucial features of human thinking—insight, intuition, understanding, and dependence upon meaning—that could be accounted for with quantum models of mind, but not with computationalist ones.

Scientifically, any quantum theory of mind would require the brain to contain a large-scale, body-temperature quantum system that underlies certain mental activities. Neural synaptic activity, especially in the retina, is known to be sensitive to single quanta, but this is not enough to underlie the unity of mental activity. Some large-scale coordinated system analogous to a superconductor or a laser beam would be necessary. Both are examples of BOSE-EINSTEIN CONDENSATION, and most current research concentrates on how there might be Bose-Einstein condensation coordinating subcellular components of neurons. Some theorists suggest that this is concentrated in the water inside neurons; others, inside the molecular membranes of neurons (see FRÖHLICH SYSTEMS); and still others, inside the neural microtubules, or cytoskeletal structure (see NANOBIOLOGY). Microtubules are a "hot" theory at the moment, because anesthetics are thought to act at these sites in the neuron.

# Quantum Tunneling

How do quantum "particles" travel in regions forbidden to classical ones? How do they jump energy hurdles and sneak through barriers that should be impossible to get beyond? Their common ability to do these things is known as tunneling and is a dramatic consequence of WAVE/PARTICLE DUALITY and HEISENBERG'S UNCERTAINTY PRINCIPLE. It has many practical applications.

In the quantum world, a "particle," which we might expect to be confined to one side of a barrier, can sometimes be found on the other

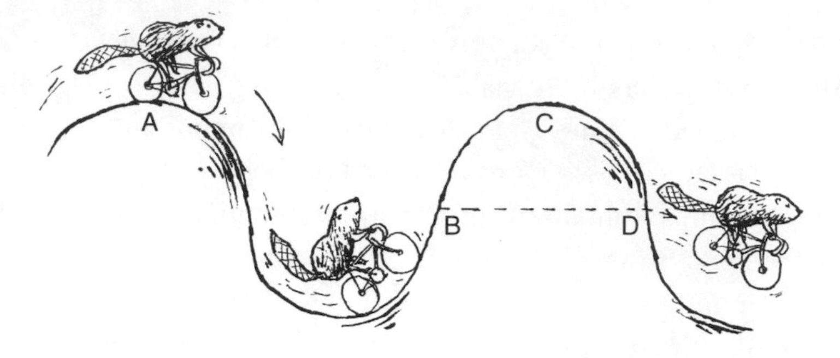


side, as if it had tunneled through like a mole. The barrier in this case is some form of energy constraint, and the likelihood of tunneling becomes less and less as the barrier is made higher or wider.

Imagine, for example, that an electron is riding a bicycle, a very tiny quantum bicycle, over a series of hills. It begins at the top of hill A and wants to get to a point *D* on the route without pedaling. In the normal course of classical events, completely discounting any effect from friction, the bicycle will roll down A's slope and have enough potential energy to climb halfway up the next hill to point *B*. At point *B*, the bicycle can climb up to C at the top of the next hill only if it is pedaled—that is, if more energy is pumped into its system. But in the quantum world, the bicycle simply tunnels through hill C and arrives directly at *D*. It goes in one side of the energy barrier and out the other, without ever going over the top. How?

There are two possible intuitive quantum models of how tunneling actually works. Both give the same mathematical predictions. One, relying on the Uncertainty Principle, calls upon us to remember that energy and time are "complementary variables." That means that when one is fixed, the other becomes fuzzy or indeterminate. So it is possible for the energy of the electron on its bicycle to fluctuate by fixing the time the journey will take. The Uncertainty Principle simply requires that the uncertainty in energy *times* the uncertainty in time remain constant ($\Delta e \bullet \Delta t \geq \hbar$, where $\hbar$ is Planck's constant). So one can increase at the expense of the other. In this case, the electron borrows enough energy, for a correspondingly short time, to increase its energy sufficiently to cross the energy barrier.

The other possible model for tunneling relies on an electron's ability to behave sometimes like a particle and at other times like a wave.

In this scenario, the electron travels up to its energy barrier as a particle, becomes a wave long enough to "wave" through the barrier (waves *can* wave through barriers, as, for instance, sound travels through walls), and then completes its journey as a particle. This, too, is completely possible in the quantum world.

Tunneling effects are common in nature. They include chemical reactions, radioactive decay (the decay particles tunnel through the attractive energy barrier that would keep them within the nucleus), and the processes by which stars generate energy. Technological applications include a special electrical switch called a tunnel diode; the scanning tunneling microscope, which can magnify up to 100 million times; and the Josephson junction, a superconducting ring that magnifies quantum effects and has endless uses from medicine to geology. (See SUPERCONDUCTORS.)

# The Quantum Vacuum

In QUANTUM FIELD THEORY, things existing in the universe are conceived of as patterns of dynamic energy. The ground state of energy in the universe, the lowest possible level, is known as the quantum vacuum. It is called a vacuum because it cannot be perceived or measured directly; it is empty of "things." When we *try* to perceive the vacuum directly, we are confronted by a "void," a background without features that therefore *seems* to be empty. In fact, the vacuum is filled with every potentiality of everything in the universe.

We can see particles, and we can see waves, but we know that neither of these is primary or permanent. Quantum reality consists of an inaccessible wave-particle dualism, and the waves and particles themselves can transmute one into the other. At high energies, one particle can transmute into another. At the level of perceived existence, everything has a kind of impermanence.

To make sense of this cosmic dance of temporary realities, physicists had to understand what lay beneath it. If particles and waves are only manifestations, what are they manifestations *of*? Seeking the answer to this question gave rise to quantum field theory, according to which everything that exists, all waves and particles that we can see and measure, literally *ex-ist*, or "stand out from," an underlying sea of potential that physicists named the vacuum. Waves and particles (and people!) "stand out from" or "wave on" the underlying vacuum, just as waves undulate on the sea.

Physicists' first motivation to look for something like the vacuum arose in response to relativity theory. (See SPECIAL RELATIVITY.) Einstein proved that the once-famous ether did not exist. The universe is not filled with a material jellylike substance. In that case, since light can be a wave, what is it a wave *on* or *in*? The later discoveries of particle physics raised the same kind of question. Since particles can appear and disappear at random, what do they emerge *from* and where do they go *to*?

Experiments in particle physics showed that existing particles, as well as coming and going from "nowhere," are slightly moved or deflected from their predicted paths, as though something were acting on them. The greatly expanded mathematical framework of quantum field theory attributed such effects to an all-pervasive, underlying field of potential—the vacuum. Unseen and not directly measurable, the vacuum exerts a subtle push on the surface of existence, like water pushing against things immersed in it. (In quantum field theory, this is known as the Casimir effect.) It is as though all surface things are in constant interaction with a tenuous background of evanescent reality. This background reality, the vacuum, replaces the material, jellylike ether. The universe is not "filled" with the vacuum. Rather, it is "written on" it or emerges out of it.

Like the Buddhist Void or the concept of Sunyata, to which it is often compared, the quantum vacuum is not "empty"; it is replete with potentiality. As the Buddhists say of the Void, "To call it being is wrong, because only concrete things exist. To call it non-being is equally wrong. It is best to avoid all description. . . . It is the basis of all."

# Quarks

Quarks are now believed to be the most fundamental constituents of large subatomic particles previously thought to be "elementary."

By the early 1960s, particle accelerators had produced hundreds of short-lived "elementary" particles, and it began to seem likely that these were, in fact, composites of some more fundamental constituent. Physicists first sought to classify the many different particles, just as Mendeleyev's periodic table had classified the chemical elements. A successful classification, nicknamed whimsically "The Eightfold Way" after the Buddhist path to enlightenment, was worked out by the American Nobel laureate Murray Gell-Mann and the Israeli physicist Yuval Ne'eman in the 1960s.

In 1964, Gell-Mann and George Zweig showed that the patterns of the then-known HADRONS (e.g., protons, neutrons, and mesons) could be explained if some were composed of three quarks and others of a quark-antiquark pair, and if there were just three kinds ("flavors") of quarks: up, down, and "strange." (The word *quark* itself was borrowed whimsically from James Joyce's *Finnegans Wake*.) Later experiments in particle accelerators bore out the scheme, until higher-energy accelerators became available. They created more exotic particles not composed only of the original three quarks. There are now six known flavors of quarks: up, down, strange, charmed, bottom, and top. (See QUANTUM CHROMODYNAMICS.) There are reasons to believe that these are all the quarks. They parallel the six (three pairs of) LEPTONS. Each flavor of quark comes in three colors and three anticolors, making thirty-six altogether.

# Quasars

Quasars, discovered in the 1960s, are small but extremely luminous objects usually found at very large distances from us—up to $10^{10}$ light-years away. They appear pointlike but can radiate a thousand times as much energy as our whole galaxy. The radiation may be at X ray, optical, or radio wavelengths. Quasars may be the bright nuclei of active galaxies otherwise too faint for us to see.

Since some quasars vary their brightness within a few hours, their diameter must be less than a few light-hours, or coordination would not be possible. And as the intense radiation pressure does not blow the source apart, its gravitational pull must be huge. The most promising candidate to explain this is a supermassive black hole, perhaps $10^9$ times the mass of the sun, onto which gas from nearby stars is accreted. (See BLACK HOLES.) Such gas would be violently accelerated and heated as it approached the black hole, generating the observed radiation. The black hole would have to swallow gas equivalent to nearly ten stars per year to produce its radiation.

For some reason, quasars were much more common when the universe was only about a quarter of its present age. At that epoch, quasars were as common as other bright galaxies, whereas now they are a hundred or a thousand times rarer. It is possible that newly formed galaxies have much more interstellar gas available for accretion onto a black hole at the nucleus. Cosmological theory about this is incomplete. What is clear is that the universe contains almost unimaginably violent processes by comparison to our rather peaceful planet.

# Reductionism

Reductionism is the belief that any complex set of phenomena or patterns of behavior can be defined or explained in terms of a relatively few simple or primitive ones. Any whole can be broken down into the sum of its parts, and a reductionist would argue that nations can be explained in terms of the behavior of individuals, individuals in terms of the biological instincts or traits they share with lower animals, and these, in turn, in terms of the principles of chemistry and physics.

The atomism of ancient physics holds that everything in the universe can be broken down into types of a few simple entities and the geometrical relations among them. Classical physics is essentially reductionist, holding that all physical reality can be reduced to a few particles and the laws and forces acting among them. Modern chemistry reduces chemical properties to ninety or so basic elements (kinds of atoms) and their rules of combination, while mathematicians have

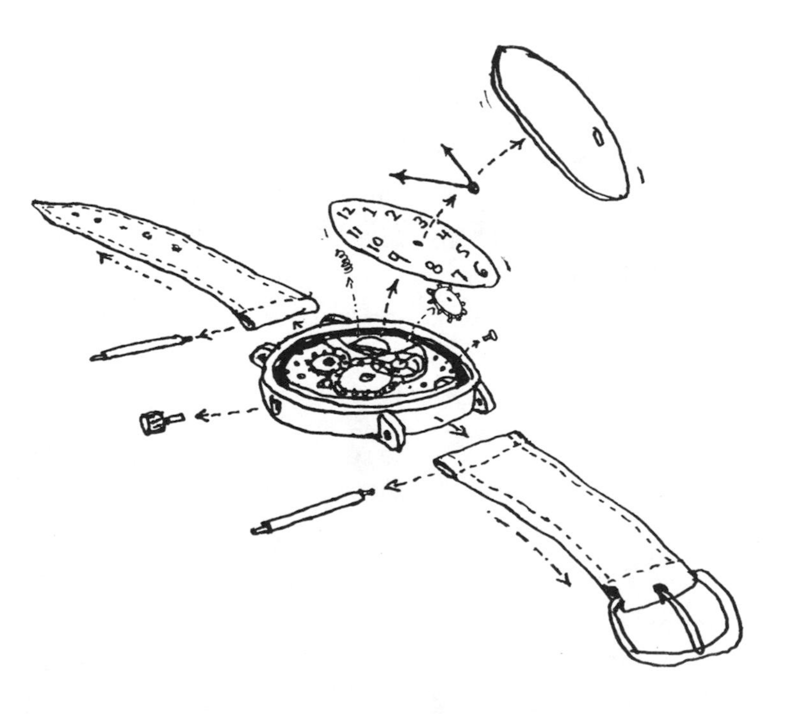

tried to assert that all mathematical truth can be expressed in terms of a universal set of axioms and principles, although GÖDEL'S THEOREM later showed that this is impossible. (See also SOCIOBIOLOGY; STRUCTURALISM.)

In spirit, reductionism is very similar to the principle of Occam's razor—that, other things being equal, we always prefer the simplest explanation or model for any event or behavior. This kind of thinking often seems to work and has enabled humans to structure their experience and understanding of the world's complexity into at least a good approximation of the way things are. But simple explanations and approximations must remain provisional, as the EMERGENCE, relational HOLISM, and COMPLEXITY that feature so widely in the physics of the twentieth century dramatically illustrate. An emotional preference for neat simple schemes can lead to wishful thinking—as can, of course, a preference for rich and "messy" world views.

The power and beauty of Newton's three laws of motion and the overwhelming success of modern science have led to a wider reductionist paradigm, derived from science. This has led from science to scientism, the view that everything can and should be modeled on scientific thinking. It has also contributed to materialism, the view that ultimately everything can be reduced to the properties of matter, as mind can be reduced to the brain. And it has encouraged the view that emotion, aesthetics, and religious experience can be reduced to biological instinct, chemical imbalances in the brain, or the rules governing genetics and physics. A twentieth-century reaction against this view is relativism; science is perhaps a middle way. (See RELATIVITY AND RELATIVISM.)

# Relativistic Cosmology

Relativistic cosmology is the result of applying Einstein's GENERAL RELATIVITY, formulated in 1915, to the universe as a whole. On this scale, gravitation is by far the most significant of the four fundamental forces. (The strong and weak nuclear forces are very short range, and most large objects are electrically neutral and thus not subject to electromagnetic force.) On this vast scale, the curvature of space-time described by Einstein marks an important difference from older cosmology. Relativistic cosmology was the beginning of modern cosmology. Solutions to Einstein's gravitational equations, coupled with the observed EXPANDING UNIVERSE, produced the widely accepted BIG BANG hypothesis.

If we consider a more or less homogeneous universe filled with gas or dust, expanding from a point (the Big Bang), then General Relativity offers three outcomes for its future, each depending upon the average density of matter in the universe. In all cases, gravitation gradually slows the expansion of the universe. If the average density is greater than a certain critical figure, the expansion will halt and then become a contraction, ending in a Big Crunch. If the average density is less than this critical figure, the expansion will continue forever. And if the average density is delicately poised at the critical figure, the universal expansion will slow down, and the universe will approach a limiting density. In each scenario, the average density of matter imposes a distinctive geometry of space: closed (rounded like a complete sphere), open (like the surface of a saddle), or flat, respectively. The first two possibilities were described by Alexander Friedmann and Abbé Georges Lemaître, the third by Einstein and Willem de Sitter, all in the early 1920s.

At this point we don't know which possibility applies to our actual universe. We do know that the critical density is not far from the delicate equilibrium point in the part of the universe that we can observe. The mass contained in the stars and gas clouds is a fair pro-

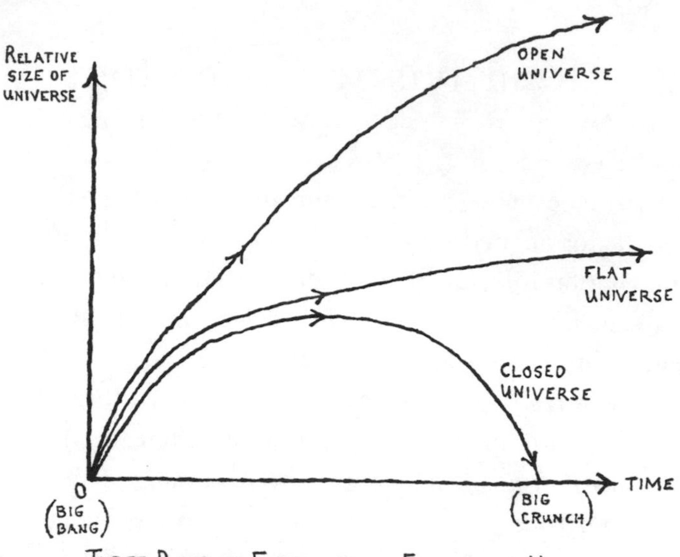


THREE POSSIBLE FATES OF AN EXPANDING UNIVERSE

portion of this, and we know from the gravitational pull of galaxies on stars that there is as much again of DARK MATTER.

We also know that our universe is about $10^{10}$ years old. If the critical density had been much different from the equilibrium point in the first place, this difference would by now have been increased enormously. This is lucky for us. Otherwise there would already have been a Big Crunch, or the universe would have become too diluted for any galaxies and stars to form. In either case, we wouldn't be here. (See THE ANTHROPIC PRINCIPLE.)

General Relativity makes definite statements about the large-scale structure of the universe, without which modern cosmology could not have gotten started. It has been confirmed by more than a dozen kinds of tests. Nonetheless, when we try to apply it to the universe as a whole, rather than to just the *observable* universe, there is tremendous need for extrapolation, and we suddenly find ourselves at the border of science and philosophy. One case in which such ambiguity applies is the question of whether we live in a closed, open, or flat universe. Given our present powers of observation, we just do not know.

# Relativity and Relativism

Einstein's SPECIAL RELATIVITY is often invoked as an intellectual support for cultural and moral relativism. Where Einstein says that we can never see beyond our own space-time framework, the relativist argues that whatever individuals or cultures happen to think is right, *is* right, for them. Everyone has a right to his or her own opinion or practice. There are no objective standards, no "correct" way of thinking or acting. There is hostility to REDUCTIONISM.

This kind of skepticism was originally expressed by the ancient Greek Sophists. It is implicit in Nietzsche's and Freud's views that a person's beliefs are often those that he or she finds most convenient in the pursuit of individual impulses. Twentieth-century anthropologists have reported very different, often conflicting beliefs in different tribes or cultures. (But there do seem to be some underlying universals; nearly all cultures prohibit cheating and murder, for example.)

Relativism is deeply embedded in the twentieth century's thinking. Philosophers of science and philosophers in general have stressed the limiting boundaries of our "language game" (Wittgenstein), our "situation" (Continental existentialists), or our "culture" (philosophical anthropologists). Much of postmodern philosophy ("deconstructive postmodernism") is a rebellion against any idea of objective standards or criteria for judging, any fixed viewpoint, any conviction, and even against reason itself. All are felt to be without foundation and constricting for the free spirit.

Within science itself, the American philospher of science Thomas Kuhn has pointed out that scientific knowledge does not always proceed through the gradual accumulation of knowledge. Sometimes there is a revolution, or a paradigm shift to a very different sort of vision. (See INTRODUCTION: THE NEW SCIENCE AND THE NEW THINKING.) The development of relativity theory and quantum mechanics is such a shift from the paradigm of classical physics. Indeed, the whole of the "new science" is such a shift. Kuhn holds that this process cannot be described or justified rationally. It just happens, and when it does, it takes us to a new way of looking at things.

Special Relativity itself does not support the view that "everything is relative." In Einstein's theory, there *are* objective descriptions. There is a unique, four-dimensional space-time desription of the "real" world in abstract terms, which contains as *aspects* the perspectives of all possible observers. These aspects, only one of which can be seen by any one observer, are related by the abstract description of the whole. Thus there is a God's-eye view, available only to God. The best that we can do is to formulate it mathematically.

A close analogy to what Special Relativity is saying about the relation between the perspective and the whole is the picture we get by looking at different perspectives on a given visual scene. If we can describe all the individual objects and their spatial locations, we can calculate how the whole scene would look from any one point of view. The different perspectives are not random or subjective; each is a partial aspect of some larger whole, and we can get some *sense* of that whole through a knowledge of several of its various viewpoints.

The same process occurs through empathy—we may not *have* the other person's emotions or experience, but we can feel what it is *like* to be him or her. And by knowing what it is like to feel or be someone else, we can get some sense of what it is like to be human. Similarly, in Kuhn's scientific paradigm shifts, it is possible to *understand* both how previous generations of scientists saw the world and how our technological progress is itself objective.

The radically new notion of a multifaceted objective reality of which we can experience only one aspect at a time was introduced into fundamental physics by Special Relativity. It was later extended in quantum mechanics. (See PERSPECTIVE AND INTERACTION.) And it has endless applications in our familiar everyday lives. It helps us understand how we might simultaneously hold both personal and impersonal points of view; how we might cling to our own views of the world and at the same time accept that they are partial and that there are other, equally valid points of view. It might help us understand, when we are angry or depressed, that at other times and in other situations we might feel different, and thus get a better perspective on the future.

The same is true of cultural attitudes. We can see that our own are valid while accepting the validity of others, and simultaneously understanding that all are aspects of some larger cultural pattern that we can never see directly. Reflecting on this might cultivate a more mature attitude toward our own beliefs and those of others. We get a hint of this two-level awareness from the opening sentence of the *Tao Te Ching*: "The way that can be expressed in words is not the eternal way."

## Resonance

Resonance occurs whenever two structures vibrate in tune. Strike a tuning fork, and an identical fork on the same table will begin to vibrate. Energy is continuously exchanged between the two forks, which are in resonance. The music from a radio comes in loud and clear when the frequency of the circuitry inside the tuner is in resonance with the incoming signal. Soldiers marching across a suspension bridge break stride in case their coordinated marching should resonate with the natural vibrations of the bridge. If this occurred, energy would be rapidly absorbed by the bridge, whose structure would oscillate out of control.

A form of resonance also occurs in the quantum world. Molecules based on benzene, with its three double bonds, often have two separate forms with identical energies. Quantum theory dictates that a single such molecule exists as a superposition of both forms. The molecule is said to be in a resonance state. It has a lower energy than either of its two component forms.

The nucleus of the hydrogen atom with its single proton acts like a tiny magnet. Exposed to a magnetic field, the nucleus spins like a top. A weak radio wave, exactly in tune with the energy of this spinning, makes the nucleus absorb or give up energy through resonance and causes the nucleus to flip into a different energy state. This is known as nuclear magnetic resonance. Because the resonance frequency of the hydrogen atom is very sensitive to its chemical environment, nuclear magnetic resonance can be used as a form of chemical analysis, proving important information about solids and liquids. Under the name of magnetic resonance imaging (MRI), it is used in hospitals as a diagnostic tool.

Resonance is the phenomenon in a microwave oven. In the oven, radiation is produced at the exactly correct frequency to resonate with vibration of water molecules in food. The molecules absorb energy, increasing their agitation and making the foodstuff heat.

Microwave ovens and magnetic resonance imaging involve what is termed nonionizing radiation, which has a frequency so low that it cannot rip electrons from their atoms. X rays and nuclear radiation, which are called ionizing radiation, can, at sufficient intensity, cause cellular damage.

Microwave and magnetic resonance imaging radiation has nothing to do with radioactivity. Like light from a bulb, the microwave radiation in an oven switches off automatically when the cooking cycle has ended or the door is opened, and no residual radiation remains in the oven.

# The Second Law of Thermodynamics

THE FIRST LAW OF THERMODYNAMICS indicates that heat and work are mutually interconvertible. The Second Law limits the ways in which this can be carried out. It prevents, for example, the total conversion of heat into work.

Much of the impetus for the study of thermodynamics was the result of the industrial revolution with its need for efficient machines. The Second Law demonstrates that absolute limits on the efficiency of machines are an aspect of nature, not the results of defects on the part of engineers and designers. It also illustrates the important role played by ENTROPY in thermodynamic change. The Second Law can be stated in several different ways, each equally important. The novelist and scientist C. P. Snow, in his famous essay "The Two Cultures," argued that every educated person should have an understanding of the Second Law.

If heat could be converted into work with 100 percent efficiency, it would be possible to build a perpetual motion machine. Driven by a bath of heat, this machine would produce work whose friction would heat up the bath, which in turn would drive the machine to produce work, which . . . and so on. It is a fundamental property of nature that such machines cannot work in this way. In every cycle, only part of the heat supply is converted into useful work; the rest is lost in the environment. One way of stating the Second Law is "Perpetual motion machines are impossible." Another is "No process is possible whose only result is to transfer heat from a lower to a higher temperature."

If you unplug your refrigerator, the contents don't get any cooler. Switch it on, and work is done, thanks to electrical energy. This allows heat to be pumped from the food inside the refrigerator until it emerges at a higher temperature out of the coils at the back. But the transfer of heat, like making water run uphill, can be achieved only at the expense of work.

Transforming energy from a useful form into wasted heat is associated with an increase in entropy. With each cycle of an engine or machine, some useful energy is lost, and entropy increases in the surrounding environment. Another way of stating the Second Law is "In every process, the total entropy (system plus environment) either increases or remains the same."

It is always possible to reduce entropy locally—for example, by making ice in a refrigerator. The random motions of water molecules slow down and form a more regular, ordered structure. But this can be done only at the expense of running a refrigerator. Statement B of the Second Law tells us that the heat from making the ice freeze cannot be pumped into the environment without doing work that generates additional heat, more heat than is extracted from the water. An increase in heat always means an increase in entropy, so more entropy is produced by the refrigerator than was extracted from the ice.

A corollary of the Second Law is that the most efficient machines are those that involve the minimum production of entropy—i.e., the minimum dissipation of useful heat. A hypothetical device in which the various parts of its cycle are so gentle and frictionless as to be reversible is called a Carnot cycle, after the French engineer Sadi Carnot. The Carnot cycle is the most efficient engine possible.

We can see why C. P. Snow felt the Second Law was so important. It dominates many of the processes around us, not only the behavior of engines and machines but even that of biological systems.

# Self-energy

The way an electron interacts with its own electrical field gives rise to one of the major headaches of modern physics. Calcuating the energy of a substance is a standard physics exercise. The energy of a cup of hot coffee is greater than that of a cold cup (heat energy), and the energy of a speeding train is greater than that of a train at rest (kinetic energy). Calculating the energy of an electron requires adding together several contributions. First, there is its kinetic energy by virtue of its speed. Second, there is the energy ($E$) arising from its mass ($m$), given by Einstein's famous equation, $E = mc^2$ (where $c$ is the velocity of light). There is also a third and more troublesome contribution to the electron's energy.

Electrons have a negative charge and are therefore surrounded by an electric field. The electron interacts with its own field, like a boat moving through the ripples it creates, and the energy involved in this must be calculated according to the laws of QUANTUM ELECTRODYNAMICS. It turns out that this "self-energy" must be calculated through an infinite series of approximations, adding corrections at each stage. The final sum turns out to be infinite!

Using a procedure called renormalization, one can determine finite results. This involves making further corrections at every stage—in effect, subtracting an infinite amount to produce a finite result. Physicists are not generally happy with this compromise and believe that the present theory must be modified or replaced by deeper ideas.

If the electron is a mathematical point but has a finite electric charge, the electric field becomes infinite as we approach it. This would also be true in classical electromagnetic theory, although there it can be ignored. On the other hand, if the electron is of finite size, it is necessary to provide some new model for it, compatible with larger-scale observations and with SPECIAL RELATIVITY. One such model is SUPERSTRINGS.

# Serial Processing

Serial processing is computation done according to a step-by-step, ordered procedure. Each operation in the sequence is done one at a time, according to a set of rules that make up the program. The theory and technology of computers using this kind of processing are more developed and familiar than any other. Any standard PC is a serial processor composed of electronic elements connected one-on-one. (See FORMAL COMPUTATION; NEURAL NETWORKS; TURING MACHINES.)

# Sociobiology

Sociobiology is based on the assumption that all aspects of animal and human behavior can be accounted for in genetic and instinctual terms. Its founder, E. O. Wilson, had done brilliant work on ants and other social insects. Then, in the 1970s, he applied the same approach to human and higher animal societies. Language, gesture, group structure, the nature of work, and the structuring of space and time are all assumed to be reducible to a single explanatory level. (See REDUCTIONISM.) Thus the complexity and ambiguity of a Hamlet can be understood ultimately in strictly biological terms. To many, this view appears overstated as well as politically controversial.

Sociobiology finds resonances in STRUCTURALISM. In each case, while the ability to gather a wide range of facts and observations within a single umbrella theory is attractive to some, the philosophy is opposed by those who believe that new levels of organization require further concepts and explanatory principles. (See EMERGENCE.)

The ideas of sociobiology inspired a great deal of research and have found a wide audience through popular books and magazine articles. These have described human beings as "naked apes" and offer explanations for human preferences, "body language," and the like, based on concepts from animal behavior. However, while it may be good to remind ourselves of our animal origins to counter too-lofty notions of ourselves, it is not clear how adequate sociobiological explanations for our higher abilities will prove.

# Solitons

Solitons are sustained waves that persist over great distances for long periods of time. Normal waves are not coherent; some parts move slightly faster than others and the wave breaks up and dissipates. Under special conditions, however, a system of waves can interact in a nonlinear way, close to the bed of a river or canal, for example. The result is a form of stabilizing FEEDBACK, an internal coherence that slows down the faster parts of a wave and speeds up the slower. This is called a soliton. It can travel for miles as a complete unit. When, for example, an incoming tide meets a flowing river at its estuary, the result can be a tidal bore, a soliton many feet high, which travels the length of the river. Tidal bores on the Amazon River reach 25 feet in height and persist up to 500 miles inland.

Solitons can be generated far out in the ocean by earthquakes. Although the wave itself may be only a few inches high, it is many hundreds of yards long and takes as long as an hour to pass a given point. Such a wave travels unchanged across the width of the Pacific or Atlantic Ocean. As it approaches the shallow continental shelf, the wave's length shortens and it grows to a 100-foot mountain of water, a tsunami or tidal wave, which can cause great destruction.

Solitons occur as stable waves in a variety of other media. In the atmosphere, stable temperature inversions can travel for hundreds of miles. Such an atmospheric tidal bore has also been observed on Mars. The famous Red Spot on Jupiter, a highly stable vortex of air in the planet's upper atmosphere, was first recorded in 1664. From time to time other atmospheric disturbances have been seen to approach and pass through the Red Spot, in exactly the same fashion as solitons in water.

Solitons can exit as stable bursts of energy in solids. Biologists speculate that electrical signals travel as stable solitons along nerve pathways. There is even a theory that the elementary particles themselves could be solitons, stable disturbances in some nonlinear quantum background.

# Special Relativity

Special Relativity was the first and most groundbreaking theory to introduce the new paradigm of twentieth-century science. Its implications reach out in every direction, radically affecting our notions of space and time, matter and energy. Perhaps more fundamentally still, it calls into question our most basic concepts of objectivity and our age-old assumption that we can have a God's-eye view on any situation.

Special Relativity was formulated by Albert Einstein in 1905. It was a new mechanics of material objects, motion, and light. In 1916, Einstein proposed the more complex GENERAL RELATIVITY, which incorporates gravitation as well. Both are supported by many experiments and observations.

In classical Newtonian mechanics, there is an absolute space and an absolute time. In Special Relativity, both space and time are aspects of a larger whole, space-time, and can be interconverted to a certain limited extent. Events on two distant stars may appear simultaneous to one observer but successive to a differently situated observer. In General Relativity, the interconversion is more far-reaching still. In classical theory, particles are indestructible. In Special Relativity, they can be converted into pure energy, and vice versa. Hence nuclear power and the atomic bomb. In classical mechanics, particles are primary and waves are secondary. Waves are undulations in some particular substance, like sound waves in air or ripples on water. In Special Relativity, and later in QUANTUM PHYSICS, light waves and particles are equally fundamental.

In the nineteenth century, light and radio waves were established as wavelike phenomena, traveling at a high but finite speed through apparently "empty" space. It was supposed that the waves were ripples in an underlying jellylike substance, the ether, which permeated the entire universe. Following this idea, Albert Michelson and Edward Morley in 1887 tried to measure the velocity of the earth's motion through the ether. The assumption was that such a motion should affect the observed speed of light in different directions, just as ripples on a river travel faster downstream than up. Light going *with* the "current" of the ether should travel faster than light going *against* the current. But no such measurable effect was found. On the contrary, the experiment showed that THE SPEED OF LIGHT is constant in all directions.

Since the earth is presumably not at rest in the universe, the classical picture of a material ether whose vibrations are light waves is somehow wrong. Light waves do not consist of a material substance "waving," but are rather a kind of physical reality, as fundamental in their own right as Newton's particles or forces. The notion of "material" has been enlarged. (See QUANTUM FIELD THEORY; THE QUANTUM VACUUM; WAVE/PARTICLE DUALITY.) The speed of light has been established as a universal constant.

But if the speed of light is the same in all directions for us, it should be so for all other observers, too, or at least for any observer moving with constant speed in a constant direction. (In a closed, soundproof train, there is no way of telling whether you are at rest or in uniform motion. If the train changes speed or goes around a bend, the acceleration can, of course, be felt.) Accelerated motions are the subject of GENERAL RELATIVITY, whereas Special Relativity is concerned with constant motions—what physicists call INERTIAL FRAMES.

In Newton's mechanics, the measured speed of light cannot be the same for all nonaccelerated observers. If I am riding on a train at 50 miles per hour and throw a ball forward out of the window at 20 miles per hour, the ball should hit the ground at 70 miles per hour. Light, according to Einstein, does not behave like that. It arrives at a constant speed, whatever the motion of the source or receiver.

One of the most surprising upshots of experiments done to confirm Special Relativity is that motion *appears* to make material bodies become shorter in the direction in which they are traveling, make clocks run slower, and make objects become more massive (the Lorentz-FitzGerald contraction). These effects depend on speed and become noticeable only when motion approaches the speed of light. (Newton's formulas for motion are good enough at more ordinary speeds.) But these strange phenomena led to two very different interpretations.

Physicists who hoped to cling to belief in the ether as a basis for an absolute space-and-time framework, Hendrik Antoon Lorentz and George Francis FitzGerald among them, argued that the various contractions really do happen with respect to a God's-eye view of the moving system, although if we are part of the system, we can never measure them. If we are on a train approaching the speed of light, the train's carriages, ourselves, and our measuring instruments will all become shorter (and we will become more massive and age more slowly). But we will notice none of these things, because all our measuring apparatus and points of reference will have undergone the same changes. Someone standing outside, "in the ether," would witness our transformation.

Einstein's interpretation of the Lorentz-FitzGerald contraction was much more radical. He argued that any observers of the moving train are themselves in other space-time frameworks, that there is no "stationary" ether in which we can stand, and that it is impossible to say who is moving with respect to whom. Therefore, it is impossible to say who is shrinking or slowing down. All that any observer can say is that the *other* travelers, in their other frame of reference, *appear* to be undergoing strange effects. (See THE TWINS PARADOX.)

We have all had the experience of sitting in a train stopped next to another train in the station. Suddenly one train starts to move. Is it ours or the other one? The only way we can know is to look at the ground, but in Special Relativity there is no stationary "ground."

If all we can measure is "apparent" space and time, what right have we to talk of an underlying "absolute" space and time? Let us, Einstein said, just speak of what would be seen by observers in different frames of reference. No one observer is more "right" than another. Two events at different places may be simultaneous according to one observer, or either may precede the other according to observers in still other inertial frames. No one human view is truer than any other. All are relative. What we are left with is an invariant, underlying space-time carved into spatial and temporal *aspects* in different ways by different observers. Only the speed of light is constant for all.

There is, of course, a God's-eye point of view available to *God*, if He is omnipresent and omnitemporal—this is the abstract, four-dimensional space-time description of all points of view. But no human can ever see this. We are always *in* space-time. The best we can do is describe it abstractly in equations.

Special Relativity ushered in a new paradigm that suggested, as the philosopher Nietzsche put it, "We can never see round our own corner." The theory has been used as an analogy to justify cultural and moral relativism, but it is not a sound analogy. (See PERSPECTIVE AND INTERACTION; RELATIVITY AND RELATIVISM.) Moreover, NONLOCALITY shows that events can be correlated faster than the speed of light, though not by sending signals. We seem to need a deeper point of view here.

# The Speed of Light

The speed of light in a vacuum is one of nature's few universal constants. (Light is slowed down by passing through glass or water, hence refraction.) In 1676, Ole Christensen Rømer could see with his telescope that the regular eclipses of Jupiter's moons seemed to be delayed when Jupiter was far from the earth. He knew that this delay was due to the finite speed at which the light could travel from Jupiter to us, but he did not know exactly how large the solar system is, and therefore couldn't calculate accurately the speed of light.

Modern technology has alowed us to discover that light travels at 186,000 miles per second ($3 \times 10^{10}$ centimeters per second). It can travel around the earth seven times in less than a second, go from here to the moon in under one and a half seconds, go to the sun in eight minutes, and go to the nearest star in four years.

SPECIAL RELATIVITY has established that the speed of light is the universal speed limit. No material object can actually reach this speed. Since any object gains apparent mass as it goes faster, gaining an infinite amount at the speed of light, it would take an infinite amount of energy to accelerate to this speed. By the same token, since light itself has only a small but finite mass, this would become zero if the light could slow down.

# Spin and Statistics

An important feature of the quantum world and an important bridge between the large and the small is the connection between the spin of elementary particles and their statistics. (Statistics here mean the way elementary particles arrange themselves in different energy states.)

Nature divides the quantum world into two classes of particles. The FERMIONS, matter particles that include all the leptons and quarks, are associated with fractional spins, such as spin ½. (The term *spin* is something of a misnomer, since elementary particles are not exactly like tiny, spinning billiard balls. Quantum spin is more subtle than this.) The other class of particles, the BOSONS, has whole-number spin. This includes the quanta of all the forces: the strong and weak nuclear, the electromagnetic, and the gravitational. Since spins can also add together, particles with integral spin include composites of an even number of fermions, such as mesons or hydrogen atoms.

The fermions obey what are called Fermi-Dirac statistics, while bosons are governed by Bose-Einstein statistics. Fermi-Dirac statistics dictate that no two fermions can occupy the same quantum state. By contrast, an unlimited number of bosons can occupy a single quantum, or energy, state.

Nature's division into two types of statistics has enormous practical significance. Because of Fermi-Dirac constraints, electrons are forced to occupy different energy states, thus making the atoms of the various chemical elements quite distinct. In a world without Fermi-Dirac statistics, there would be no chemistry, no rich behavior at the molecular level—indeed, no life on earth. Our planet would collapse to the size of a tiny ball.

By contrast, bosons are permitted to congregate in the same energy state. This gives rise to BOSE-EINSTEIN CONDENSATION, which is characteristic of SUPERFLUIDS, SUPERCONDUCTORS, and LASERS. It also allows forces of unlimited strength, composed of many bosons acting in concert.

We can describe these statistics mathematically, but their deeper meaning is less obvious. Why, for example, should the particles that compose matter not be allowed to congregate in the same quantum state as do photons of light? It is built into the symmetry of the respective wave functions of fermions and bosons, but what has this to do with the particle's spin? The deep connection of spin, statistics, and symmetry can be understood, but only in terms of a QUANTUM FIELD THEORY and the structure of space-time in SPECIAL RELATIVITY. Discussion of this is mathematically complex.

(See also IDENTITY IN QUANTUM MECHANICS for some philosophical background.)

# Split-Brain Phenomena

Split-brain phenomena are noticed when the corpus callosum, the main section of the brain connecting its right and left hemispheres, is cut during a surgical operation. Patients undergoing this operation are sometimes noticed to have split or multiple conscious fields and associated split behavior. One such patient, seemingly in the grip of two conflicting motives, was observed hugging his wife with one arm while pushing her away with the other. Such phenomena raise deep philosophical questions about the unity of consciousness, the nature of the personality, possibly even the existence of the soul. Is each of us really only one person? Do conflicts of mood or motivation imply a deeper conflict between split selves within us? If each of us possesses one immortal soul, how can that soul be split into two through the action of surgery?

Split-brain operations were first undertaken in the 1960s as a treatment for some forms of uncontrollable epilepsy. Known as commissurotomy, the surgery leaves the brain's two hemispheres virtually unconnected to each other, although each remains separately connected to the lower brain. As a result, epileptic attacks are reduced, or confined to one hemisphere. Under normal circumstances, postoperative patients remain normal in both intelligence and personality, but subtle psychological tests reveal interesting differences.

Roger Sperry, a California neurologist, did pioneering tests on split-brain subjects in the 1970s. He found that when each of the two disconnected hemispheres belonging to the same patient is subjected to a different experience, each hemisphere behaves as though it were a separate mind. If, for instance, such a patient is asked to look straight ahead while a phrase like "key ring" is flashed on a screen for a tenth of a second, one part to the right and the other part to the left of the fixation point, the patient has two separate experiences, one of "key" with the right hemisphere and another of "ring" with the left. Since the speech area is in the left hemisphere, the patient will *say* that he or she has seen "ring." But both hemispheres *see* a word. Thus, a patient who is presented with a box of unseen objects and asked to choose those that have been named will pick out a ring with the right hand, but a key with the left. Each hemisphere saw *something*, but neither saw the combination of both words to make the connection "key ring."

Under normal circumstances, when both hemispheres are able to take in the same information, the behavior of split-brain patients is well coordinated. They can walk or even play the piano without difficulty. It is assumed that such residual coordination is made possible by the two hemispheres' common perceptions and common connection to the lower brain. But under more testing circumstances, like those in the key ring test, or in the case of the man who embraced his wife with one arm while pushing her away with the other, conflicts sometimes arise.

Psychological phenomena similar to those seen in split-brain patients have been observed in subjects who have undergone hypnosis. Subjects who have been given a hypnotic suggestion have no memory of what they have been told to do, but will nonetheless perform the suggested behavior on cue. If told to remove a jacket at a specific moment, the posthypnotic subject will do so, but may then justify the behavior by saying something like "I felt too hot." Cases like this interested Freud, who believed they supported his theories of neurotic dissociation, the view that some parts of the personality are split off from consciousness but nonetheless strive to influence conscious behavior. (See PSYCHODYNAMICS AND PSYCHOTHERAPY.) Freud thought most such dissociated strivings arise from the id, the personality's "deep basement" of instinct and unconscious motives. People influenced by dissociated parts of themselves often claim to have "behaved out of character." More commonly still, mild dissociation of consciousness can be observed in the act of talking while driving. Neither activity is fully automatic; each seems to be supervised by some different part of the mind.

Split-brain phenomena and dissociative experience raise problems for the most popular Western model of personal identity. Most Western philosophers, including Plato and Descartes, have believed each person has one enduring and unified mind, or soul.

How, then, can we account for one person's behaving as though he or she has two minds or more? One way of resolving this question is to say that one hemisphere of the brain is conscious and the other merely acts as an automaton. The problem with this approach is that the behavior of *both* hemispheres seems to be intelligent and motivated. And there are cases of people who have lost the entire left hemisphere—and hence the speech center—but who behave as though fully conscious.

A newer way to think about split-brain phenomena is to accept the fact that certain kinds of brain activity are associated with consciousness. Any one person may have zero, one, two, or more centers of consciousness active within the brain at any moment. All these may from time to time merge into a higher-level unity of consciousness, or one of them may at any moment be the subject of selective ATTENTION. The "self" at that moment is the arena of consciousness that has access to the motor, language, and long-term memory systems. Any one arena of consciousness may have a greater or lesser degree of stability over time, and "the person" is a composite "choir" of conscious experiences. We would say that a split-brain patient is one person with two separate islands of consciousness. This model is similar to those of both COMPUTATIONAL PSYCHOLOGY and Buddhism. (See also THE MIND-BODY PROBLEM; QUANTUM THEORIES OF MIND.)

# The Standard Model

The Standard Model is a way of making sense of the multiplicity of elementary particles within a single scheme. By the late 1960s, there were two radical schemes of unification within elementary particle physics. The first involved the combination of the weak nuclear force and the electromagnetic force into a single ELECTROWEAK FORCE. The second was the development of the quark model of the hadrons, in which mesons were pictured as a quark-antiquark pair, and baryons were composed of three QUARKS. The force holding the baryons together is called the color force and is carried by gluon particles. By analogy with QUANTUM ELECTRODYNAMICS, the new scheme was called QUANTUM CHROMODYNAMICS.

These two schemes, taken together, are called the Standard Model. This does not represent a complete unification, for the electroweak and color forces remain distinct, but rather the combination of two different domains of elementary particle behavior. Nevertheless, the Standard Model brings a considerable degree of order into the multiplicity of hadrons. It has led to important predictions, which have later been experimentally verified, and it has stimulated many theoretical investigations and speculations. Nevertheless, the model is not without its difficulties.

Despite a large number of different experiments, a free quark has never been observed. A variety of explanations have been offered. One suggests that, as two quarks are pulled apart, the gluon force increases in strength. This is something new in physics, since forces generally fall off with distance. Not so the color force. When two quarks are finally separated, perhaps in a violent collision, theoretical physicists suggest that the energy released by the breaking of the color bond is so great that it creates new quarks. The instant it is set free, the isolated quark acquires a neighbor and becomes a bound state again.

Another problem is that the Standard Model contains a large number of arbitrary constants. The masses of the mesons, baryons, and leptons, as well as the forces between them, are determined by fixing the values of these constants. By choosing the numbers correctly, it becomes possible to match theoretical preditions with experimental values. But this is not a particularly satisfactory state of affairs. A good theory should account for the forces and masses of the elementary particles without the need for so much external input. It should fix the fundamental constants themselves, showing how their particular values arise. It should also explain why the electric charge on the electron is equal in size to that on the proton, and why parity is not conserved. (See CPT SYMMETRY.) For these reasons and others, some physicists feel that something is missing from the Standard Model.

Yet the success of the Standard Model cannot be denied. The mesons and baryons can be grouped together in symmetric schemes. It also looks as if the hadrons are composite particles, which can be accounted for in terms of quarks that interact by exchanging gluons. Such a scheme produces the observed symmetry exactly. The idea that electromagnetism and the weak nuclear force can be unified is also attractive. Combining these two approaches gives a uniform approach to the wide range of elementary particle phenomena. But beyond this point, the theory seems to falter. It is more descriptive than predictive and fails to fulfill its initial promise of explaining the full nature of the hadons and leptons, together with the strength of their forces. Nevertheless, the success of electroweak unification and the Standard Model has encouraged physicists to seek further unification—a Grand Unified Theory of hadrons and leptons. (See GRAND UNIFIED THEORIES.)

# Stars

The ancients thought of stars as serene and godlike, made of a "celestial substance," looking down on the turbulent world of human beings from an unchanging vantage point. Today we know that this picture is completely wrong. Stars of various kinds are part of the same universe as ourselves. They, too, have lifetimes, crises, and dramas, but on a larger scale. Our own bodies are made out of stardust.

Stars are the main contents of the observable universe. There are about $10^{11}$ in our galaxy, among which the sun is quite average. Stars vary in mass, from about sixty times the mass of the sun to about 10 percent of its mass. If their mass is below this limit, the object cannot generate enough internal pressure and temperature to ignite the nuclear reactions that cause a star to shine. (Jupiter, our sun's largest planet, for instance, is only 0.001 of the sun's mass.)

As a proto-star mass of gas condenses gravitationally, its pressure and temperature increase, particularly at the center. Originally, it is composed of about 75 percent hydrogen by weight and 25 percent helium. All other elements present vary from 0.01 percent to about 3 percent. In the sun, when the core temperature reached $15 \times 10^6$ K, hydrogen was fused into helium, with the radiation of much energy. We have been trying to reproduce this fusion process on earth as a cheap source of nuclear power, but so far there is no commercially viable way.

The heat produced at the core of a star escapes very slowly through the outer layers. The photons interact with layers of matter, gradually cooling until they reach the surface—the photosphere. In the sun, this is about 30,000 years later, at a temperature of 5800 K. These are photons of yellow light, hence the yellow color of the sun. Beyond the visible sun, there is another layer of very thin, much hotter gas known as the corona, which is hot enough to emit X rays, and which gradually blows away into space as the "stellar wind."

The energy output of the sun, or any other main-sequence star, is self-regulating. Its outer layers are kept distended by the radiation pressure of photons escaping from the core. If for any reason the core rate of energy output was to decrease, the star would shrink. The core pressure and temperature would then increase again, and the rate of core energy production would increase. Similarly, a slightly increased core energy output would be decreased again by the same mechanism. The main regulator of the energy output is the mass of the star. Heavier stars are shorter-lived than the lighter ones. They are also brighter and bluer (the "main sequence") until late in their lives.

When a star has burned about 12 percent of its hydrogen into helium, it becomes unstable. The inert helium core contracts, and hydrogen is burned in a shell around the core. The outer layers expand, due to radiation pressure, and grow cooler. The star then moves off the main sequence, growing both brighter and redder, and eventually becoming a red giant. Our sun is expected to do this, engulfing the earth, some $5 \times 10^9$ years from now. During this phase of expansion, a massive star may become hot enough to burn the helium core into carbon and even heavier elements.

Eventually, the expanding star runs out of nuclear fuel. It is no longer supported by radiation pressure, and the core collapses under its own gravity. The outer layers are blown off. There are now three possible end states, depending on the mass of the star. Stars that at this stage weigh less than about 1.4 times the sun (they may have been heavier at an earlier stage) end up as white dwarfs. They are about the size of the earth, with the mass of the sun, and gradually cool to become invisible. Heavier stars suffer an enormous supernova explosion. (See SUPERNOVAS.) They then become NEUTRON STARS or, if heavy enough, BLACK HOLES.

The ejected outer layers of a star rich in heavy elements return to the interstellar medium and are eventually recycled into the next generation of stars. Our sun is a fairly recent star. The heavy elements of earth and of our own bodies were synthesized ages ago in the interiors of earlier stars. (See CHEMICAL ABUNDANCES.)

# Statistical Mechanics

Statistical mechanics explains the laws of THERMODYNAMICS in terms of atoms and molecules. Thermodynamics was formulated in the early nineteenth century as an expression of the relationship between heat and work for discussing the efficiency of machines. Its fundamental concepts—temperature, heat, internal energy and ENTROPY—were defined long before the existence of atoms had been confirmed. These concepts were taken as basic properties of nature, not subject to further analysis. THE FIRST LAW OF THERMODYNAMICS and THE SECOND LAW OF THERMODYNAMICS express the relationships between these properties.

Toward the end of the nineteenth century, the physicist Ludwig Boltzmann suggested that the laws of thermodynamics could be explained in terms of the underlying motion of hypothetical molecules. Temperature is a measure of the average speed at which molecules travel—the more agitated the molecules become, the higher the temperature. Heat is understood in terms of the total molecular energy within a substance. (See ABSOLUTE ZERO; ATOMISM.)

Boltzmann also gave a molecular explanation of entropy: It is the degree of molecular disorder. Left to itself, the internal order of a system breaks down, and its entropy spontaneously increases. The only way in which entropy can decrease is to make the molecules align in a more orderly fashion. This happens when steam condenses or water freezes, slowing down molecules until they can arrange themselves in patterns in response to attractive forces. Slowing down means a decrease in temperature, and in this way Boltzmann was able to relate entropy, temperature, heat, and internal energy in terms of molecular motions.

The number of molecules in even the smallest crumb of matter is astronomically large. It is out of the question to discuss molecular motions in detail. The best that Boltzmann could do was to work with average, statistical effects. Hence the term *statistical* mechanics.

Boltzmann's ideas were not accepted in his lifetime. The influential physicist and philosopher Ernst Mach dismissed the whole notion of molecules as an unnecessary hypothesis. Faced with this rejection, Boltzmann committed suicide. It was only after his death that the fundamental significance of his approach was understood.

# The Steady-State Hypothesis

See THE PERFECT COSMOLOGICAL PRINCIPLE.

# Structuralism

Structuralism, strongly associated with France, is based on hypothetical deep structural relationships. Sociologist Claude Lévi-Strauss argued that the complex webs of social relationships and institutions in many different societies are based on a small number of principles related to incest taboos. In psychology, Jean Piaget and others isolated basic cognitive strategies that give rise to a wide variety of human behavior. The linguist Noam Chomsky argues that the surface structures that characterize individual languages are less important than deep linguistic structures common to all human speech. Even mathematics can be given a structuralist edge. Categorical algebras are not so much about particular mathematical relationships within, say, geometry or algebra, but about the sets of universal relationships that exist between them.

Structuralism assumes that a small number of objective, universal principles—structural relationships—lie behind all human knowledge and behavior. By contrast, poststructuralism, or deconstruction, stresses subjectivity and relativism. (See RELATIVITY AND RELATIVISM.) A text does not have a single objective meaning, as structuralism would hold. It is generated in the act of reading and reflects the background and prejudices of the author. A character in a novel appears different to male and female readers and undergoes a sudden dislocation of meaning if one learns the author one had assumed to be male is, in fact, female. Clearly, the meaning does not lie exclusively in the text but is partly created in the act of reading.

While a structuralist universe is based on a few structural laws, a poststructuralist's is self-generating, with a strong subjective element. Contemporary science steers a course between these two extremes. Nonlinear systems stress principles of self-organization and context dependence, while quantum theory emphasizes the role of the observer and the context-dependent nature of complementarity. Science also assumes a certain level of objectivity and independence. The world may be a text whose meaning lies in the act of reading, but the text itself, most scientists believe, has an objective, independent existence. (See also EMERGENCE; REDUCTIONISM; SOCIOBIOLOGY.)

# Superconductors

In a superconductor, electrical currents can flow without any electrical resistance for many years. This is related to BOSE-EINSTEIN CONDENSATION, in which pairs of electrons move in a coherent and collective fashion.

Electrical currents occur when an astronomical number of electrons flow through a wire. In a normal metal, electrons collide with atoms and other obstructions. The result is a disruption of smooth flow called electrical resistance. In a superconductor, it vanishes because the river of electrons becomes coordinated, like a ballet. Individual electrons become merged into a large-scale coherent wave that can pass obstructions undeflected.

Superconductivity is caused by small but subtle attractive forces between a pair of electrons, which enable it to behave like a single (spin-1) entity. The connection between SPIN AND STATISTICS tells us that spin-1 particles can all condense into the same quantum state, called Bose-Einstein condensation. Something analogous occurs in a superconductor. The entire gas of electrons can be described by a single, coherent quantum-wave function, which extends over several feet rather than over molecular dimensions.

Normally, the effects of these attractive forces are swamped by the random, thermal movements of atoms and electrons within the metal. It is only when a suitable metal or other compound is cooled to a few degrees above ABSOLUTE ZERO that these motions become sufficiently damped for attractive forces to dominate. It is at this critical temperature that superconductivity becomes possible.

Because superconducting cables can carry a high electrical current, they are used to build electromagnets with extremely high magnetic fields. Superconducting magnets are used in a wide variety of applications, from magnetic resonance imaging in medical diagnosis to the superconducting supercolliders that investigate elementary particles. The twin phenomena of superconductivity and intense magnetic fields are also found in NEUTRON STARS.

Superconductors' ability to transmit electrical energy without resistive losses gives them great technological importance. To take another example, a significant percentage of electrical energy in transmission lines between a power station and a city is lost in the form of heat. No such loss is experienced with superconducting cables. The advantage is balanced by costs incurred in refrigerating the cables to maintain them at a sufficiently low temperature for superconductivity to operate.

Today physicists are developing high-temperature superconductors, new materials that have stronger attractive forces between pairs of electrons. In these materials, superconductivity occurs at temperatures tens of degrees higher than absolute zero, but much lower than the freezing point of water. The dream is to design a superconducting material that will operate at room temperature, avoiding the need for expensive refrigeration of superconducting cables.

# Superfluids

A superfluid has the remarkable property of flowing without any resistance. Molecules or atoms in a normal liquid behave in a chaotic way, constantly colliding and scattering. Set the liquid in motion, and these random processes produce a resistance to flow. Parts of the liquid move slower than others, creating drag, eddies, vortices, and even turbulence. Obstacles in the liquid's path also act to reduce the flow, whose energy is dissipated as heat. The result is that the initial movement soon comes to rest.

In a superfluid, by contrast, motion is totally coherent. The entire liquid exists in a single quantum state and is described by a single wave function of macroscopic dimensions. Superfluids, like LASERS and SUPERCONDUCTORS, are examples of BOSE-EINSTEIN CONDENSATION.

The longest-known superfluid is helium II, discovered (though not understood) by Heike Kamerlingh Onnes in 1908. It forms at 2.18° above ABSOLUTE ZERO. It is composed of atoms of helium-4, each of which, although made up of fermion particles, acts effectively as a boson. (See BOSONS.) Helium-4 contains a nucleus, composed of two protons and two neutrons, surrounded by two electrons. Its behavior is that of an integral-spin boson. (By contrast, the nucleus of helium-3 contains two protons but only one neutron, and, having a fractional spin of ½, behaves like a fermion. But at three thousandths of a degree above absolute zero, the helium-3 atoms pair up, like electrons in superconductors, to behave like bosons and also form a superfluid. This was discovered in 1972, using more advanced cooling methods.)

At sufficiently low temperatures, the individual helium atoms condense into a singe quantum state—a Bose-Einstein condensate—in which the entire liquid acts as a single entity. Bound together within a single wave function, the helium atoms are no longer able to scatter or move as free individuals: They are coordinated within a single movement. As a result, helium-4 or helium-3 becomes a superfluid able to flow through the narrowest capillary without any resistance at all. Placed in a beaker, the superfluid has the bizarre property of flowing up the walls in a very thin film, and then down the outside, from which it can drip onto the floor! Raise the temperature by a fraction of a degree, and the superfluid state is lost and liquid helium-4 behaves like a normal liquid.

The interiors of NEUTRON STARS are believed to contain a superfluid composed of paired neutrons. This is at a high temperature, but also at a high enough pressure to squeeze its neutrons into this state. Superfluids are one example of a large-scale state of matter that defies common-sense experience and can be understood only by quantum principles.

# Supergravity

Supergravity was one of the early (mid-1960s) attempts to formulate a Theory of Everything that could solve the inconsistencies that divide QUANTUM PHYSICS and GENERAL RELATIVITY. (See THEORIES OF EVERYTHING.) Supergravity was especially favored by Stephen Hawking.

Any Theory of Everything must be able to unite the properties of General Relativity (gravity and space-time) with those described by quantum theory (bosons, which are particles of force, and fermions, which are particles of matter). Supergravity tried to do so by uniting gravity and SUPERSYMMETRY, a theory according to which bosons and fermions (forces and particles) are interchangeable. But supersymmetry itself couldn't cope with certain experimental results, such as the non-conservation of parity in weak nuclear interactions. Experiments showed that all neutrinos, like our corkscrews, spin in the same direction relative to their forward motion—a violation of parity. (See CPT SYMMETRY.)

Because of the problems, it is now accepted that supergravity is not a successful way forward. It has since been superseded by the theory of SUPERSTRINGS, which currently remains the favored candidate for a successful Theory of Everything.

# Supernovas

A supernova is a remnant of a heavy star (see STARS) that suffered an enormous explosion as it reached the end stages of its life. For a few days, a supernova shines as brightly as a whole galaxy. Then it gradually cools down for a few months, leaving behind a neutron star or a black hole, surrounded by a cloud of heated, glowing gas. (See NEUTRON STARS; BLACK HOLES.) Supernovas are observed about once a year in each average galaxy.

The beautiful Crab Nebula in the constellation Taurus is the remnant of a very bright supernova, visible even in daylight, whose explosion in 1054 was recorded by Chinese and Japanese astronomers. A bright explosion in 1987 in a galaxy near ours was studied intensively by modern astronomers and confirmed our general picture of these events. It radiated a huge pulse of NEUTRINOS, the expected result of the conversion of ordinary matter into neutrons.

A heavy star, like other stars, begins its life by burning hydrogen into helium at its core. When the core hydrogen is exhausted, the radiation pressure diminishes. The core then shrinks and becomes hot enough to burn helium into carbon. The process is repeated to burn carbon successively into neon, oxygen, sulfur, and iron, but that is the end of the line. Iron cannot be burned into any further chemical element with release of energy. The radiation pressure now decreases further, and the star collapses under its own gravity. If the core is too massive to end up as a white dwarf, it is crushed into a neutron star or, if it is heavy enough, into a black hole.

The heavy star's collapse creates a shock wave and enormous heat and light, which blow off its outer layers. The explosion also creates many other nuclear reactions. It enriches the interstellar medium with the heavier chemical elements that become part of later-formed stars. We would not be here were it not for supernovas before our sun was formed.

# Superpositions

The fictional case of Schrödinger's alive/dead cat (see PROLOGUE) and the adventures of A QUANTUM HUSSY illustrate the fact that quantum entities can experience more than one possible reality at a time, each playing out its individual drama simultaneously with many others. In quantum language, these multiple possibilities are known as superpositions. We get one possible reality literally "on top of" another. In the quantum realm, superpositions are the norm. Schrödinger's wave function, the mathematical construct that describes any piece of quantum reality, always contains a plethora of possibilities, all equally "real" and many mutually contradictory.

Any wavelike process can be superposed onto another to form a combined process. This is true of colors, sounds, water ripples, force fields, or, for that matter, our own experiences. White light is a superposition of all the colors of the rainbow, which can be separated out with a prism. Space is filled with a superposition of different television programs that a receiver can tune in to. But particles of the classical Newtonian sort cannot be superposed to form a single large particle. Each Newtonian particle occupies its own separate place in space and time and retains its own identity, even though several particles may be stuck together by forces. In classical mechanics, all things are either waves, which can be superposed, or particles, which cannot. Nothing is ever both.

Quantum mechanics, with its WAVE/PARTICLE DUALITY, is more subtle. Every quantum entity has both a wavelike aspect and a particlelike aspect, so that, though one or the other aspect predominates, *any* quantum entity can get into a wavelike superposition. Schrödinger's equation for describing quantum reality is a wave equation. Thus even particles, like photons—or cats—can get into a superposition of two states to give a third state. In the two-slit experiment, a single photon exists as the superposition of two possibilities, each one of which goes through one of the two open slits. In the cat paradox, the possibility that the cat is alive is superposed with the possibility that the cat is dead, to give the third possibility, that the cat is *both* dead *and* alive.

Quantum mechanics is counterintuitive, or a strain on common sense, because in everyday life we never see one thing going through two slits, nor cats that are both alive and dead. In ordinary experience, things behave like waves *or* like particles. So how do we get from the quantum realm where everything exists in a state of superposition (Schrödinger's wave function) to our world, where only some things do? How do all those multiple possibilities become a single actuality? This is the outstanding problem facing quantum theory. (See COLLAPSE OF THE WAVE FUNCTION; THE MEASUREMENT PROBLEM.)

Ordinary logic, like our everyday experience of cats and photons, is an either/or logic, a logic of particles, multiple-choice forms, and lawyers who ask, "Did you see it or did you not? Just answer yes or no." But there is a newer branch of logic, known as FUZZY LOGIC, that deals with superpositions and matters of degree. This is a both/and or wavelike logic.

# Superstrings

Does space consist of three dimensions, or ten, or even twenty-six? The theory of superstrings proposes that there are hidden dimensions to our universe.

String theory began with the observation that elementary particle resonances (the different energies at which new elementary particles are produced in the colliding beams from particle accelerators) form regular patterns, not unlike the overtones from a plucked string. This led the Italian physicist Gabriele Veneziano to propose in 1968 that the HADRONS, the strongly interacting elementary particles, are in fact energy vibrations of incredibly small strings. Roughly, the QUARKS in a hadron are bound together by strings. The most elementary units of geometry are not points in space but tiny extended strings. The quantized vibrations and rotations of these strings are supposed to account for all hadrons.

Veneziano's original theory was ingenious, but soon ran into difficulties when it was discovered that the only way for its mathematics to satisfy both quantum theory and relativity was if these strings existed in a space of twenty-six dimensions (for BOSONS) or ten dimensions (for FERMIONS). At the time, physicists persisted in trying to connect string theory to the theory of quarks, suggesting that quarks are actually the ends of strings. The reason that isolated quarks have never been seen became immediately obvious—break a string in half in the hope of capturing a free end, and a new end is generated.

The creators of string theory had hoped that using strings instead of geometrical points would avoid the infinities and divergences that plague quantum field theory. As it turned out, the theory contains other unexpected difficulties. For a time it was eclipsed by SUPERSYMMETRY and SUPERGRAVITY.

In the early 1980s, the English physicist Michael Green and the American John Schwarz married the ideas of string theory to those of supersymmetry to create superstrings, an approach that, at the time, claimed to be the Theory of Everything. (See THEORIES OF EVERYTHING.) Superstrings proposed to explain not only the properties of all the elementary particles and their interactions, but even the nature of space-time. The theory was hailed as a true unification of quantum theory and General Relativity (gravity). This time the superstrings were closed loops, only $10^{-33}$ centimeter in size.

In this "heterotic" ("two-way," applying to both bosons and fermions) superstring theory, there are twenty-six dimensions. Waves traveling counterclockwise around the closed string are the bosons; waves traveling clockwise and using only ten dimensions of space-time are fermions. Strings can interact with each other, or with the space-time background. As a model for reconciling bosons, fermions, and space-time, the theory is mathematically ingenious, not to say beautiful. But we are not certain that it describes the real world.

If space-time has so many dimensions, why do we sense only three of space and one of time? Theory has it that all the dimensions were created at the instant of the Big Bang, when the size of the entire cosmos was far smaller than that of an elementary particle. In the period of rapid expansion that followed, three of these dimensions expanded or unrolled, while the remaining dimensions remained tightly curled up. Today three of the spatial dimensions can be measured in billions of light-years, while the others are much smaller than the radius of an elementary particle. The short dimensions are effectively invisible, yet make their effects felt through the forces of nature.

For a time superstring theory looked as if it could indeed serve as a true Theory of Everything, until physicists began to find problems. The process by which just three spatial dimensions expanded was mysterious. Fashions changed and less interest was expressed in the theory, although it had given rise to a number of insights and some intriguing mathematics, such as American physicist Ed Witten's axiomatic field theory. Today many theoretical physicists persist in pushing ahead with Theories of Everything, although the mathematics proves to be enormously difficult, and the concrete predictions so far are few. This has led some physicists to wonder whether what may be required is not simply more "ideas" and new mathematics, but fundamental insights into some much deeper theory that lies below both relativity and quantum theory.

# Supersymmetry

When the fundamental symmetries of nature are discovered, elementary particles that once appeared to be markedly different turn out to be relatives reflected in a mirror of symmetry. Later, if these symmetries do not turn out to be exact, the deviations can be ascribed to processes of SYMMETRY BREAKING.

The LEPTONS became unified under THE ELECTROWEAK FORCE, the HADRONS under QUANTUM CHROMODYNAMICS, and both in GRAND UNIFIED THEORIES, yet one absolute division remained, that between FERMIONS and BOSONS. In many ways fermions and bosons represent quite different orders of matter. Fermions, with their fractional spin and obedience to Fermi-Dirac statistics, are the particles out of which the material world is composed—nuclei, atoms, and molecules. Bosons, by contrast, are the quantized excitations of the various force fields that bind matter together. These two very different types of particles do not, at first sight, appear to be manifestations of a single underlying symmetry.

Nevertheless, physicists have proposed the new supersymmetry, under whose transformations bosons become fermions and vice versa. This symmetry itself is quite unlike anything previously used in physics; it mixes together abstract and space-time symmetries. The transformations of supersymmetry include what is called the Poincaré group—all the transformations found in SPECIAL RELATIVITY—plus an abstract transformation that converts bosons into fermions. The totality of all these transformations is supposed to unify all the elementary particles, as well as proposing a whole range of supersymmetric partners.

Corresponding to the spin-1 photon, there should now be a fermion particle called a photino—a sort of hybrid of a photon and a neutrino. The supersymmetric companion of the graviton is a gravitino; that of the gluon, a gluino. The W and Z particles of the electroweak force have supersymmetric wino and zino partners. Sleptons partner leptons, and squarks reflect into quarks. None of these hypothetical new particles has been discovered, which implies either that they do not exist or that their masses are extremely high.

A significant theoretical advance occurred in 1976 when supersymmetry was combined with earlier string theories to create SUPERSTRINGS. Not only did superstring theory attempt to explain all the elementary particles and forces in terms of the geometry and topology of extended stringlike structures, it also sought to provide a foundation for the structure of space-time. The theory is extremely elegant mathematically, but too complex to solve its equations in general.

It is difficult to tell at this stage if supersymmetries represent profound new insights into the nature of the physical world, or are no more than another expression of the ingenuity of theoretical physicists in coming up with novel mathematical schemes. Physics in the earlier years of the century was often guided by philosophical considerations and deep physical insight. Mathematics came later, as a tool for the expression of physical ideas. Today it is often mathematics that drives physics, and novel schemes that do not always have a clear underlying physical meaning are proposed.

## Symmetry

See overview essay D, THE COSMIC CANOPY; ANTIMATTER; CONTINUOUS SYMMETRIES; CPT SYMMETRY; SUPERSYMMETRY; SYMMETRY BREAKING.

## Symmetry Breaking

Symmetry breaking indicates that nature evolves through the unpredictable breaking of its most fundamental symmetries. (See overview essay D, THE COSMIC CANOPY.) Physicists believe that the most fundamental laws of nature are also the most simple and, mathematically speaking, the most elegant. For example, it is believed that just after the instant of the BIG BANG creation of the universe, the cosmos was totally symmetric. All the forces of nature had equal strength, the masses of the elementary particles were all identical, and space was everywhere the same. (See GRAND UNIFIED THEORIES.) Yet when we look around, we realize that most of these symmetries have been broken. The universe itself is far from uniform, with stars clustering into galaxies, and galaxies forming clusters and superclusters, not to mention such ordinary—to us—things as chemical elements and compounds and biological species.

To take other examples, a magnet can easily pick up a paper clip, and a comb brushed through the hair will cause tiny pieces of paper to jump into the air, indicating that the electrical and magnetic forces are now much stronger than the force of gravity. The very different masses of the elementary particles also show that quantum matter is very far from being symmetric. But how did such an asymmetric universe grow from a symmetric birth? And why don't the symmetric laws of nature have symmetric solutions? The answer is symmetry breaking, one of the cornerstones of modern physics.

Think of a new penny; aside from the markings on front and back, the two sides are totally symmetric. Toss the coin in the air, and there is an equal chance of its landing heads or tails. The laws that govern the tossing of a coin are symmetric with respect to heads or tails. Yet when the coin lands on the floor, only one out of two events is possible; it has to land on one side or the other. Coin tossing is a matter of pure chance, and the result must always be one of two possibilities, either of which breaks the basic symmetry of the situation.

In a similar way, at the quantum level, a symmetric law has several different solutions, all equally probable. Taken together, these solutions reflect the original symmetry, just as petals on a rose, taken all together, display the flower's symmetry. One petal by itself breaks this symmetry. Likewise, in the real world, only one outcome or solution from a range of equal possibilities can be made manifest. The result is called a broken symmetry. The physical outcome has a much lower symmetry than the laws that determine its particular manifestation. Evolution to greater structure and more information is via symmetry breaking. A blank page has more symmetry than one with writing.

A magnet is an example of symmetry breaking. But is the original isotropic symmetry of each spinning electron broken, or merely hidden? Physicists have proved that whenever a symmetry has been broken, there will be an associated field of massless particles that, in a sense, hold on to the hidden symmetry—these are called Goldstone bosons. In the case of a ferromagnet, these take the forms of "spin waves," fluctuating waves of magnetism that pass along the magnet. When more and more of these waves are excited, the original ferromagnetism is lost, individual spinning electrons point in arbitrary directions, and the isotropic symmetry of space is restored.

A single atom can be placed anywhere in empty space. With no markers around, it could sit anywhere. Gather together a collection of similar atoms, and they will form a regular pattern—a crystal lattice. Although the crystal has its own degree of symmetry, it has broken the total homogeneity of space. As before, whenever symmetry breaking occurs, physicists expect to find a new field with massless particles. In this case, the particles associated with symmetry breaking are called phonons and turn out to be the waves of vibration that pass along a crystal lattice. Excite enough phonons, and the crystal vibrations increase to such a point that the entire lattice breaks down and the original homogeneous symmetry of space is restored.

Symmetry breaking is equally important when it comes to the world of elementary particles. Physicists believe that just after the original Big Bang, all the forces of nature were identical, and all elementary particle masses were the same (zero). But, as with the tossed coin, particular solutions occurred by chance, and in the process symmetry was broken. The first instants of our universe therefore involved a series of symmetry breakings. Initially the color force between QUARKS broke away from the electroweak force, and the hadrons developed quite different masses from the leptons. Next, the electroweak force fragmented into two parts—electromagnetism and the weak force. At this stage of symmetry breaking, the leptons themselves gained different masses. Our universe is an evolving one.

At each stage of symmetry breaking, massless particles, Goldstone bosons, appear. But things turn out to be even more complicated. Other types of symmetry breaking create new massless particles called Higgs bosons. These massless particles interact and, in the process, gain mass. While the particle that carries the electromagnetic force—the photon—is massless, the vector bosons that carry the weak nuclear force all have mass. (See GAUGE FIELDS.)

In the case of a crystal lattice or a ferromagnet, it is not so much that symmetries are irreversibly broken as that they become hidden. Excite the associated massless bosons (lattice vibrations or spin waves) and shadows of the original symmetry appear. Finally, as ferromagnetism vanishes or a crystal lattice breaks apart, the original symmetry reappears. Something similar occurs with elementary particles. Go to high enough temperatures using high-energy particle collisions or, at times, close to the creation of the universe, and the electromagnetic and weak nuclear forces become indistinguishable. Physicists speculate that, at energies far beyond what can be produced in the largest particle accelerators, the color force would again become indistinguishable from the electroweak force. In this region, conditions would approach those that applied at the Big Bang origin of the universe.

# Systems Theory

Using systems theory, it is possible to describe the functioning of a corporation, a national railway, a computer, or even the human body in terms of a set of general rules. The overall approach is to analyze highly complicated organizations into interconnected subsystems that operate in relatively autonomous ways. Some are interconnected by arrows (stimulation or inhibition). The same rules apply to a wide variety of different systems.

The human body contains heart, lungs, kidneys, liver, and brain. Each contributes to the correct functioning of the whole body and cannot (normally) exist independently of it. While it is possible to analyze the structure and functioning of an organ at ever finer levels—in terms of its biochemical and cellular composition, for example—from the perspective of systems theory it need be considered only a BLACK BOX, a subsystem that generates a specific output in response to a variety of inputs. Some organs secrete biochemical substances that act on other organs and in turn receive chemical messages that indicate when more or less of this secretion is required. According to systems theory, it is not necessary to understand the detailed cellular makeup of the organ in question to describe the approximate dynamics of the overall process. Scientists need know only qualitatively how the organ responds to messages it receives.

Systems theory has wide applications in a variety of fields. According to it, everything from machines to human organizations, ecologies, plants, and animals can be modeled as a dynamic system. It offers a powerful approach to management theory. A visitor to an organization sees physically distinct buildings and offices. The same organization could be viewed as the input of raw materials, the manufacture of components, and the final marketing of a product. The organization can also be seen in terms of the way information and decisions flow along its telephone lines, computer links, and interdepartmental mail, during boardroom meetings, and in verbal instructions given by supervisors.

Systems theory analyzes the internal dynamics of an organization in terms of information flows or the various sequences involved in the manufacture of a product. Subsystems and their boundaries are identified together with their connections to other subsystems. Having discovered the different black boxes, or suborganizations, one can build a dynamic model of the overall organization. Once this has been set down diagrammatically as a series of interconnected boxes (subsystems), it is immediately apparent where lines of communication are missing, or FEEDBACK loops need to be added. Viewing an organization in a diagrammatic fashion gives a clear picture of its internal structure and the way information, decisions, and commands are flowing.

General systems theory enables the dynamics of organizations and natural systems to be modeled on a computer, which can be used to improve efficiency and productivity as well as to expose defects in management structure. Negative feedback stabilizes a system through small corrective signals issued each time behavior deviates from a prearranged norm. These signals enable the system to swing back on track. (See CYBERNETICS.) But what if there is a noticeable delay between the sending of a corrective signal and its reception? In this case, behavior and correction may become out of phase, and, rather than stabilizing the system, feedback can push the dynamics into a series of oscillations. Dynamic modeling is able to identify these potentially dangerous situations and modify an organization's internal structure accordingly.

The simplest systems consist of two boxes interconnected by two arrows, representing enhancing effects (positive feedback) or diminishing effects (negative feedback). There are three possible systems in this simple case. If both arrows are positive, the two boxes are cooperative. (See COEVOLUTION.) If one arrow is positive and the other negative, we have a PREDATOR-PREY system. Two negative arrows give a competitive (Darwinian) system. (See DARWINIAN EVOLUTION.) With more boxes, the number of possible types of system increases rapidly.

General systems, with their networks of subsystems, inputs, outputs, and feedback loops, have the capacity to exhibit all the characteristics of the nonlinear systems that are discussed in this book—stability, amplification of fluctuations, oscillations, THE BUTTERFLY EFFECT, and even chaos. (See CHAOS AND SELF-ORGANIZATION.) Effective control does not, therefore, lie in old-fashioned mechanical causality, such as rigid top-down commands, but rather in analyzing an organization into its various subsystems and then determining the range and character of its dynamic behavior. By identifying potentially troublesome regions of behavior it becomes possible to modify the organization's structure accordingly. Effective organizations can be considered dynamic wholes in which the management structure is only one aspect or subsystem, rather than as fixed, rigid structures in which commands come down from on high.

Creative managers look toward the sciences to supply their metaphors and images. In a world of change, new technologies develop rapidly, the nature of work is changing, and national and economic boundaries are in flux. It is important for organizations to be specifically designed for adaptation to change. Even systems theory can be too rigid an approach if the boundaries around subsystems are rigidly fixed or inapplicable, if its overall system boundaries are insufficiently defined. Rather, organizations should be thought of as OPEN SYSTEMS, in constant communication with the environment and capable of self-organization. The concept of learning systems has also been applied to organizations to describe the way in which they respond intelligently to a changing world.

The whole notion of the meaning of a corporation is undergoing a revolution, a change of perspective as profound as that which accompanied the emergence of the first banking and business houses in Renaissance Europe. In many organizations, creative managers are as likely to be learning about semiotics, contemporary art, and theoretical physics as they are about economics and tax laws. Where once concerns were exclusively for profits, efficiency, and productivity, today they may also be about improving human, environmental, and aesthetic values. (See also overview essay B, ORDER IN SCIENCE AND THOUGHT.)

# Tachyons

Can anything move faster than THE SPEED OF LIGHT? Some physicists have proposed the existence of faster-than-light particles. Einstein's relativity shows that the faster a particle moves, the more its mass increases. This means that a very rapidly moving particle requires more force to accelerate to a higher speed than a slowly moving particle. As speeds become very close to that of light, it takes an infinite amount of force to increase the particle's speed. At the speed of light, a particle's mass would become infinite and its length (in the direction of motion) would shrink to zero.

For a variety of reasons, the velocity of light represents, in Einstein's theory, an absolute limit to the speed at which a material particle or signal of information can travel. Nevertheless some physicists have proposed a bizarre mirror world located beyond the speed of light, a world inhabited by elementary particles called tachyons. Just as the speed of light represents a barrier in our world, it would be a barrier within the domain of tachyons. Apply force to an electron, and it accelerates to a higher speed. Apply a force to a tachyon, and it slows down. Just as infinite force would be needed to accelerate an electron to the speed of light, so, too, it would require an infinite force to slow down a tachyon to the speed of light. The tachyon world, if it existed, would be a curious mirror reflection of our own subluminal world.

Not many physicists take the tachyon proposal seriously today, not so much because tachyons have never been observed, or because they break a fundamental law of physics, but simply because assuming their existence would not help physicists resolve the many difficulties and anomalies that face them. There seems to be no good reason to complicate matters further with a new set of particles. Therefore, when theories, such as early versions of string theory, predict the existence of tachyons, this is assumed to indicate errors in the theory.

# Teleology

Teleology, from the Greek *telos* ("end") and *logos* ("reason"), is the concept that things are directed by virtue of their ends and ultimate purposes. Early philosophers believed that living beings have an inbuilt goal toward which they move. Aristotle taught that everything from rocks to planets strives to attain its proper place and motion. Teleology, the final cause, was one of several causes believed to be operating within nature and took its place alongside material, efficient, and formal causes. Thus individual human behavior was thought to be influenced as much by the relative mixture of four elements, or humors, and astronomical harmonies as by accidental encounters with the external world or personal history. (See CAUSALITY.)

Galileo, Descartes, and Newton gave pride of place to material or efficient cause—mechanical pushes and pulls and the attraction of gravitation—and formal cause: the form of the laws of physics. When the fall of an apple or the motion of the moon could be explained by the force of gravity, there was no longer a need to invoke teleology. The human soul was exempted from this mechanical view. But what applied to inanimate matter must, later nonreligious scientists believed, be equally true of life and human society.

In such an environment, teleology was soon discounted. For a time it lingered on in VITALISM, which held that living systems contain an inner life force and a directedness that can never be reduced to purely mechanistic explanations. While Darwin's theory argued that evolution is a chance matter, teleology suggested that life moves toward predetermined goals. In political theory, teleology found expression in nineteenth-century utilitarianism, in which every action is (or should be) taken to further the goal of "greatest good (or happiness) of the greatest number." Law and ethics seemed to need an underlying belief in human freedom and purpose. (See GAMES, THEORY OF.)

Today, despite the great success of Western science and technology, teleology survives within science under a variety of guises. SYSTEMS THEORY describes the complex nonlinear behavior of artificial and natural systems in terms of goals and ends. Biological behavior and cognition are often described as goal oriented. THE ANTHROPIC PRINCIPLE in physics and cosmology argues that the present structure of the universe, with its delicate adjustment of nature's constants, did not happen by chance. Versions of the Anthropic Principle suggest that this process is directed toward the evolution of life and consciousness. Earlier in the century, Pierre Teilhard de Chardin and others suggested that nature's goal is the creation of the noosphere (the domain of the mind) through the intermediate stages of geosphere and biosphere.

The theoretical physicists John Wheeler and Fred Hoyle have independently proposed a highly speculative notion, that teleology is not preexistent or innate, but the goal is (or will be) created in the future and then acts back on the past. The result of a quantum event that took place in the past may not be perceived (or registered) by a human agent until the present moment. This action, Wheeler argues, then has a direct effect on the past, collapsing the wave function of that earlier time period. Because some past quantum states collapse only when we observe their fossil traces today, human beings can, through their observations, influence events millions of years ago. Thus, the past can be restructured by the present.

# Theories of Everything

A Theory of Everything is a longed-for, all-embracing theory of the universe that will unify matter (FERMIONS), forces (BOSONS), and curved space-time in one grand picture that applies from the first split second after THE BIG BANG. Such a theory, which would also have to include an adequate theory of quantum gravity, has not been achieved so far.

For the GUT era, before the universe was $10^{-36}$ second old, our physics can successfully describe one kind of matter, one kind of force, and curved space-time. (See GRAND UNIFIED THEORIES.) But the three things are described separately and cannot be combined. Even the separate concepts fail in the still earlier—and still more speculative—PLANCK ERA, before the universe was $10^{-43}$ second old. A basic difficulty is that quantum theory is formulated in a "flat" (Euclidean) space, whereas GENERAL RELATIVITY involves a curved space. The two theories do not mix well (see QUANTUM GRAVITY; THE QUANTUM VACUUM), and the difficulty becomes acute at the Planck scale. This is where physicists feel the need for a new concept or new kind of theory.

The most promising candidate so far is a new type of symmetry called SUPERSYMMETRY, according to which every boson (particle of force) or fermion (particle of matter) has a supersymmetric partner, hitherto undiscovered because it is too heavy—quark/squark, lepton/slepton, boson/bosino, and so on. In these terms, the theory of SUPERGRAVITY works better, though it is not yet fully consistent. Since 1984, a new type of theory called SUPERSTRINGS has held out even greater promise than supersymmetry. If the enormous difficulties raised by trying to calculate superstring theory could be overcome, and useful predictions could be made with it, it might be the sought-after Theory of Everything.

Of course, it remains the physicists' assumption that a Theory of Everything actually exists and that scientists merely have to look for it. Armed with such a theory, physicists could calculate in principle all the properties and processes of elementary particles. Physics itself would not necessarily come to an end; interesting and difficult problems connected with COMPLEXITY, NONLINEARITY, and consciousness (see CONSCIOUSNESS, TOWARD A SCIENCE OF) still exist. But if a Theory of Everything was achieved, there would be a feeling that physics had at last touched the ultimate level of matter.

At this stage, however, a Theory of Everything remains an act of speculation. Reputable physicists have often rashly claimed that such a theory was at hand, only to be disappointed. There is no compelling scientific or philosophical reason why such a fundamental theory should exist. It might equally well be that any finitely describable theory is only approximately true. Perhaps beneath the quantum domain there is a world of pure chaos, without any fixed laws or symmetries. What appear, at the level of elementary particles, to be unifying theories may be no more than the averaging out of underlying fluctuations. After all, the eighteenth-century laws that govern the behavior of gases were later found to be the average effects of random collisions of molecules. Just as, in chemical, social, and economic systems, order can emerge out of chaos, so, too, a Theory of Everything might conceivably be no more than a statistical effect of chance processes originating at an underlying level.

# Thermodynamics

Literally *thermodynamics* means the study of the movement of heat. It is the science of the way energy moves in a system and how it is transformed into useful work or heat.

Thermodynamics arose in the wake of the industrial revolution through attempts to improve the efficiency of engines. THE FIRST LAW OF THERMODYNAMICS deals with transformations of heat and various other forms of energy, under the principle that the total energy within the universe is constant.

THE SECOND LAW OF THERMODYNAMICS places additional limits on the nature of these energy transfers. Cycles of energy transformation are a little like what happens to money during international travel. At the border, dollars are converted into German marks, marks into francs, francs into lira, or lira into pounds. Each transaction in governed by laws of exchange. The individual bank notes look different, but the same concept of money (like energy) applies in every case. Just as with energy in the First Law of Thermodynamics, the total amount of money remains constant. But this does not apply to the cash in your pocket.

Suppose you begin with a thousand dollars and convert it into a variety of different currencies and then back into dollars, without ever spending a cent. As any experienced traveler knows, you will end up with slightly less than your original thousand dollars. Some of your original money is no longer available for purchasing goods. But, like energy, it has not vanished; it has simply been diverted from your pocket to pay for a variety of bank charges.

In a similar way, when energy moves around a circuit, some of it is used to pay nature's bank charge—in this case, heat. In every energy circuit, a percentage of useful energy—energy that can be used to do work—is lost as heat, which dissipates into the environment.

If this did not happen, it would be possible to build a perpetual motion machine. Electrical energy could be used to drive a dynamo that rotated and drove a generator to produce the original electricity that drove the dynamo! The Second Law of Thermodynamics demands, however, that within each cycle a certain percentage of energy must be dissipated as heat. (Exceptions are SUPERCONDUCTORS and SUPERFLUIDS.)

The dissipation of potentially useful work into heat is also associated with an increase in ENTROPY. Entropy can be thought of as a measure of disorder within a system. A highly ordered system can be put to work, but as that order is lost, the ability to do work decreases. Increasing entropy is associated with a loss in the potential to perform work. The Second Law of Thermodynamics dictates that in an energy cycle the total entropy always increases, or remains the same. It never decreases.

One consequence of the Second Law is that you have to switch on your refrigerator if you want the milk to cool. Left to itself, a cup of hot coffee will cool. But if you want the coffee to keep cooling below room temperature, the Second Law dictates that you must use some energy—the electrical energy used to power a refrigerator. Conversely, if you want to heat the coffee above room temperature, energy must be used—electrical or gas energy in a stove.

In the nineteenth century, thermodynamics was formulated using macroscopic concepts of heat, work, temperature, and entropy. But it is also possible to give an explanation of these laws in terms of the underlying behavior of molecules. This is called STATISTICAL MECHANICS. Newton's laws of motion are used to describe the collisions of molecules. By averaging out billions of such collisions, it becomes possible to reproduce the conventional laws of thermodynamics. To take an example, thermodynamics absolutely forbids the spontaneous movement of heat from a cold to a warm body. Statistical mechanics does not impose an absolute restriction on this; instead it shows that such a process is highly improbable. Just as it is unlikely that a shuffled deck of cards will be dealt in exact order by value and suit, so it is even more improbable that the entropy of a system will ever decrease spontaneously.

The Second Law applies to closed systems, in which matter and energy can neither enter nor leave. But there are many OPEN SYSTEMS: animals, plants, cities, whirlpools, the weather. In these, order can be produced in a part of the system, even if entropy increases elsewhere. This is a new area of thermodynamics. (See CHAOS AND SELF-ORGANIZATION.)

# Thinking

What do we mean by *thinking*? Does the word cover all our conscious and semiconscious mental activity—specific ideas, problem solving, concentrating, remembering, dreamy reflection, insight, and some dream activity? Or is *thinking* a more limited concept with specific boundaries? To common sense, the broader definition seems to apply, but among cognitive scientists and computational psychologists, *thinking* generally refers to problem solving—to calculation, data analysis, and pattern recognition. There are two clear models for these things, one based on SERIAL PROCESSING, the other on parallel processing (see NEURAL NETWORKS.)

Things like mental arithmetic or planning a journey are done slowly, in step-by-step stages, following known rules, like multiplication tables or train timetables. Serial computers are excellent at such tasks. (See FORMAL COMPUTATION.) But a quite different style of computation is needed to recognize a voice or a kind of tree, or to play a successful game of tennis. In these cases, many different kinds of data must be rapidly combined into a whole. Neural networks are good at such tasks.

Serial and parallel computation are complementary. Serial computers are good at doing precise tasks that are "rational," i.e., rule following or algorithmic. These rules must be defined and stated in a language, the program. Serial computers are very bad, however, at recognizing complex patterns and executing complex activities, such as driving a car. Parallel processors are, by contrast, good at pattern recognition but have no language and no ability to follow explicit rules. They operate by association, not by "reason."

Human beings often use both methods of computation in tandem. A good chess player immediately and "intuitively" will recognize the general type of a chess position and select a small number of moves for further analysis by serial processing. This is different from the method used by a chess-playing computer, which misses the intuitive stage and simply analyzes *all* possible moves, no matter how absurd. At the other end of the spectrum, children and lower animals survive quite well by "intuitive" thinking, but are less good at rational calculation, which seems to be associated with a later stage of evolutionary development.

The human brain appears to have structures corresponding to both types of processors, serial and parallel. Parallel processors require rich networks of more or less randomly interconnected elements. The brain contains many such elements; indeed, parallel computers are more commonly known as neural networks. Serial processing, by contrast, requires precise, point-to-point connections resembling telephone cables or personal computers. The brain also has this kind of "wiring"—for example, in the optic nerve. The cerebellum is also very precisely structured.

Parallel processors learn gradually by repetition, like Pavlov's conditioned reflexes on a more complex scale, whereas serial processors store information after one input. The brain has both kinds of memory. The parallel memory system is spread throughout the cortex, as are the neural networks, whereas the "one-off" memory system involves the hippocampus, a part of the primitive forebrain. Hippocampal damage or deterioration, common in older people, leaves individuals able to learn things slowly, not always able to remember things seen only once. The two systems are also biochemically different.

Apparently, human beings use both serial and parallel processing in an integrated way. This is compatible with many everyday observations. People who are intoxicated or half asleep often regress to a more associative type of thinking, but seem able to be quite rational when fully in possession of their faculties. Freud's two parts of the mind—the rational, conscious ego and the associative, less conscious id—compare quite well to serial and parallel processing abilities. Freud, of course, thought that these two systems are destined to be at war, whereas many modern therapists feel that an integration of the two is at least possible.

Do serial and parallel processing systems in the brain exhaust all we have to say about thinking? Both everyday experience and data from psychologists argue not. Neither type of computation has anything to say about consciousness, responsibility, meaning, or creativity. Neither casts any light on how we create new ideas or learn a language embodying new ideas. Parallel processing uses no language at all. Serial processing has a program language, like that of a personal computer, but it cannot be modified. Yet we know that children, as an obvious example, expand their language effortlessly.

Many philosophers argue that there is more to thinking than just computation, and thus more underlying its capacity than just serial and parallel processing. This feeling lies at the heart of the whole ARTIFICIAL INTELLIGENCE debate and is the motive behind seeking new physical theories of mind. (See CHAOS THEORIES OF MIND; PENROSE ON NONCOMPUTABILITY; QUANTUM THEORIES OF MIND.)

# The Three-Body Problem

A French mathematician's attempt to solve the famous "three-body problem" at the end of the nineteenth century opened the doors to chaos in science. Henri Poincaré demonstrated that instability and infinite sensitivity lie at the heart of Newton's clockwork.

In the eighteenth century, Isaac Newton demonstrated that, under the influence of the earth's gravitational attraction alone, the moon would move in a closed, elliptical orbit. Newton showed that the same force of gravity that causes an apple to fall from a tree also acts to pull the moon toward the earth. Under any other law of force, the moon's orbit would not be closed and the periodicity of the moon's phases and eclipses would not be maintained. Newton's triumph was to apply the principle of universal gravitation and his three laws of motion to the entire solar system, demonstrating that each planet (neglecting the small influences of the other planets) has a fixed, periodic orbit, returning to its exact starting point at the end of that planet's "year."

This Newtonian clock, once set in motion, continues its precise cyclic motions for millions of years. Nature at heart was regular, and the future could always be predicted, manipulated, and controlled.

The fly in the ointment was that, while Newton could fully solve a two-body problem (the earth-moon or earth-sun system), he was unable to work out precisely what would happen when the effects of an additional third body were added. The three-body system, earth-moon-sun, does not repeat exactly.

The Three-Body Problem cannot be completely solved. The best mathematicians and astronomers have been able to do is make a series of approximations. The sun exerts a relatively small gravitational pull, called a perturbation, on the moon orbiting the earth. Astronomers first determine the orbit of the moon in the absence of the sun, then add an approximate correction for the sun's gravitational pull. The result is a slight modification of the moon's orbit. In the second stage of the approximation, a further small correction is added to the new orbit. In this fashion, successive approximations, each tending to be smaller than the last, are added together, and the same procedure is applied to the whole solar system.

But is the solar system stable in the long term? In the last years of the nineteenth century, Poincaré tried to resolve the question by attempting to classify the various possible solutions to the Three-Body Problem. He confirmed that, in most cases, the tiny perturbations produced by a distant third body have a negligible effect on a final orbit. But in certain exceptional circumstances, very tiny perturbations accumulate, feeding back into an asteroid's orbit and causing it to behave erratically. There were even cases in which, under cumulative perturbations, one body would begin to behave chaotically, changing its orbit wildly or even flying out of the solar system altogether.

Poincaré's result came as a tremendous shock. It suggested that chaos sleeps within our solar system. For over 200 years Newton's clockwork universe had been the paradigm of order and predictability. Now it was found to be infected by chaos and uncertainty. Did this mean that the earth's orbit would be stable into the far distant future, or would it eventually behave chaotically and fly away from the sun? Poincaré's result was the first surfacing of chaos theory, an intimation of the very rich behavior inherent in nonlinear systems in which tiny effects can feed back and amplify. (See FEEDBACK.)

Poincaré's analysis of the Three-Body Problem was overshadowed by the scientific revolutions of relativity and quantum theory, and the inherent difficulties in solving the nonlinear differential equations that describe the Three-Body Problem appeared to be insurmountable. (See NONLINEARITY.) It was not until 1954 that the mathematicians A. N. Kolmogorov, V. Arnold, and J. Moser developed new techniques for solving nonlinear differential equations. Their result, the KAM theorem, made Poincaré's result more precise. It proved that the solar system is stable—providing all perturbations are small and the "years" of the planets do not fall into simple ratios like 1:2 or 2:3. In these cases, even the smallest perturbations will accumulate whenever the planet orbits, like an adult regularly pushing a child on a swing. The result is a form of RESONANCE or positive feedback, what happens when a public-address system begins to screech. Instead of remaining in a regular orbit, the planet's motion becomes increasingly erratic and chaotic. It is currently thought that our solar system is probably stable.

Today, thanks to the development of high-speed computers, it is possible to work out the details of these complex planetary motions. Astronomers speculate that regions of potential chaos may explain the gaps in the rings of Saturn. Place a rock in one of those regions, and its orbit becomes wildly unstable, causing it to fly off into empty space. This could be why an asteroid orbits the sun in a regular way for millions of years, until one day its motion becomes erratic, causing it to leave the asteroid belt and wander through the solar system until it hits the earth as a meteorite.

# Time

Long ago, Saint Augustine said that time is both very familiar and very hard to think about. It is a rich concept, with several layers of meaning. It can be merely a marker for an appointment, or an experience of flowing and changing. Time is irreversible in our lives, but physics does not tell us why or how. We can distinguish at least three of its meanings and discuss some of the speculative problems they raise.

### *Spacelike Time*

We can say where an event took place, in three spatial dimensions, and when it was, in one time dimension. All four dimensions are described by ordinary numbers and together form space-time. They are treated differently by the laws of physics, but are to some degree interconvertible. (See SPECIAL RELATIVITY.)

There are some questions we can't seem to help asking. If the time dimension of our universe arose with THE BIG BANG, what happened "before" that? If it will rain tomorrow, is that fact "already" in time? Can we travel in time? (See TIME TRAVEL.) Such questions are based on thinking of ourselves and our consciousness as outside physical time, yet experiencing the flow of time. But flowing time is not the same as spacelike time.

### *The Arrow of Time*

There are only five or six known phenomena in the physical universe that are time-irreversible: ENTROPY increase, THE EXPANDING UNIVERSE, consciousness (see CONSCIOUSNESS, TOWARD A SCIENCE OF), quantum COLLAPSE OF THE WAVE FUNCTION, and a few particle phenomena (see CPT SYMMETRY). The relation of these phenomena to each other and to time-symmetric phenomena in other areas of physics remains an unsolved problem. (See THE ARROW OF TIME.)

### *Flowing Time*

We are not consciously aware of any single instant of time, but rather of a "slice" of time perhaps a few seconds long. We hear things like tunes or sentences as a unity, in the period of time known as the specious present. This is the length of each of our experienced "moments." But this is not the way time is treated in Newtonian physics, where it is pointlike. Despite this, time spreads over a finite region in some systems, for instance the COHERENCE time of LASERS (often one second or more). If our consciousness was associated with some such coherent physical system, there might be less conflict between physical time and our experience of a specious present. (See QUANTUM THEORIES OF MIND.)

Within the region of the specious present, we experience a flow of time. New events begin while others fade into the past. Can we find any parallel for this in physics that could underlie its representation in consciousness? We appear to need a framework in which past, present, and future would be qualitatively different—future events are mere possibilities until they happen. This framework does seem to appear in modern physics, for example in the link between ACTUALITY AND POTENTIALITY IN QUANTUM MECHANICS. Again, if consciousness was associated with a collapse of the quantum wave function in the brain, as some have suggested, the paradox of flowing time would be nearer resolution. At this stage, however, such ideas are tentative.

# Time Travel

The idea of being able to travel backward or forward in TIME has fascinated generations of science fiction writers and is a serious preoccupation of a few scientists. Since one aspect of time in physics is that it is a dimension not unlike the three spatial dimensions, why shouldn't we be able to travel along it as we do through space? The issues of travel into the future and the past, however, raise very different scientific and philosophical problems.

It may be just possible, given known science, that I—that is, my body and my consciousness—could travel into some future time. At the simplest level, my body could be deep-frozen and then thawed back to life at some future date. Some institutions in California already offer this service to those who may be suffering from a fatal disease for which there is not as yet a cure. Another more theoretical possibility is that I might take a high-speed journey on a spaceship. On my return, I would be younger than my former contemporaries. (See THE TWINS PARADOX.)

There is philosophical difficulty about travel into the past, some imagined examples of which give rise to embarrassing paradoxes. Suppose I travel into the past, where I murder my own grandfather-to-be. Then neither my father nor I would ever have existed. So I *couldn't* have done the murder. (The self-contradictory logic of this example is reminiscent of the "I am lying" paradox, and of GÖDEL'S THEOREM.) To exclude the paradox requires still further mental gymnastics, such as the supposition that there are many worlds, and that I have performed the murder only in some alternate branch of reality. (See THE MEASUREMENT PROBLEM.)

The equations of GENERAL RELATIVITY do allow limited forms of theoretical time travel. On an ultramicroscopic scale (the Planck length, $10^{-33}$ centimeter, many times smaller than the size of a proton), HEISENBERG'S UNCERTAINTY PRINCIPLE means that such a well-localized portion of space-time is not smooth, perhaps not even continuous. Space-time has a shifting, foamlike structure. Mini–BLACK HOLES may constantly form and dissolve. A baby universe may branch off; connected at first by an umbilical cord (a "wormhole") to our universe, it then launches on a career of its own. Or a wormhole may momentarily connect two widely separated space-time points within our own universe.

A kind of evanescent time travel via wormholes seems possible, but it is of no practical use. A macroscopic body would be crushed out of existence by the energies in a wormhole. In any case, it would be a

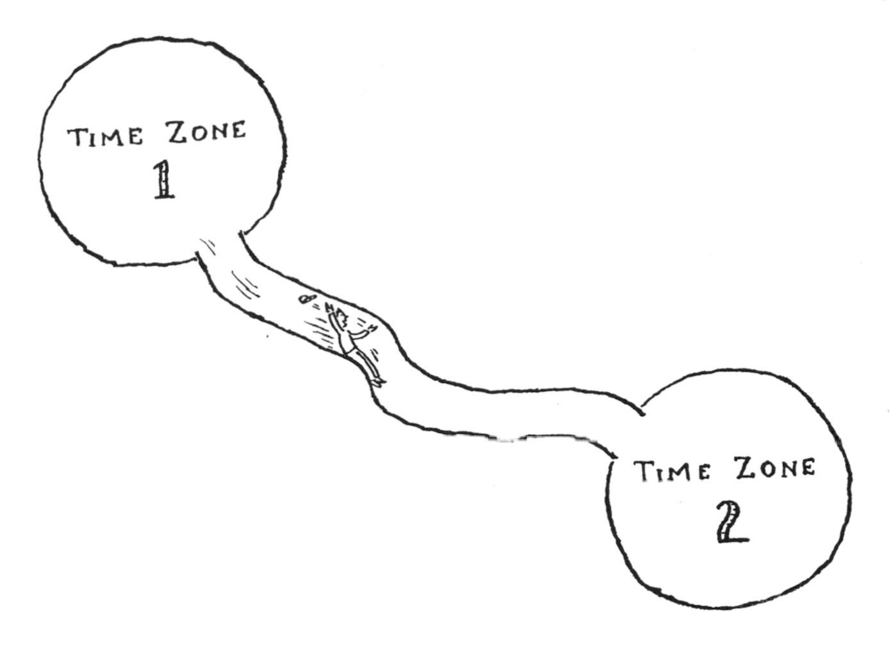


one-way ticket to an unpredictable destination. It is more realistic to conclude that our concept of time, and therefore of time travel, breaks down at the Planck length.

Even on a macroscopic scale, General Relativity holds out further possibilities of time travel. There might, for instance, be a large wormhole at the center of a large black hole, though this would still probably crush any large body entering it, and one would never return to describe the adventure. Slightly more feasibly, General Relativity predicts that a large, heavy, rotating cylinder, perhaps 1 light-year long, would so distort space-time around it that circumnavigating it would shift a person forward or backward in time. In practice, this is out of the question, but it does raise the philosophical problem mentioned earlier. Even the theoretical possibility of time travel and its paradoxes means that some assumptions must change, either about General Relativity or about our notions of time. At this stage, it is unclear which.

# Transpersonal Psychology

The post-Jungian analyst James Hillman has described the self as a mansion with many rooms. Each branch of psychology has concerned itself with some of these rooms—the rat cellar of the behaviorists' reductive vision, Freud's dark and tempestuous id, the higher mental functions of the Gestalt and cognitive psychologists (see GESTALT AND COGNITIVE PSYCHOLOGY), the Freudian ego (see PSYCHODYNAMICS AND PSYCHOTHERAPY), or the "drawing room" of interpersonal relationships and ego-centered self-development described by HUMANISTIC PSYCHOLOGY. But what lies beyond the ego and its day-to-day relationships? What function or room of the self accommodates those dimensions of human experience that transcend the instinctive, the personal, or the ego-bound? What accounts for our religious sense or the higher inspirations of creative artists? Can any branch of psychology address these questions? Can it be "scientific"?

Inspired by the foundational work of William James and Carl Jung, transpersonal psychology concerns itself with these further realms of experience. Its disciplines rest on the belief that there are higher states of consciousness and higher spiritual values beyond the personal ego. Such further experiences, which can occur spontaneously or through participation in rituals, prayer, or meditation, are thought to be transformative. They can lead to personal growth. Where humanistic psychology goes one step beyond the individual to consider his or her personal relations, transpersonal psychology focuses on a more spiritual or cosmic level—the individual's relation to God or unity, to beauty, to nature, or to "archetypes"—shared, recurring patterns of psychic energy or imagery—of the unconscious.

Descriptions of transpersonal experiences and views about them are found in all the world's major religions, East and West. More than half of the ordinary people in the West, according to various surveys, report having occasionally experienced a cosmic sense of oneness, peace, beauty, or contact with "something beyond the self." Most have valued these experiences.

Some psychologists have always been interested in the transpersonal dimension of experience, which they recognize as existing independently of, or without commitment to, any particular religious belief system. They see the transpersonal as a universal human capacity that may, in some individuals, become focused within a given belief system. William James's *The Varieties of Religious Experience* and Jung's work on psychological archetypes have provided a basic foundation for study.

Transpersonal psychology as an organized branch of psychology dates from the 1950s as a "fourth force" after behaviorism, psychodynamics, and humanistic psychology. Its academic study has been paralleled by a greatly increased interest in Eastern religions and westernized forms of them. The decline of traditional Christianity in the West left a vacuum of meaning, which many felt might be filled by cultivating transpersonal awareness or Eastern insights. There is a broad general agreement on basic principles among those who practice transpersonal psychology or psychotherapy, but there is considerable variation about details.

Fundamental to all transpersonal psychology is the view that our familiar, ego-bound state of consciousness is limited. "I" simply cannot encompass all of conscious reality. Higher states of consciousness are held to exist, and it is thought that these can be experienced through a variety of practices, including meditation on complex subjects or images, simple awareness meditation (such as counting one's breaths or repeating a simple mantra), prayer, ritual, rhythmic chanting or drumming, and the use of psychedelic drugs.

Most of these practices have been borrowed from one or another of the great religious traditions, where they were used by committed people within the context of a whole religiously based lifestyle, and with aid from a supportive community. Without this inclusive context, such practices may be ineffectual or even, to some personalities, dangerous. Jung himself believed that the whole transpersonal dimension was best explored by older people who had already rooted themselves successfully in everyday life. Many of the great religious traditions have counseled the same restraint. Nonetheless, there clearly are transpersonal phenomena associated with higher or altered states of consciousness that must form part of any comprehensive paradigm in psychology. (See MEDITATION.)

What might lie in some transpersonal domain beyond the individual ego? Answers to this question have traditionally depended upon religious teachings, but today some people draw them from new interpretations of scientific insight. For Christians, the transpersonal is envisioned as a union of the immortal soul with God, analogous to human love. In the Hindu Vedanta it is envisioned as a reabsorption of the self into its source, resembling sleep. For Buddhists, it is the awareness of an inexpressible something (the Void) that transcends our intellectually formulated pairs of opposites and the details of everyday life, but highlights these details with great clarity. This special awareness is similar to being fully awake. For those who look to science, the transpersonal may be a resonance between qualities of THE QUANTUM VACUUM, the underlying ground-energy state of all that is in the universe, and related qualities of the human mind.

The spontaneous transpersonal experiences of ordinary Western people are of different types—overwhelming love, a feeling of oneness with everyone and everything, an overwhelming sense that the whole universe is alive, or the sense of a conscious presence that pervades everything. Are these various experiences simply different ways of approaching a single, indescribable state? Is the underlying state, or states, purely subjective, or does it have some kind of objective existence beyond the individual experience? These basic philosophical questions are still unresolved, yet the experiences themselves remain of psychological importance because people invest them with meaning and value.

# Turing Machines

A Turing machine is an idealized "digital computer" defined by Alan Turing in 1938. A digital computer is a machine for processing definite, discrete symbols, like dots, dashes, numbers, or letters, that it is fed as input. Such machines are very exact. Turing's machine was a crucial breakthrough in the thinking about and development of all electronic computers. It was a mental model, or mathematical construct, rather than an actual machine.

A Turing machine is a computer that has unlimited potential memory storage and only a few very simple kinds of processing operations. It is equivalent to any other digital computer ever defined, except that it is terribly slow because its basic operations are so simple. It is this simplicity that makes the Turing machine so theoretically important. Though it can do (slowly) what any other machine can do, its operations are easier to think about clearly.

THE CHURCH-TURING THESIS states that any mathematically definable or physically constructible computer will have the same basic powers as a Turing machine. Thus in understanding Turing machines, we are understanding the fundamental nature of all digital computing machines.

A Turing machine has a "language" consisting of a finite "alphabet" of symbols, one of which may be printed in any square of an infinitely long tape. This tape is the machine's "memory." When the machine is started up, its tape is blank except for a finite number of squares—its "input data." Its "program" is located in a reading/writing head as a finite set of rules. If, for example, we suppose there are four symbols in the machine's alphabet and five states the head could be in—five ways the head's program could manipulate the symbols of the alphabet—then there are twenty possible combinations of symbol and head state. Thus there are twenty different actions the machine might perform.

Each rule in the Turing machine's program, located in the head, is "if this, then do that." It might say, "If the head is in state *s* and sees a tape square on which a *y* is printed, *then* replace *y* by *y/*, go from state *s* to state *s/*, and move one square to the left (or to the right, or stop)." If the machine eventually stops, what is printed on its tape is called the output.

All known mathematical and logical operations can be reproduced by simple Turing machines. More important still, there is a "universal Turing machine"—one simple Turing machine capable of *every* logical or mathematical operation because it can simulate *any* other computing machine's function. Thus, the alphabet and program (set of rules for the reading/writing head) of the universal Turing machine can imitate the behavior of any other Turing machine if the universal machine's input contains information about the other machine's construction.

The universal Turing machine was a fundamental development in the mathematics of computing and crucial to the whole computer revolution. Until Turing's work, people like Charles Babbage were designing a different machine to meet the needs of each new computing task. After Turing—and his discovery that if we can prove something about a universal Turing machine, we can prove it about *any* computing machine—computers became quite versatile. The "Stone Age" computer described in FORMAL COMPUTATION is another design for a universal computing machine. Any computation it can perform can be translated into one performable by a universal Turing machine, and vice versa, just as French and English can be translated into one another.

Some workers in ARTIFICIAL INTELLIGENCE have argued that all human thinking is reducible to the operations of a Turing machine or formal computation in general. The whole argument about the "computability" of human thinking centers on this claim. To say that something is computable is to say that it is, in principle, calculable by a Turing machine, that it can be calculated with a program operating according to a set of definite rules known as an algorithm. Thus, the thought processes in our heads are equivalent to the linear *click, click* of the Turing machine's tape. This argument has, however, met with opposition.

GÖDEL'S THEOREM proved that the set of results provable by any one Turing machine will always be incomplete—there will always be some *extra* step outside the formal language of the machine's program needed to "understand," or to prove, or to sum up the rules of the program itself. Turing showed this in a different way. We now know that human abilities and the structure of the brain suggest that we use a kind of parallel processing (see NEURAL NETWORKS), at least in practice, though this may not be necessary in theory. Parallel processing is a very different form of information processing from that done by a Turing machine. Many people working on the capacities of the human mind think that broader models still, going beyond both serial and parallel processing, may be needed to encompass qualities like consciousness, free will, or creativity. (See also CONSCIOUSNESS, TOWARD A SCIENCE OF; PENROSE ON NONCOMPUTABILITY; QUANTUM THEORIES OF MIND; THINKING.)

# The Turing Test

The facility to build "intelligent" machines has sparked the larger question of whether such machines should be considered conscious. Is a high-powered computer different in principle from a human mind? How can we know if an animal or an alien is conscious? In 1950, mathematician Alan Turing proposed a simple behavioral test for deciding these questions. The principle underlying the test is that if a system *behaves* as if it is conscious, we must assume for all practical purposes that it *is*. The Turing criterion is accepted by many cognitive scientists and workers in AI, but it remains controversial.

The Turing test itself is very simple. Imagine yourself seated before a typewriter or computer screen that is connected to another unit in a different room. At the second unit, there may be either a human being or a computer standing in for one. You can type in questions or messages from the console, and the second unit will answer them. According to Turing, if you cannot *tell* from the nature of the answers you get whether you are communicating with a human being or a computer, then the system at the other end *must* be regarded as conscious.

You may want to reject Turing's criterion on the basis of common sense. If, by analogy, you hear music over a telephone wire and cannot *tell* whether the music is being made by a live performer or a recording, it does not follow that it *must* be a live performer. But a supporter of the Turing test would reject such an analogy. With the music, he would say, it is in principle possible to go to the source of the music and observe whether it is live. By contrast, we can never directly observe another's consciousness. All we have to go by is behavior. This argument carries some weight but places too much emphasis on verbal behavior. We usually regard infants, dogs, speakers of foreign languages, and paralyzed people as conscious, but none of them could pass the Turing test.

There is a still deeper criticism of the test. We presume that consciousness in human beings is associated with the activity of certain brain systems, not just with external behavior. Any given behavior can be simulated by a computer or a robot, even though it lacks the relevant brain structures. Equally, the correct brain structures might be functioning even though there is an absence of telling behavior—as in paralysis or partly anesthetized people. Such considerations lie outside the scope of BEHAVIORISM in psychology, yet behaviorist assumptions are the basis of the Turing test. (See also CONSCIOUSNESS, TOWARD A SCIENCE OF; FUNCTIONALISM; PENROSE ON NONCOMPUTABILITY.)

# The Twins Paradox

In SPECIAL RELATIVITY, the mass, length, and time measurements of phenomena depend on the relative motion of the observer and the observed, each in its own inertial frame. (See INERTIAL FRAMES.) The effect becomes significantly large only at relative speeds approaching THE SPEED OF LIGHT. Experiments confirm this. Fast-moving, short-lived particles from particle accelerators or cosmic rays appear to us to live longer when their speeds are greater.

The paradox in all this arises when *two* inertial frames are moving near the speed of light with respect to each other. If two observers in spaceships pass one another, each will regard the other's "time" as slowed down relative to his or her own. But what happens if one of two twins travels away in a spaceship at a great speed and returns much later? Each twin cannot be younger than the other.

This "twins paradox" ceases to be a paradox in GENERAL RELATIVITY. It is seen that the traveling twin must have accelerated when turning around to come back again; that twin was not in an inertial frame throughout the journey. A recent experiment has confirmed this. Two extremely accurate atomic clocks were the "twins." One was flown around the world in a jet plane. On its return, it was found to be slightly slow by comparison with its nonaccelerating twin.

# Twistors

Twistors suggest a way of describing space-time and the elementary particles without having to resort to point particles and continuous space. They also reveal a link between QUANTUM PHYSICS and GENERAL RELATIVITY.

Most physicists believe that the infinities encountered when calculations are made in QUANTUM FIELD THEORY have their origin in the assumption that space can be defined down to the level of infinitesimal points. Yet, as physicists like John Wheeler point out, quantum theory indicates that space-time must break down into a foamlike structure at very short distances. (See THE PLANCK ERA.) This use of continuous space-time coordinates and mathematical points troubled the English mathematician Roger Penrose, who attempted a different way of building up space-time.

Penrose had demonstrated earlier how many of the properties of three-dimensional space can be created out of networks of spinors, the simplest possible quantum mechanical objects, used to define the two possible values of an electron's spin. This suggests that space-time is not a passive backdrop against which the quantum theory is played, but rather that space-time and the elementary particles may emerge from the same ground.

Penrose's next step was to generalize the spinor into what he called a twistor. Because the mathematics of complex numbers is used, it is difficult to visualize a twistor, but it is a little like a moving corkscrew of infinite length, or, as Penrose jokingly put it, "the square root of a light ray." In conventional geometry, the point becomes a more complicated secondary notion, defined by a conjunction of many individual twistors. This is a little like the way a tiny region of space is defined experimentally through the intersection of many elementary particles coming from a particle accelerator.

Penrose's original goal was to define space-time and the elementary particles in terms of twistors rather than points, and at the same time to derive the various elementary particles and their fundamental symmetries. The mathematical difficulties involved are daunting; nevertheless, the theory has already produced a number of provocative insights. A twistor and its complex-number partner behave a little like position and momentum in quantum theory. When the structure of twistor space is disrupted, as if a wave of gravity had passed through the space, the identities of twistors mix. The result is the same as if a quantum event involving position and momentum had taken place. Conversely, "quantum events" in twistor space look a little like waves of gravity, suggesting that some deeper sort of connection has been made between General Relativity (gravity) and quantum theory.

Penrose has been able to write down a number of fundamental field theories using the twistor language, and connections between twistors and SUPERSTRINGS exist. However, the goal of Penrose's original program remains to be achieved, and the technical difficulties barring further progress are severe.

# Virtual Particles

Virtual particles are the elementary particles that never quite come into existence. In our everyday world, an object is defined and occupies a specific location in space. Not so in the domain of elementary particles, where energy can be borrowed and paid back from the vacuum field. In this way sufficient energy can sometimes be taken on loan in order to create temporary virtual particles. Though not directly observable, they make possible QUANTUM TUNNELING and other effects.

The more you try to pin down the position of an electron, the more uncertain its momentum becomes. Something analogous occurs with time and energy. If you define an elementary particle over shorter and shorter time intervals, its energy becomes more and more uncertain. And, according to Einstein's famous $E = mc^2$, some of this fluctuating energy can be used to create particles.

However you look at it, every elementary particle is surrounded by a fluctuating cloud of virtual particles that blink in and out of existence. The heavier the virtual particle, the more energy has to be borrowed, and the shorter its life. These ephemera are called virtual particles because they are never free, never able to venture out on their own and leave their traces in bubble chamber detectors, in Geiger counters, or on photographic plates.

To take one example, the proton surrounds itself with virtual mesons. (See QUANTUM CHROMODYNAMICS.) It is constantly borrowing energy from the vacuum field to create and emit a meson and then reabsorb it. Put another way, over very short time intervals, its energy becomes so uncertain that it is able to create a meson. When two protons approach closely, they are able to exchange virtual mesons, producing an attractive interaction, the "strong nuclear force." As before, the energy accounting is strict, and net energy is never lost or gained by the vacuum field.

Normally, these virtual particles cannot be observed. Observation would violate the law of conservation of mass-energy, since a virtual particle registered in the laboratory would have extracted net energy from the vacuum field. But Stephen Hawking has proposed a case in which isolated virtual particles could become real near a black hole, by taking energy from it. (See BLACK HOLES.)

Like other particles, virtual particles obey conservation laws of charge, spin, and momentum. Thus, when a virtual electron is produced, it is always accompanied by a virtual positron. Normally, these two particles will disappear almost as soon as they have been created, returning energy to the vacuum field. If one of them happens to stray across the event horizon of a black hole, the point of no return, it vanishes forever from our observable universe. A virtual particle has suddenly been made real. Its energy can no longer be paid back to the vacuum state and must therefore be extracted from the black hole itself. The net result, as predicted by Hawking, is that the black holes appear to radiate elementary particles and, at the same time, begin to lose equal amounts of mass. According to this mechanism, they will slowly evaporate into space, becoming smaller as they do so. But only a tiny black hole could have evaporated completely in the $10^{10}$ or so years since THE BIG BANG. (See also VIRTUAL TRANSITIONS.)

# Virtual Transitions

A virtual transition is the movement of an "impossibility" through time, an illicit sort of "feeler into the future" that a quantum event can make. Virtual transitions are the Cinderellas of the quantum world. They can travel briefly into forbidden or impossible zones of reality, but they have to get back again before their coaches turn into pumpkins.

"Real" or "really possible" processes in physics are those that conserve energy and momentum. Virtual processes are able to break these fundamental constraints of nature for a brief time. A virtual particle, such as a virtual photon or virtual electron, can break through an energy barrier, as in QUANTUM TUNNELING, and appear on the other side. In radioactive decay, for instance, the attractive forces of the nucleus are so strong that nothing should be able to escape, yet entities like alpha and beta particles get out by making a virtual passage through the nuclear energy barrier.

Other virtual transitions occur in high-energy particle physics. The collision and mutual annihilation of two particles may give rise to virtual intermediate states that then promptly turn into other real particles that conserve energy and momentum. The likelihood of the total process in these cases depends upon the existence and number of the virtual pathways.

According to David Bohm, virtual transitions are badly named because they aren't illusions. They have *real* effects. Though only possibilities, quantum possibilities have an existential reality of their own. They evolve, and they can interact with each other and with the environment. Bohm compares them to possible "missing links" in biology, trial runs made in the course of nature's evolution that, though they don't themselves survive, possibly give rise to new species that do. We might also compare them to acts of the human imagination. We can imagine any number of future scenarios for a given situation, some of which may be totally "crazy" associations. These, though "unreal," might cause us to change our characters or our behavior.

Another set of virtual transitions occurs continually on the "surface" of THE QUANTUM VACUUM, the underlying sea of potentiality out of which arises everything that exists. Particles can emerge from the vacuum only if they have sufficient energy to do so, but VIRTUAL PARTICLES continually spring into existence for a brief time before disappearing back again. They are like a cloud of impossibilities that cluster on the edge of possibility. And though "unreal," their presence exerts a subtle pressure on all material existing things. This pressure, which can actually be measured with sensitive instruments, is known as the Casimir effect.

Virtual particles can also become real and stay on this side of reality if they "steal" energy from another real particle through collision. When this happens, the particle that has been "robbed" may disappear back into the vacuum. In this case, the virtual particles are behaving like quantum vampires, drawing from others that have it the energy they need to exist.

The purpose of a quantum entity's virtual transitions is to explore all possible futures with a view to finding the best one. Usually, the best one will be the one of highest probability. When the entity has sufficiently tested the waters, a "real" event will be actualized. An electron will be seen actually to have traveled to a certain destination via a particular path, and at that point all its other possible paths disappear (see COLLAPSE OF THE WAVE FUNCTION), but not without sometimes leaving their effects. As physicist Richard Feynman describes the process, "The electron does anything it likes. It just goes in any direction at any speed, forward or backward in time, however it likes, and then you add up the amplitudes [of these virtual paths], and it gives you the wave function." (See THE WAVE FUNCTION AND SCHRÖDINGER'S EQUATION.)

It is because of their simultaneous virtual transitions that quantum systems can evolve so quickly. Instead of trying out each possibility one by one, they can try out all possibilities at once. (See A QUANTUM HUSSY.) Some mathematicians believe a technology might be conceived to employ them in superfast quantum computers, which could explore all possible options at once instead of serially.

# Visual Perception

Vision is one of the easiest to study and also the most complex of our sensory abilities. Because it is simple to correlate measured neuron activity with visual data shown to animals and humans, scientists can make a direct link between brain activity and visual experience. They know when a part of the brain responds to green, or another part to roundness, and this is useful for understanding not just vision but the brain itself. Some scientists, like Francis Crick, believe that a better understanding of visual perception may lead to a rudimentary understanding of the brain's link to our conscious experience. (See CONSCIOUSNESS, TOWARD A SCIENCE OF; CRICK'S HYPOTHESIS.) Western philosophers have often seen vision as a proper subject for speculating on the nature of the mind.

There are more than twenty areas of the brain's cortex that specialize in some aspect of vision. The retina itself is a part of the brain. In human beings and other primates, the cortex is the most important part of the conscious visual system, although there remain connections to lower, more primitive parts of the brain that dominated earlier in the history of evolution. Because these lower brain connections are still relevant to human vision, some physiologists concentrate their research on things like the frog's visual system, which has structures similar to those in the human lower brain.

When light falls on the eye, it is focused on the retina, a sheet of neurons located at the back. Unlike a passive camera film, the retina does more than simply record the visual pattern. It is actively involved in the first stage of its processing. The retina contains over 100 million light-sensitive elements—many "rods," which are activated by dim light, and fewer "cones," which respond to color but require brighter light. Visual information is encoded and summarized all over the retina by a mechanism of parallel processing (see NEURAL NETWORKS) and is sufficiently processed to pass on to the optic nerve, which contains only one million neurons.

In 1981, David Hubel and Torsten Wiesel received the Nobel Prize for crucial work on a cat's cortex, two decades before, that had shown key features of our visual processing. By measuring the sensitivity of cells in the primary visual cortex to signals coming from the retina, they discovered that these cells respond to a light spot on a dark background, and vice versa, or to something like a red spot on a green background. But the brain does not respond to stimuli from scenes of uniform brightness and color. If presented with a wholly uniform scene, we see nothing. The brain needs diversity for its sensory apparatus to function. (See GESTALT AND COGNITIVE PSYCHOLOGY.)

In the first instance, axons of the optic nerve originating in the retina end in the thalamus, a primitive part of the forebrain. The cerebral cortex is a specialized outgrowth of the thalamus that is large in the higher mammals, and the two are connected by a two-way point-to-point neural mapping. While still in the thalamus, the visual signals coming along the optic nerve undergo some form of parallel processing, a stage still not well understood.

Once the signals have passed from the thalamus, they go to the first visual areas in the cortex, at the back of the brain. At this point, individual neurons called feature detectors respond to specific aspects of the visual scene—some to spots, lines, or edges, others to orientation or movement, still others to corresponding parts of the two eyes' retinas, which are used for depth perception.

From this stage on, visual processing becomes increasingly complex. Output from the primary visual area goes to several other areas, then on to several more, in a rough hierarchy. At each stage, the visual features detected by individual neurons are more complex and specialized. Color, shape, and movement are processed in different areas. Some neurons respond best to faces or to other patterns. Thus damage to some localized part of the brain, through illness or injury, may result in the loss of one aspect of vision—the ability to discriminate color or movement, the ability to recognize faces or to pay attention to some area of the visual field. The broad function of each of these localized areas for processing is determined genetically, though the details of what individual neurons may respond to are usually built up through experience, as we would expect in the parallel processing done by neural networks. This is why a rich variety of visual experience is crucial for an infant's later development.

The overall current model of visual perception is of an interconnected hierarchy of visual areas in the brain, each functioning as a neural network for parallel processing. Each higher level of the hierarchy detects more complex features in the visual field. In addition, there are further brain areas in which information from the visual field is combined with information from the other senses, and all this combined information is fed into still other areas, where it can be associated with things like memory, emotion, and action.

This model is soundly supported by experiment, but it has obvious gaps that leave it unable to explain our visual experience fully. Different areas of the brain are known to process different aspects of a visual scene, but there is no one area of the brain where all this information comes together. How, then, do we perceive a whole, unified visual scene with all its colors, movements, and orientations? How do we bring all these features together as perceived objects? These as-yet-unanswered questions are known as THE BINDING PROBLEM. More fundamental still, what have all the known patterns of neural firing associated with vision got to do with our visual *experience*, our *lived sense* of seeing things? Unconscious visual perception is possible. (See BLINDSIGHT.) Are the neurons firing to process visual data different from those that represent sounds or thoughts? These questions are crucial to our attempts to link our understanding of vision with that of consciousness. They remain at the cutting edge of experimental mind science.

# Vitalism

Is life the result of molecular processes, or is a hidden "life force" at work? Vitalism was the philosophical and scientific movement of the nineteenth century that argued that a drive to life lies behind all biological systems. At the time, it seemed impossible to some people to explain life in terms of mechanistic Newtonian physics.

The philosopher Hegel argued that the World Soul evolves through a dialectical movement, one example of which is the transformation of Being into Becoming. Related streams of thought persisted throughout the nineteenth century. The Romantic movement celebrated it. Goethe's "eternal feminine" drew humanity ever upward, Beethoven's music shook its fist at God, Caspar David Friedrich painted lone trees or single figures standing defiant in vast and terrifying landscapes, and Byron's poetry celebrated the heroic journey of Childe Harold. The Romantics' dream was also reinforced by dramatic social change, social revolutions in France and the United States, and Germany's dream of unification.

There were similar ideas in philosophy. Henri Bergson wrote of the *élan vital*, which finds its expression in the material domain as plant and animal life. Evolution, he suggested, is creative rather than mechanistic; nature is constantly seeking new forms. In biology, these ideas became the basis of vitalism, which held that the essence of biological systems cannot be reduced to a collection of molecules and their reactions. What distinguishes a living cell from a mere, albeit complex, collection of molecular reactions is the life force, the *élan vital*. (See also HOLISM.)

According to vitalism, the life force acts on inanimate matter to create life. It is the animating principle of the eternal struggle whereby nature raises itself into consciousness and humanity takes on the aspect of the gods. Versions of vitalism found their way into the plays of George Bernard Shaw, including his epic *Back to Methuselah*. For Pierre Teilhard de Chardin, the goal of evolution's climb is the global planetary consciousness he termed the noosphere.

Vitalism was an account of evolution alternative to that of Darwin. The biologist Lamarck (see LAMARCKISM) proposed that evolution takes place through nature's efforts to improve. When environments change, species attempt to better themselves in a purposeful way rather than through a series of blind mutations. These habits and adaptations are in turn, passed on to future generations.

At first sight, vitalism and Lamarck's notions of evolutionary advance do not appear unreasonable. Since the complex behaviors of living systems are infinitely richer than the reactions of ordinary chemical systems, it is natural to propose new principles of organization. But the more biologists discovered about living systems, the more they realized that everything can be explained in terms of processes at the molecular level. For example, living systems appear to defy THE SECOND LAW OF THERMODYNAMICS, which demands that entropy should spontaneously increase. Does this mean some new vitalistic principle is involved? Chemists also discovered self-organizing (nonlinear) inorganic chemical reactions in which entropy decreases. The Second Law is not violated; order within the system *does* increase, at the expense of pumping excess entropy into the environment. (See CHAOS AND SELF-ORGANIZATION.)

Throughout the nineteenth century, vitalism had many opponents. One of the most powerful was the German scientist Hermann Helmholtz. Vitalism held that, since life is governed by nonmechanistic principles, its essential processes cannot be reduced to laboratory measurements. Life is a holistic process; analysis into components destroys its essential principles. Helmholtz demonstrated that, on the contrary, a wide variety of processes, such as the way in which impulses travel along nerves and the neurological basis of sensations, could all be measured and studied in the laboratory. Another blow to the theory had already come in 1828 when the German chemist Friedrich Wöhler synthesized urea, an organic compound found in the urine of animals. The fact that a compound previously associated only with the life process could be produced by artificial chemical means was seen as the denial of any special principle of life.

Today, vitalism has been largely discredited in the biological sciences. The principle of Occam's razor dictates that scientific theories should not contain unnecessary assumptions, no matter how seductive. Since biological functions all seem to be explainable in terms of molecular reactions, there is no need to invoke the assumption of a life force. Nevertheless, echoes of vitalism occur in many branches of yoga, acupuncture, and alternative medicine in general; it is generally unclear whether the life energies referred to are reducible to forms of energy known to physics. (See also EMERGENCE.)

# The Wave Function and Schrödinger's Equation

Quantum systems have many indeterminate aspects (variables) because those features of the system remain unfixed, or unrealized. They are possibilities rather than actualities, things that might be or might happen, rather than things that are. The quantum wave function is a mathematical description of the possibilities associated with a system at any moment.

Consider the situation of Schrödinger's alive/dead cat. Before we open the box to look at the cat, he has two equally "real" possibilities—the possibility that he is alive and the possibility that he is dead.

The cat himself exists in a superposition of these two states. (See PROLOGUE; SUPERPOSITIONS.) The cat's wave function is a mathematical representation of these two possibilities. Graphically, we could draw it as a wave with two humps, each hump representing one possibility. Alternatively, we could draw the wave function of the top card on a full deck of shuffled but as yet unobserved (quantum) cards. This card has fifty-two possibilities, so its wave function would have fifty-two humps.

The wave function is called that for two reasons. If we describe the possibilities associated with a quantum system mathematically, they look like the mathematical description of a set of waves undulating. But more "tangibly," a wave function describes the momentary state of a system that really does consist of some kind of waves. These could be light, sound, or water waves. The quantum wave function describes the wave aspect of matter—that is, its indeterminate aspect, the aspect that is spread out (as waves) all across space and time.

Today physicists even know what the wave aspect of matter is waving in—each wave is an undulation or excitation of the underlying QUANTUM VACUUM, the underlying ground state of physical reality. (See QUANTUM FIELD THEORY.) Mathematically the wave function can be thought of as a menu of possible meals that one might eat, or a list of horses running in a race, each of which might possibly win.

In quantum theory, all events are possible (because the initial state of the system is indeterminate), but some are more likely than others. While the quantum physicist can say very little about the likelihood of any single quantum event's happening, quantum physics works as a science that can make predictions because patterns of probability emerge in large numbers of events. It is more likely that some events will happen than others, and over an average of many events, a given pattern of outcome is predictable. Thus, to make their science work for them, quantum physicists assign a probability to each of the possibilities represented in a wave function. How likely is it that, of the fifty-two possibilities existing, the top card on a fully shuffled deck will be the queen of hearts? How likely is it that when we open the box, we will find Schrödinger's cat alive and well? The answer to these questions is called a probability function, and is arrived at mathematically by squaring the amplitude of each possibility's hump on the wave function.

$$i\hbar \frac{\partial \Psi}{\partial t} = \left( \hat{p}^2 / 2m + V \right) \Psi$$

SCHRÖDINGER'S EQUATION

Both the wave function (also known as Schrödinger's wave function) and the probability function tell us the possibilities and probabilities associated with a quantum system *at any moment*. They are like still snapshots, which catch a segment of action. But all physics is concerned with how things change over time, with how they evolve. To calculate how the wave function evolves through time, quantum theorists use Schrödinger's wave *equation*. (In very high energy situations, a slightly different wave equation must be used, to take account of SPECIAL RELATIVITY effects.)

Schrödinger's equation describes the dynamic unfolding of a set of possibilities over time and tells us the probability of finding any one possibility actualized in a given experimental situation. This is the equation with which physicists can accurately predict the outcome of a large run of quantum events. And while the evolution of a single event is always indeterminate, Schrödinger's equation describes a fully determinate situation—at any given time, as the wave function evolves, the probabilities associated with any possibility are fixed. For a large number of events, these can tell us exactly what to expect.

Schrödinger's equation is like a set of bookmaker's odds, and we can accurately use it as such with two important provisos. First, the equation is calculated on an undisturbed or unmeasured quantum system that exists as an array of possibilities. The moment any one of these possibilities is actualized—through measurement or observation, for instance—the whole equation must be recalculated to give a new set of odds. If there is a one-in-fifty-two chance of the queen of hearts' being drawn from a full deck of cards, the odds change every time a card is drawn that is not the queen of hearts. Once this card is drawn, the possibility of its being drawn again is zero. If there is a 50 percent chance of finding Schrödinger's cat alive before we open the box, this reduces to zero if we open it and find him dead. (See COLLAPSE OF THE WAVE FUNCTION.)

The other limitation on thinking of Schrödinger's equation as a set of bookmaker's odds is ontological; it has to do with the kind of existence found in quantum reality. The events that Schrödinger's equation describes are *more* than mere probabilities. The wave function really is waving in something. In the absence of observation or measurement, its possibilities evolve and interfere (interact) with each other. They have a real effect on the real world. The possibilities are not yet actualities, but they are more than mere mathematical entities. They are a different kind of being—*potentialities*—described for the first time in modern science by quantum physics. (See ACTUALITY AND POTENTIALITY IN QUANTUM MECHANICS.)

# Wave/Particle Duality

One of quantum theory's most revolutionary ideas is that all the constituents of matter and light are *both* wavelike *and* particlelike *at the same time*. This is known as the wave/particle duality. Neither aspect of the duality—the wavelike or the particlelike—is more primary or more real. The two complement each other, and both are necessary for any full description of what light and matter really are. (See COMPLEMENTARITY.)

The oldest concept of matter is that it is made up of particles, individual pointlike entities, with a few simple properties such as position, movement, mass, and charge. The early Greeks attributed other properties, such as color, to them, but these were not considered basic in classical physics. For purposes of calculation, particles can be very large, like apples or planets, or very small, like atoms or electrons. Any bulk substance, like a pile of sand or a jug of water, can be seen as composed of myriads of tiny particles. (See ATOMISM.) Bulk properties like weight, pressure, and volume are considered the sum of the properties of the parts.

Waves are almost as familiar as particles, although very different.

There are sound waves, ripples on the surface of water, electromagnetic waves (light, radio, X rays, and so on), and the vibrations of a guitar string. In classical physics, waves were treated as disturbances or excitations of some material medium, itself composed of particles—air, water, strings, or "the ether," the universal medium believed by classical physicists to fill empty space. But since Einstein proposed SPECIAL RELATIVITY, it has been recognized that electromagnetic waves are not like this (because there is no ether) but have a reality of their own, as fundamental as that of particles.

A wave has a succession of peaks and troughs, a wavelength (the distance between successive peaks), a frequency (the number of peaks per second), and a velocity. Waves carry energy, which depends upon their amplitude (the average height of the peaks in any region). At any given point as it passes, a wave has a phase, which is where, in the cycle of peaks and troughs, it is at that moment.

Particles behave quite differently from waves. They are localized at one point in space and time, and when two particles meet, they bump into each other, clash, and go their separate ways. Waves are not localized; they can be spread out across vast regions of space and time. When two meet, they can overlap and pass through each other. The resulting disturbance at the point of meeting can be increased or decreased as the two disturbances add or subtract, depending on their phase. Two waves meeting produce interference patterns, a patchwork of crisscrossing peaks and troughs where their disturbances add and subtract. Particles are always individuals, but since any two wave patterns add up to make a third, waves are not. Particles are ultimately discrete or irreducible, but a wave can be regarded as the sum of various "components" in an infinity of ways.

Both waves and particles have well-established, quite distinct mathematical descriptions in classical physics. Trying to combine them in some mathematically coherent way is like trying to marry fire and water, yet it was the genius, and the necessity, of quantum physics to do so. The major impetus came from a renewed attempt to understand the nature of light in the face of certain experimental anomalies.

Newton firmly believed that light was a stream of particles, but his view was overridden by nineteenth-century physicists trying to understand how light could bend around corners (diffraction) and give rise to interference patterns. They decided it was a wave undulating in the newly proposed background ether. Light's supposed wavelike nature could not, however, explain the new experimental results associated with black-body radiation: why a body disposed to emit nonpreferentially the full spectrum of colors does so in an observed bell-shaped pattern of intensities, rather than trailing off to infinity as predicted by the wavelike equations of classical theory. Planck's theory that light is absorbed or radiated, not continuously but in little packets called photons each associated with a quantum of action, accurately explained the observations (see QUANTUM), but left physicists with the conclusion that light was behaving like a stream of little "lumps," i.e., *particles*.

Building on Planck's theory, Einstein was able in 1905 to explain another experimental puzzle, the photoelectric effect, by proposing that light is a stream of photon particles, each carrying one quantum of action. His theory that electrons are knocked out of a metal surface by the photons, like coconuts being knocked off a fairground shelf by balls, was verified by the appearance of a photographic plate exposed to a very weak beam of light. The plate shows a patchwork of black spots where each photon has knocked out an electron, rather than a uniform gray exposure as would be expected if the light were a series of continuous waves.

These particlelike interpretations of light's behavior led to a paradox. Unquestionably, light can behave like a stream of particles. There are other times when it behaves like a series of waves (interference and diffraction, for example). This paradox led to confusion until the 1920s, when it was shown to make mathematical, if not common, sense to say that light *sometimes* behaves like a particle, and *sometimes* like a wave. Light itself cannot be said to be either; it must instead be seen as a potentiality to be *both* at any given time, depending upon the circumstances and experimental surroundings in which it finds itself. (See ACTUALITY AND POTENTIALITY IN QUANTUM MECHANICS; CONTEXTUALISM.)

> "Light seems to behave like a wave on Mondays, Wednesdays and Fridays and like a particle on Tuesdays, Thursdays and Saturdays."
>
> —WILLIAM BRAGG

The opposite paradox arose with solid matter, which can usually be interpreted as consisting of particles, but sometimes behaves as though these had a wavelength. Streams of photons and electrons can, in some experiments, give rise to interference patterns. Even large objects like apples and ourselves have a wavelength, although this is so infinitesimally small as to be of no practical consequence. The wave nature of electrons is the physical basis of the electron microscope, which uses beams of electrons, whose wavelengths are millions of times shorter than those of photons, to view objects too tiny for examination by a light microscope.

Though both light and matter have wave and particle aspects, the two are not identical. Matter is more "solid" than light. (See BOSONS; FERMIONS.) Mathematically, light is described by Maxwell's equation, and solid matter by Schrödinger's equation (see THE WAVE FUNCTION AND SCHRÖDINGER'S EQUATION), or by its relativistic refinements in QUANTUM FIELD THEORY. The analogies are very close.

The both/and, rather than the either/or, nature of light and matter is one of the outcomes of quantum physics that has the most profound philosophical implications. Viewed from within the old paradigm, it seems utterly paradoxical, but taken into our way of thinking and extended through metaphor to things like the individual and relational aspects of the self or society, it gives us a new way of looking at our own experience.

> "A paradox is not a conflict within reality. It is a conflict between reality and your feeling of what reality should be like."
>
> —RICHARD FEYNMAN

# Wormholes

See TIME TRAVEL.

# Wrinkles in the Microwave Background

COSMIC BACKGROUND RADIATION provides a snapshot of the universe when it was 300,000 years old. At that point, the primeval "soup" became transparent to radiation. It was almost perfectly isotropic—i.e., it looked the same in all directions. But we would not expect it to be *completely* isotropic. At that epoch, the primordial gas should have already begun to form slight condensations, which later were to become galaxies, clusters, and superclusters of galaxies.

In the 1970s, George Smoot and his colleagues in the United States set out to measure any faint irregularities in the microwave background. To do so, they had to invent new and more sensitive instruments that could be taken above most of the earth's atmosphere by being placed on mountains or in balloons, planes, or satellites. This was because the earth's atmosphere itself emits microwave noise.

In 1977, Smoot's team made an unexpected discovery. The background radiation showed a slight Doppler shift—a reduction in the wavelength that occurs because something is moving away from us. This slight shift is known as the cosmic dipole. It showed that our whole galaxy and our neighboring galaxies are moving in the direction of the constellation Leo, at about 400 miles per second. This can be explained only by the presence of a very large, concentrated mass exerting a gravitational pull on the whole group of galaxies. The distribution of matter in our universe is lumpy on the largest scales that we can observe, and the particular concentrated structure pulling us toward Leo has been named THE GREAT ATTRACTOR. (See THE COSMOLOGICAL PRINCIPLE.)

In 1989, after many more improvements in technology, a satellite known as Cosmic Background Explorer (COBE) was launched by rocket. By 1992, enough data had been accumulated for a pattern to emerge. There definitely are slight irregularities ("wrinkles") in the intensity of the microwave background, to 1 part in 100,000. These wrinkles are of all spatial sizes, small and large, and the distribution of sizes is that predicted by INFLATION THEORY (a variant of BIG BANG theory).

The confirmed presence of wrinkles in the background radiation has led to another clue about the makeup of the universe. Since the high-energy radiation present at the time the universe was only 300,000 years old would have prevented visible matter from clustering at all, it is thought that the original lumpiness must consist of DARK MATTER. Unaffected by the radiation, this dark matter would already have become slightly lumpy. In a metaphor suggested by Smoot, the visible matter would then cluster around it like dewdrops on an unseen cobweb. This would lead to the observed wrinkles in the radiation.

The detection of Smoot's "wrinkles in time" is one of the five main pieces of evidence supporting Big Bang theory. It explains how galaxies and even larger structures have had time to form, as well as supporting inflation theory and the existence of some kind of dark matter.

# INDEX